Design Technology of Synthetic Aperture Radar

# Design Technology of Synthetic Aperture Radar

*Jiaguo Lu*
East China Research Institute of Electronic Engineering
Hefei, China

This edition first published 2019


Published by John Wiley & Sons Limited, The Atrium, Southern Gate, Chichester, West Sussex, PO19 8SQ, United Kingdom, under exclusive license granted by NDIP for all media and languages excluding Simplified and Traditional Chinese and throughout the world excluding Mainland China, and with non-exclusive license for electronic versions in Mainland China.



*Registered Offices*
John Wiley & Sons, Inc., 111 River Street, Hoboken, NJ 07030, USA
John Wiley & Sons Ltd, The Atrium, Southern Gate, Chichester, West Sussex, PO19 8SQ, UK

*Editorial Office*
The Atrium, Southern Gate, Chichester, West Sussex, PO19 8SQ, UK

For details of our global editorial offices, customer services, and more information about Wiley products visit us at www.wiley.com.

Wiley also publishes its books in a variety of electronic formats and by print-on-demand. Some content that appears in standard print versions of this book may not be available in other formats.



*Library of Congress Cataloging-in-Publication Data*

Names: Lu, Jiaguo, 1964- author.
Title: Design Technology of Synthetic Aperture Radar / Jiaguo Lu.
Description: Hoboken, NJ, USA : Wiley-IEEE Press, 2019. | Includes bibliographical references and index. |
Identifiers: LCCN 2019011087 (print) | LCCN 2019017006 (ebook) | ISBN 9781119564638 (Adobe PDF) | ISBN 9781119564676 (ePub) | ISBN 9781119564546 (hardcover)
Subjects: LCSH: Synthetic aperture radar.
Classification: LCC TK6592.S95 (ebook) | LCC TK6592.S95 L83 2019 (print) | DDC 621.3848/5–dc23
LC record available at https://lccn.loc.gov/2019011087

Cover design: Wiley
Cover image: © imaginima/Getty Images

Set in 10/12pt WarnockPro by SPi Global, Chennai, India
Printed and bound in Singapore by Markono Print Media Pte Ltd

10 9 8 7 6 5 4 3 2 1

# Contents

# About the Book

This book is a technical monograph on the Synthetic Aperture Radar (SAR) system design and technologies with focuses on high resolution imaging, moving target indication, and system engineering technologies. All the contents presented in the book are a summary of the research and development activities conducted by the author and his team over the past 20 years.

This book introduces the theory, applications, and new techniques of SAR. The signal characteristics, system parameters, and critical design factors of SAR imaging especially in the moving target mode are discussed. The SAR antenna design, array analysis, antenna element, and antenna structure, etc are elaborated. The basic requirements of transmitters and T/R modules are discussed with an emphasis on the fundamental function, design, fabrication, and application of the T/R modules. The book also discusses the wide band receiver technologies, including direct demodulation receiver (analog demodulation and digital demodulation), de-chirp receiver, monolithic hybrid integration and multi-band receiver. In addition, the integrated phase-locked frequency synthesizer and direct digital frequency synthesizer are described in details along with the wideband waveform signal generation method based on DDS direct or parallel structure intermediate frequency generation, digital baseband multiplexing and sub-band concurrency. It is followed by the signal processing methods and research achievements of high resolution imaging, SAR ground moving target indicator (SAR-GMTI), marine moving target indication (MMTI) and airborne moving target indication (AMTI) are discussed and analyzed. We also elaborate the SAR image interpretation, processing and intelligence capture, with a focus on the intelligence acquisition from SAR images.

This book is intended for engineers and technicians who work in SAR system research and development, and it can be served as a reference book for postgraduates majoring in SAR system design, microwave antenna, signal and information processing.

# Preface

Synthetic aperture radar (SAR) was firstly presented more than 60 years ago. After more than 60 years of rapid development of microwave and digital technology, SAR has evolved to a technology that can achieve target information acquisition with high resolution, multi-band, multi-polarization, and multi-mode operations, and has become one of the most important sensing technologies. Harnessing the development of aviation, aerospace and missile technology, SAR has also been applied extensively in different platforms such as satellite, aircraft, and missile. Moreover, SAR has been developed from two-dimensional imaging to three-dimensional imaging with a resolution at sub-meter level. In all, SAR plays an irreplaceable role in the field of military surveillance and civil microwave remote sensing.

In China, the research of SAR originated in 1970s and flourished in 1990s. During the past 20 years, Chinese SAR technology has made great progress in many aspects. As a participant, I am honored to witness this historical process. However, I feel a lack of suitable SAR reference book for practical engineering. Thus, this book, based on the research and experiment I have been involved in, systematically describing the SAR design including the latest research methods, the research progress, and results, may serve as a reference for readers.

This book is divided into seven chapters. In Chapter 1, the basic concept of SAR is introduced firstly. Based on that, the SAR application on different platforms is discussed, and new technologies that could impact the future development of SAR are proposed. The general design technology of SAR, the conventional radar and SAR equation, and the operation mode of SAR are discussed in Chapter 2. Considering the engineering design issues, the difficulties and keys of designing a radar system on different platforms are discussed. In Chapter 3, the planar array antenna is analyzed, based on the analysis of basic antenna parameters. It focuses on antenna radiation elements such as microstrip patch, dipole, and slotted waveguide. The corresponding latest research achievements are presented as well. The fundamental function and classification of T/R module, as well as some typical T/R modules are analyzed in Chapter 4, in which, the communication design, architecture design, electromagnetic compatibility and the environmental adaptability of T/R module are also discussed in detail. In addition new technologies and applications of the T/R module are presented. In Chapter 5, three kinds of frequency synthesizer are analyzed in detail, namely, analog direct frequency synthesizer, phase locked frequency synthesizer, and digital direct frequency synthesizer. This chapter also describes five types of wideband waveform generation methods, including DDS (direct digital synthesizer) based direct waveform generation, parallel DDS intermediate frequency waveform generation, digital baseband waveform generation, multiplex waveform generation and sub-band parallel wideband waveform generation. Two kinds of signal processing algorithms of SAR echo, namely time domain and frequency domain, are introduced in Chapter 6. The characteristics and process of RD (Range Doppler) algorithm, CS (chirp scaling) algorithm, $\omega$-$k$ algorithm, and SPEAN (spectral analysis) algorithm are emphasized. In this chapter, to improve the imaging resolution, the accurate estimation of Doppler parameters based on the raw echo data, and the motion compensation based on sensor, echo data, and image data, are analyzed. This chapter also presents the research achievements of the author and his research team in high resolution imaging, SAR-GMTI (ground moving target indicator), MMTI (maritime moving target indicator), and AMTI (airborne moving target indicator). In Chapter 7, from the perspective of the intelligence application requirements, SAR image target detection, target change detection, target recognition, and multiple SAR image fusion of the intelligence extraction system are discussed. The latest achievements in intelligence extraction based on domestic SAR image are presented. From the purpose of the intelligence application, this chapter emphasizes on the technical process and processing results. In view of the difficulties in the engineering application of image intelligence processing, intelligent target recognition, and image intelligence application system, the book gives the perspectives on the technology.

With the development of modern science and technology, one can predict that the further efforts of SAR technology will be put in the aspects of ultra-high resolution, wide swath, moving target detection, and moving target imaging. I believe this book will enable the readers to further think and understand the development of SAR.

During the writing of this book, I paid equal attention to the theory and the engineering technology. The SAR theory, the technology and design of SAR system and sub-systems are discussed from the engineer's point of view. In the meantime, image processing and intelligence extraction are studied and introduced.

Finally, I would like to expresses my gratitude to the SAR system research team of East China Research Institute of Electronic Engineering (ECRIEE), whose hard work and significant contributions have greatly enriched the contents of this book. I am grateful to Academician of Chinese Academy of Engineering (CAE) Wu Manqing, who established and developed this research team. I sincerely appreciate Mr. Zhang Changyao, Mr. Ge Jialong, Mr. Zhang Weihua, and Mr. Jiang Kai as well as other professional researchers for their pioneering work in the field of SAR. In the process of writing this book, I received the support and useful discussion of Researcher Zhu Qingming, Dr. Li Tong and Dr. Liu Zongang. Researcher Wang Chuansheng has done a lot of detailed work for the translation of this book. Academician of CAE Wang Xiaomo and Academician of CAE Lu Jun gave me encouragement and careful guidance during the writing process, and I would like to express my deep gratitude. In addition, due to my limited knowledge, there might be some mistakes in this book, please do not hesitate to let me know.

February 2018 Jiaguo Lu

## List of Acronyms

| | |
|---|---|
| 1D | One-dimensional |
| 2D | Two-dimensional |
| 3D | Three-dimensional |
| 3DT-STAP | Three Doppler transform STAP |
| A/D | Analog to digital |
| AASR | Azimuth ambiguity-to-signal ratio |
| ADC | Analog to digital converter |
| AMTD | Airborne moving target detection |
| AMTI | Airborne moving target indication |
| APC | Antenna phase center |
| ATI | Along-track interferometry |
| BITE | Built in test equipment |
| BJT | Bipolar junction transistor |
| BP | Back projection |
| CAD | Computer aided design |
| CF | Carbon fiber |
| CFAR | Constant false alarm rate |
| CMC | Commercial mezzanine card |
| CPCI | Compact peripheral component interconnect |
| CPI | Coherent pulse interval |
| CPU | Central processing unit |
| CSI | Clutter-suppressing interferometry |
| D/A | Digital to analog |
| DAC | Digital to analog converter |
| DAR | Digital array radar |
| DBF | Digital beam forming |
| DDR | Double data rate |
| DDS | Direct digital synthesis/synthesizer |
| DLR | Deutsches Zentrum fur Luftund Raumfahrt |
| DMA | Direct memory access |
| DPC-MB | Displaced-phase center multibeam |
| DPCA | Displaced phase center antenna |
| DSP | Digital signal processor |
| EMC | Electromagnetic compatibility |
| EMI | Electromagnetic interference |

| | |
|---|---|
| ERIM | Environmental Research Institute of Michigan |
| ESD | Electrostatic discharge |
| FET | Field-effect transistor |
| FFBP | Fast factorized back projection |
| FFT | Fast Fourier transform |
| FIR | Finite impulse response |
| FM | Frequency modulation |
| FPGA | Field-programmable gate array |
| GMTD | Ground moving target detection |
| GMTI | Ground moving target indicator |
| GPIB | General purpose interface bus |
| GPS | Global Positioning System |
| GPU | Graphics processing unit |
| HEMT | High-electron-mobility transistor |
| HIRF | High intensity radiated field |
| HRWS | High-resolution wide-swath |
| HTCC | High temperature co-fired ceramic |
| ICA | Independent component analysis |
| IF | Intermediate frequency |
| IFFT | Inverse fast Fourier transform |
| IMU | Inertial measurement unit |
| INS | Inertial navigation system |
| INSAR | Interferometric synthetic aperture radar |
| ISAR | Inverse synthetic aperture radar |
| JPDA | Joint probabilistic data association |
| LAN | Local area network |
| LDMOS | Longitudinally diffused metal–oxide–semiconductor |
| LFM | Linear frequency modulation |
| LNA | Low noise amplifier |
| LTCC | Low temperature co-fired ceramics |
| LTCF | Low temperature co-firing ferrite |
| LPF | Low-pass filter |
| MCM | Multi-chip module |
| MCM-C | Ceramic thick-film multi-chip module |
| MDV | Minimum detectable velocity |
| MEMS | Micro-electromechanical Systems |
| MIMO | Multiple-input, multiple-output |
| MMIC | Microwave monolithic integrated circuit |
| MMTI | Maritime/Marine moving target indicator/indication |
| MRF | Markov random field |
| MTD | Moving target detection |
| MTI | Moving target indication |
| MTT | Multiple-target tracking |
| MUX | Multiplexer |
| NCO | Numerically controlled oscillator |
| OCXO | Oven-controlled crystal oscillator |

| | |
|---|---|
| PCA | Principal component analysis |
| PCB | Printed circuit board |
| PD | Pulsed-Doppler |
| PGA | Phase gradient autofocus |
| PHEMT | Pseudomorphic high electron mobility transistor |
| PLL | Phase-locked loop |
| PRF | Pulse repetition frequency |
| PRI | Pulse repetition interval |
| PRT | Pulse repetition time |
| RAM | Random-access memory |
| RASR | Range ambiguity-to-signal ratio |
| RD | Range Doppler |
| RF | Radio frequency |
| ROI | Region of interest |
| ROM | Read-only memory |
| SAIP | Semi-Automated Image Intelligence Processing |
| SAM | Scalable array module |
| SAR | Synthetic aperture radar |
| SCNR | Signal-to-clutter noise ratio |
| SDRAM | Synchronous dynamic random-access memory |
| SIMD | Single-instruction, multiple-data |
| SIMO | Single-input, multiple-output |
| SISO | Single-input, single-output |
| SLC | Single look complex |
| SMD | Surface mount device |
| SNR | Signal-to-noise ratio |
| SOA | Service-oriented architecture |
| SOI | Silicon on insulator |
| SPCMB | Single phase center multiple beams |
| SPCMEB | Single phase center multiple elevation beams |
| SPECAN | Spectral analysis |
| SRAM | Static Random-access memory |
| STAP | Space-time adaptive processing |
| STT | Single-target tracking |
| SVD | Singular value decomposition |
| SVM | Support vector machine |
| SVR | Support vector regression |
| T/R | Transmit/receive |
| TAS | Tracking and searching |
| TBD | Track before detect |
| TOPS | Terrain observation by progressive scans |
| TOPSAR | Terrain observation by progressive scans synthetic aperture radar |
| TT | Target tracking |
| TTL | Transistor–transistor logic |
| TWS | Tracking while scan |
| UART | Universal asynchronous receiver/transmitter |

| | |
|---|---|
| UAV | Unmanned aerial vehicle |
| UHF | Ultra-high frequency |
| VDMOS | Vertically diffused metal–oxide–semiconductor |
| VSWR | Voltage standing wave ratio |
| VCO | Voltage-controlled oscillator |
| VME | VersaModule Eurocard |

# 1

# Introduction

## 1.1 Overview

Synthetic aperture radar (SAR) is a type of imaging radar that receives wideband echoes at different positions through relative movement between the loading platform and the target. Within a certain accumulating period, SAR obtains a two-dimensional (2D) image of the target after coherently processing the received echoes, which allows high-resolution observation of the real image of the target.

The resolution of the 2D image of the target is normally expressed as range resolution and azimuth resolution. In the range direction, high resolution can be achieved by employing the pulse compression technique in the transmitted linear frequency modulation signal, which has a large time-width and bandwidth product. In other words, similar to conventional radar, SAR improves the range resolution by employing the pulse compression technique on the transmitted and the received linear frequency modulation signal. In the azimuth direction, one radar sensor of uniform motion in a straight line is used to transmit and receive the pulse signal at interval positions with a certain pulse repetition frequency, and then a high azimuth resolution is obtained by a coherent processing of the received echoes. To achieve a high resolution, a synthetic aperture equivalent to an actual huge but unrealizable array aperture is constructed based on the track produced from the relative movement between the target and the radar.

As shown in Figure 1.1, the range resolution of the radar is normally defined as the minimum distance between two points that can be distinguished. If the arrival time of the leading edge of the pulse echo at a distant point is later than the arrival time of the trailing edge of the pulse echo at a near point, then these two points are considered distinguishable. Therefore, the range resolution of two distinguishable points is given by

$$\Delta R_g = \frac{\Delta R_s}{\sin \eta} = \frac{c\tau_p}{2 \sin \eta} \tag{1.1}$$

where $\tau_p$ is the signal pulse width, $c$ is the light speed, and $\eta$ is the incident angle between the antenna and the ground.

It can be seen that the pulse width is required to be very narrow to achieve a high-range resolution. However this will dramatically reduce the average power level of the radar system and will make the signal-to-noise ratio (SNR) of the echo of the observed target too low. A pulse compression technique is commonly employed in SAR to avoid

*Design Technology of Synthetic Aperture Radar,* First Edition. Jiaguo Lu.

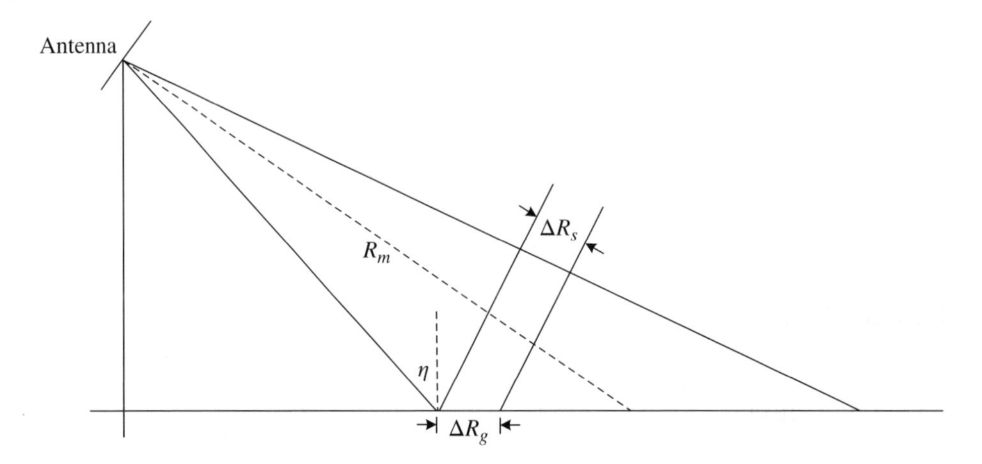


**Figure 1.1** Sketch of range resolution.

this problem. Consequently, a relatively wider pulse can be used to realize both high resolution and high SNR.

In SAR, the linear frequency modulated signal transmitted can be expressed as

$$\mu(t) = rect\left(\frac{t}{\tau_p}\right) e^{j2\pi\left(f_c t + \frac{1}{2}kt^2\right)} \tag{1.2}$$

where $\tau_p$ is the signal pulse width, $f_c$ is the center frequency of the carrier, and $k$ is the slope of the linear modulated frequency. The bandwidth of the signal is then $B = k\tau_p$. After matched filtering, the pulse width $\tau_p$ is compressed into *1/B,* and the resolution is improved to

$$\Delta R_g = \frac{c}{2B\sin\eta} \tag{1.3}$$

For azimuth resolution, the motion of the radar antenna with the platform forms a SAR virtual antenna array, while the physical antenna serves as a unit of the array. During the motion, the echo of the target is sequentially received and recorded, and the phase difference caused by the wave path difference of multiple antenna positions is compensated by signal processing. Therefore, the echoes of the same point target can be stacked in phase, and azimuth compression is performed to realize high azimuth resolution.

As shown in Figure 1.2, suppose the radar moving distance is $L_s$ (the distance of the antenna beam beginning to illuminate the target to leaving the target), and the size of the antenna in the azimuth direction is $D$. If $\lambda$ is the wavelength of the radar wave, the maximum length of synthetic aperture is determined by the actual antenna beam width $\theta_B$, the distance of the target $R$, and the wavelength of the system, as expressed by Eq. (1.4).

$$L_s = \theta_B R = \frac{\lambda R}{D} \tag{1.4}$$

For synthetic aperture, transmission and reception of the radar pulse are considered as two-way propagation. The phase difference between two arbitrary array elements to the target is twice that of the single-way transmission, hence the equivalent bandwidth

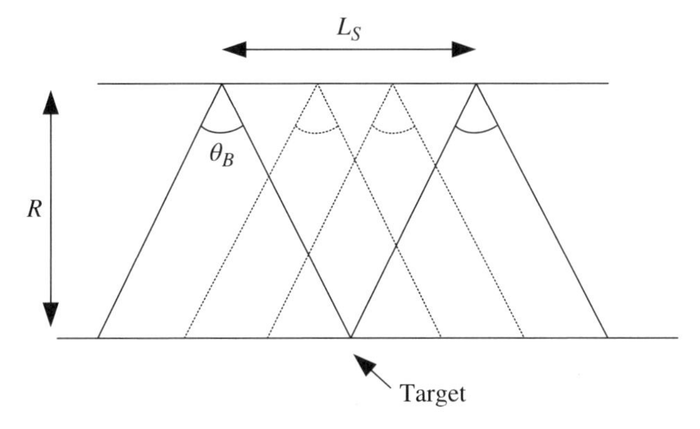


**Figure 1.2** Diagram of synthetic aperture.

of synthetic aperture is given by

$$\theta_s = \lambda/2L_s \tag{1.5}$$

After synthetic aperture processing, the azimuth resolution can be improved to

$$\Delta R_{az} = \theta_s R = \frac{\lambda R}{2L_s} = \frac{\lambda R}{2} \cdot \frac{D}{\lambda R} = \frac{D}{2} \tag{1.6}$$

It is evident that the azimuth resolution of the synthetic aperture is irrelevant to the target distance because the further the distance, the longer the effective synthetic aperture and the narrower the formed beam. This is balanced out by the widening of the azimuth resolution unit caused by the increase of the distance, which keeps the size of the azimuth resolution unit unchanged.

From the previous discussion, SAR features a high resolution because the length of the synthetic aperture is equivalent to that of a large antenna aperture. To obtain a high-resolution 2D image, the echoes within the coverage area of the radar beam are collected and coherently processed. Broadly speaking, SAR is also distributed radar that sequentially samples a synthetic aperture to replace the distributed antenna aperture.

## 1.2 SAR Applications

### 1.2.1 Military Applications

As SAR system technology moves forward, the resolution increases from tens of meters to meters to submeters; the polarization modes vary from single polarization to multipolarization to full polarization; the radar waveband ranges from meter-wave band to microwave band to millimeter-wave band to submillimeter-wave band. Furthermore, SAR has covered military, civil, and scientific applications, where the platforms include unmanned aerial vehicles (UAVs), helicopters, fixed-wing manned aircraft, satellites, and missiles. To fulfill different requirements, different operation modes are realized by the antenna beam scheduling, for instance, strip-map SAR, spotlight SAR, scan SAR, moving target detection (MTD), inverse SAR (ISAR), and interferometric SAR (INSAR). At different operation modes, useful information is interpreted and extracted from

SAR images, including military intelligence, mapping, marine meteorology, hydrology, geology, forestry, and object deformation. SAR plays an important role in military applications and is an indispensable and irreplaceable sensor for mapping the surface of the earth, due to its availability for high resolution, in all types of weather, and at all times.

#### 1.2.1.1 Military Intelligence

SAR plays a very important role in obtaining military intelligence. SAR mounted on different platforms has different characteristics in military applications.

By providing reliable intelligence gathering and surveillance, spaceborne SAR performs reconnaissance missions, including military maneuvering, military target and enemy troops surveillance, and military assessment. Spaceborne SAR with different wavebands, polarizations, and resolutions has many functions in military applications [1].

Due to high mobility and time continuity of the platforms, manned aircraft and UAV SAR can make up for disadvantages in satellite observation. Airborne SAR therefore has a wide application in intelligence reconnaissance [2]. Furthermore, the features of airborne SAR include a short development cycle, rapid implementation of the most advanced techniques, ultra-high resolution, and large-scale search. All of these features drive the application of airborne SAR for tactical reconnaissance [3].

Missile-borne SAR can obtain real-time images of ground scenery in flight or real-time images of the earth and objects along the track of the missile flying to the target [4, 5]. By comparing with the reference map installed on the missile, the longitudinal or lateral deviation of the missile from the target or the predetermined track can be obtained, which further guides the missile. Missile-borne SAR has become a hot topic because it can effectively improve the terminal guidance precision of medium- and long-range attack weapons, such as camouflaged underground missile launch silos, ground objects in fog-covered areas, etc.

#### 1.2.1.2 Moving Target Detection

Battlefield information includes not only the precise location of the enemy, the facilities, and the deployment of the enemy troops but also the enemy tanks, armored vehicles, mobile artillery, mobile missile launchers, helicopters, cruise missiles, and the activities of mobilization and supply of the enemy troops. SAR can extract the information of moving targets from geographical objects [6]. Different from a static target, a moving target has a certain radial velocity and Doppler frequency. During imaging, it can possibly move out of the image. To retain the moving target in the image, a key mission of MTD is to remove the stationary clutter, or clutter of the geographical object. By combining SAR imaging techniques with features of MTD including airborne moving target detection (AMTD) and ground moving target detection (GMTD), SAR can obtain high-resolution images of the scene and detect the moving target within the scene as well, which is of particular importance for practical applications.

#### 1.2.1.3 Military Topography and Mapping

SAR imaging also offers an accurate and fast means of military mapping. Through processing of SAR images, military mapping can be provided for a variety of purposes (topographic scales 1 : 10,000 and 1 : 50,000), such as quickly drawing and repairing overseas areas, basic mapping of combat command in hot-spot areas and surrounding areas, providing a target location for modern weaponry and precision-guided weapons, and radar image reference mapping for target matching guidance.

#### 1.2.1.4 Detection of Marine Meteorology and Hydrology

A radar image is very sensitive to marine structure. Therefore, qualitative and quantitative analysis of the interaction, the mechanism, the results of the wind field, the marine wave, and the ocean current can be performed to detect marine meteorology and hydrology. Different types of sea have different marine features and coastline. An electromagnetic wave emitted by radar is very sensitive to such parameters as the geometric structures of the marine wave and marine roughness. Based on different backscattered waves, SAR can be applied in medium-scale sea feature detection and large-scale sea feature recognition, such as water mass and peak area. Marine changes caused by moving targets under the water (for instance, a submarine) can also lead to changes in the radar images.

### 1.2.2 Civil Applications

Incident radar waves of SAR at different wavebands and polarization modes have different effects on detection of the same geometric object. More accurate target features can be detected on the echoes from geometric objects at different wavebands and the copolarization and cross-polarization information in linearly polarized status.

SAR can be widely used in resource exploration and research; environmental monitoring and research; disaster monitoring; agricultural yield estimation; hydrological and geological exploration; engineering surveying; marine monitoring, for example, detecting floods, drought, storm surges, and other major natural disasters; and marine research.

#### 1.2.2.1 Geological Exploration

SAR images provide very fruitful geological and mineral information, such as geological structure; lithology; and the presence of a concealed geological body, which is especially useful in detecting the geological structure of a volcano, a meteorite, or a fault and in detecting metal deposits in the structural belts. With the development of polarization and coherence technology, SAR can be used in measuring and researching crustal deformation, seismic inoculation, plate movement, and ground subsidence.

#### 1.2.2.2 Oceanographic Research

SAR can obtain continuous data and images of the marine and Arctic sea ice during all types of weather and at all times, so marine transportation and worldwide climate change (including the prediction of major climate change) can be surveyed [7]. The intensity of backscattered waves is reduced because of the decrease in marine roughness caused by an oil spill on the surface of a sea. Due to this, SAR is the best effective method to survey oil pollution and natural oil spill film on the water and to detect marine oil and gas.

#### 1.2.2.3 Forestry Research

SAR normally works in the microwave band with different frequencies. Backscattering and penetration of microwave signals differ for different geological surfaces and plants.

SAR can provide abundant vegetation and soil information; it also provides estimations of the vegetation height, forest stock volume, and forest biomass; it can identify forest types, forest disasters, forest density, forest age, and health status as well as monitor logging activities; moreover, it can estimate the biomass of forests in cloudy, rainy, and foggy tropical, subtropical, and high-latitude regions.

Soil moisture is an essential condition for plant growth. Therefore, the estimation of soil moisture is of great concern. SAR can also be used for soil moisture estimation and soil classification. The penetrating characteristics of SAR provide a certain penetration capacity for vegetation and soil, and radar's backscatter coefficient is very sensitive to soil moisture and soil type.

#### 1.2.2.4 Deformation Monitoring

The differential SAR interference for monitoring the deformation has been widely used, for example, for high-rate dynamic monitoring of the stability of manmade buildings like dams, bridges, and tall towers and long-time monitoring and early warning of natural disasters like surface subsidence, landslides, avalanches, glacier displacement, and volcanic activity [8, 9]. Airborne, spaceborne, and ground-based SAR can be used for monitoring and early warning. Spaceborne SAR commonly employs repeat track interferometry to monitor the deformation of the ground surface. However, due to the long retracking time, it is difficult for spaceborne SAR to continuously monitor a fixed point in the deformed area. Double antenna interference in airborne SAR is commonly used for monitoring the deformation of the ground surface; however, the precision is not good enough for deformation. Ground-based SAR is commonly used to monitor the regional and continuous deformation of a fixed point. Ground-based SAR features good flexibility and feasibility, is contact free, and has high resolution and low cost.

## 1.3 Features of SAR

SAR is similar to other types of radar systems regarding the composition of the radar equipment. However, SAR has unique features in terms of the platform loading mode, radar system, and information processing.

### 1.3.1 Radar Loading Platforms

Many types of platforms, such as satellites, aircraft, and missiles, are used to load SAR. SAR is usually used to observe the two sides of the platform (except the satellite platform) to remotely map and image the ground (or sea). In working mode, the radar beam with the wave-propagation direction has a substantial component perpendicular to the flight-path direction and points to the sides of the platform.

In some operating modes, such as spotlight and squint modes, the radar beam deviates from the perpendicular direction of the flight path. To meet the requirements of multimodes, small SAR antennas are normally mounted on UAV and missile platforms, less than one square meter in size. Large SAR antennas, which are often mounted under the sides of an aircraft or underneath the belly to minimize the influence of the belly on the antenna radiation performance and to enlarge the ground observation angle, have dimensions of less than tens of square meters in size. Most satellite platforms will prefer large SAR, in which antenna size varies from tens to dozens of square meters. The radar antenna is folded during transmission and is unfolded when operating in orbit.

The basic function of all kinds of platforms is to provide a necessary resource to ensure the normal operation of radar and to transfer the processed results of the radar back to the ground. The five elements of the radar platform commonly referred to are the geometric size, weight, power consumption, motion parameters, and data communication rate.

### 1.3.2 Radar System

The movement of the SAR platform is used to create a virtually large aperture antenna to obtain a high azimuth resolution, while the azimuth resolution of conventional radar (especially ground-based radar) is the physical aperture. Their mechanism difference is mainly caused by the difference of the radar system. Conventional radar transmits a series of pulses, which then are bounced back by the target, and the target can be detected by signal processing. However, SAR needs to transmit and receive pulses many times until an expected aperture signal is collected to perform the signal processing to obtain a full SAR image. In the meantime, the SAR subsystems have their own characters.

For antennas, the instantaneous bandwidth of conventional radar and SAR are different due to their different range resolutions. Conventional radar normally has to work with a hopping frequency within a broader band range to improve the electronic countermeasure capability; therefore, demands for conventional radar and SAR on the antenna bandwidth are identical. For conventional radar, to reduce the interference power entering into the receiver in the side lobe region, the antenna side lobe is required to be less than −30 dB. For airborne early warning detection radar, to reduce the influence of strong ground clutter on the performance of the detecting target, the antenna side lobe is required to be less than −35 dB. For SAR, though a relatively high antenna side lobe has an effect on the ambiguity of the images and the detection ability on the moving target, the antenna side lobe is normally controlled to be around −20 dB. Generally, it is preferable to further lower the antenna side lobe.

Conventional radar normally works at a narrow instantaneous bandwidth (usually 2.5~10 MHz), whereas SAR normally works at a broad instantaneous bandwidth (normally more than 400 MHz). Due to the signal bandwidth difference of the receiver, the dynamic range of the SAR receiver (the noise level of the receiver is taken as a benchmark) is 20~30 dB lower than that of a conventional radar receiver. However, given the target characteristics and pulse compression features, the transient behavior of the SAR receiver and conventional radar receiver are comparable.

Regarding waveform design, linear frequency modulation, nonlinear frequency modulation, phase encoding, and step frequency are used in signal waveforms of conventional radar. To obtain a high-range resolution, the bandwidth characteristics and the pulse compression effect of the system are specially emphasized in SAR, which normally employs a linear frequency modulated signal, or sometimes the stepped frequency pulse signal. To achieve a higher resolution at the centimeter level, waveform sub-band synthesis is a good choice. It can effectively reduce the difficulty of wideband waveform design.

For a frequency source, to improve the SNR of the radar system, especially the detection of a moving target in strong clutter, coherent processing is normally applied in conventional radars. Similarly, coherent processing is the foundation and prerequisite of SAR. Operation of high azimuth resolution results from coherent accumulation, closely related to the stability of the frequency source and coherent accumulation. In other words, the imaging resolution is closely linked with the stability of the frequency source. The frequency error caused by the random fluctuation has an effect on the phase characteristics of the echoes and thus deteriorates the azimuth resolution of SAR. The frequency error caused by the sine wave results in the appearance of paired echoes after pulse compression and has an impact on the imaging quality. In contrast with conventional radars, a long-time pulse accumulation is required for SAR system. In the meantime, the loading platform of SAR offers strong random characteristics, thus the frequency source of a SAR system focuses on frequency antivibration characteristics and long-time phase stability.

For signal processing, conventional radar and a SAR system both need to perform a range direction pulse compression. They differ greatly in azimuth. Azimuth accumulation of conventional radar is used to distinguish the target and the clutter in the frequency domain and to obtain a certain gain. Coherent accumulation of SAR is used to obtain a high-angle resolution along the range dimension. Furthermore, in a SAR system, different algorithm systems are used at different operation modes, such as SAR, ISAR, INSAR, GMTI, etc. For signal processing of a moving target, conventional radars and SAR are similar in starting pulse compression in the range direction, clutter suppression, and detection and information extraction of the moving target. The difference is that the SAR system usually adopts phase-parallel synthesis to obtain the angle high resolution, and the conventional radar coherent synthesis obtains the frequency resolution. In measuring the azimuth angle of the target, multibaseline interferometry is often used in SAR radar, while a narrow-beam characteristic or sum-difference beam method is used in conventional radar.

### 1.3.3 Information and Intelligence Processing

The basic operations of conventional radar (except the weather radar and other special radars) are almost similar to SAR. Conventional radar mainly measures the range, altitude, and azimuth information of the target and forms the target track. Stable target tracking is of great importance.

A 2D image is produced from the SAR signal processing. The image processing transforms the complex SAR image into usable intelligence. The targets can be quickly interpreted, detected, and identified from SAR images, with particular interest in the realization of small target detection and extraction in images with strong interference, strong scattering, and high-density electromagnetic signals.

Not only is the fusion of radar images obtained from different platforms included in fusion with other sensors but also the fusion of information obtained from passive detectors and other sensors, such as infrared, optical, and acoustic sensors.

## 1.4 New Technologies of SAR

More and more, applications show an urgent need for a wide swath, high resolution, high accuracy, real-time acquisition and processing, and timely update of SAR observation data. Some new technologies and methods that are appearing and will appear will give new vitality to the traditional SAR system. These new technologies and methods have been widely studied.

### 1.4.1 Digital Array Technology

Digital array technology (i.e. receiver/transmitter digital beam-forming [DBF] technology) provides such multidimensional information as spatial domain, time domain, and frequency domain, which can enhance the performance and simplify the structure of a radar system. Digital array technology has been successfully applied in target detection radar. If it is combined with SAR technology, it will significantly enhance the performance of the imaging system [10].

Digital array technology can realize flexible allocation of radio-frequency (RF) signal power in the observation zone. Compared to multichannel technology with single transmitting and multireceiving, or the receiver DBF technology only in the range direction, it can realize a new adaptive SAR imaging mode so that the SAR system can be switched between different modes, including strip-map, spotlight, scanning, GMTI, and interference. It can also produce image results of different modes based on the same group of echo data, enabling the system to operate under multimode simultaneously. The improvement of the observation efficiency has a greater significance for a spaceborne SAR system, which is restricted by the revisit period. It is not difficult to predict that the digital array technology is one important trend of future SAR.

Digital array technology can effectively alleviate the contradiction between high azimuth resolution and the wide swath of traditional satellite SAR, realizing high-resolution wide-swath imaging. It is mainly realized by azimuth digital array technology and range direction digital array technology as well as a combination of these two technologies to further improve the system performance. Azimuth digital array technology can greatly alleviate contradiction between minimum detectable speed and blind speed in conventional a SAR/moving target indication (MTI) system and can improve the performance of SAR/MTI. By using digital array technology, multimode and multimission operation can be realized simultaneously. In addition, the digital array technology provides simultaneous multibeam adaptive capability, which can resist many kinds of electronic jamming. When a disturbance signal is detected, the interference can be weakened in the signal processing stage by zero setting the antenna pattern in an appropriate suitable direction. The digital array technology can also vary the transmitted waveform during beam scanning, which lowers the probability of interception. Therefore, digital array SAR has stronger anti-interference ability and battlefield survivability.

Both the theory and technology of digital array SAR have evolved from receiving DBF SAR and transmitting DBF SAR to DBF multiple-input, multiple-output (MIMO) SAR, namely, the gradual integration of the digitalization technology, software, the transmitting and receiving system, and the distributed system architecture. The distributed SAR system based on digital array technology with respective transmitting and receiving cannot only complete missions that are difficult for a single platform but also can obtain more and more diversified target information. After integrating the transmitting and receiving system with the distributed system architecture, the radar system has more reliability and more flexibility and is suitable for many different applications, such as continuous monitoring, wide-swath, and high-resolution imaging.

Compared with a conventional SAR system, a digital array SAR system is simplified and has obvious performance and functional advantages. However, to make a full use of these advantages, there are still some technical issues to be resolved, mainly including high-density integration of radar system and broadband data transmission and processing. As these problems are solved, the performance and function of digital array SAR systems will be improved.

### 1.4.2 MIMO Technology

MIMO radar is capable of obtaining equivalent observation channels whose number is far more than the actual number of the antennas, through simultaneous multiple antennas transmitting and receiving. For target detection, parameter estimation, and radar imaging, it is better than those of the traditional radar system. It has been widely researched by experts and scholars in the field of radar. The combination of MIMO technology and SAR provides a new way to solve the conflict between high resolution and wide swath and the problem of detecting slow moving targets, which commonly exists in conventional SAR.

There are four primary characteristics in the concept of MIMO-SAR: (1) Multiple transmitting/receiving antennas are distributed on the motion platform; (2) the transmitting antennas simultaneously transmit multiple waveforms independently, and the waveforms can be mutually orthogonal or uncorrelated; (3) the receiving antennas simultaneously receive the echoes of the swath independently, and the echoes of each transmitted signal can be separated by a group of filters; and (4) the performance of SAR system is improved by joint processing the echo data of multiple observation channels during signal processing.

According to the classification of MIMO radar, MIMO-SAR can be divided into two categories, common platform MIMO-SAR and distributed MIMO-SAR. For common platform MIMO-SAR, all transmitting and receiving antennas are installed on one platform, while for distributed MIMO-SAR, the transmit and receive antennas are placed on different platforms, and they are connected by network, except for bistatic SAR.

After years of development, much progress has been made in the theory and application of MIMO radar. But there are few for MIMO-SAR. A complete theoretical system is seldom seen as far as system concept, theoretical model, imaging strategy and method, performance evaluation, and so on. Many critical technologies have yet to be further addressed and improved, such as orthogonal waveform optimization, array configuration optimization, integrated imaging processing, etc. The special transceiver mode of MIMO-SAR and its diverse waveforms make the existing SAR imaging algorithm difficult to apply directly. Therefore, it is important to explore the imaging processing method and imaging method that are suitable for MIMO-SAR.

Current research on MIMO-SAR imaging technology remains in the initial stage of theoretical exploration, but its potential advantages have been widely noted by researchers. It is expected that MIMO-SAR will play an important role in SAR and GMTI applications, UAV multibase SAR imaging applications, sensing radar imaging applications, radar communication integration applications, and so on.

### 1.4.3 Microwave Photonic Technology

Microwave photonic technology is an emerging technology accompanied by the development of semiconductor lasers, integrated optics, optical fiber waveguides, and microwave monolithic integrated circuits. It is the combination of microwave and photon technology. It has potential applications in microwave signal generation, transmission, processing, and so on. To overcome the inadequacies in conventional SAR systems, researchers began to explore the application of microwave photonic technology in SAR. The microwave photonic technology in SAR applications will become a new frontier.

The microwave signal is modulated onto the optical carrier, and the long-distance transmission of the microwave signal by using the optical fiber has been applied in communication. The advantages of optical fiber have been used in the radar research field. Designers introduced microwave optoelectronic technology in phased array systems, such as optical fiber being used as the transmission line of radar data and signal, the true optical delay line being used to form optical beam, and so on. Metal waveguide or coaxial cable is usually used as the microwave signal transmission line, and an electronic phase shifter is used to form and control the antenna beams in conventional radar systems. They are not only bulky, heavy, and easily disturbed but also very expensive. To realize beam forming and control, eliminate the phenomenon of crowding on the back of the array through metal waveguides and coaxial cable, which can greatly reduce the system size and weight. Optical fiber instead of coaxial cable is used as the transmission medium, and real-time delay is used to replace an electronic phase shifter.

The optical phased array technology comes from the combination of microwave photonic technology and phased array technology. In an optical phased array, the RF signal is modulated on an optical carrier and is transmitted to the antenna units through different optical paths. Beam forming and control in the optical domain will produce many anticipated effects. It is well known that optical domain is nondispersive, and the propagation time will not change with respect to the frequency in a very wide microwave frequency band. The optical device is a kind of nonconductive transmission medium, and the light in the optical device offers good anti-interference capability. At the same time, the crosstalk to the connector is almost zero. These remarkable advantages cannot be realized by microwave devices. In addition, optical devices are not sensitive to interference and are durable, which makes them especially suitable for airborne and spaceborne SAR applications.

The application of microwave photonics technology in SAR has evolved from realizing remote control by using an optical link to control amplitude and phase of the antenna array elements. One of important research directions is to distribute the RF signal of the antenna unit by using the optical system. At present, most research focuses on realizing the true time-delay line of the large instantaneous bandwidth by microwave optoelectronic technology. The delay line is characterized by small size, light weight, low cost, and wide bandwidth. More attention will be paid to the improvement of performance and environmental adaptability of optical delay line in the future.

### 1.4.4 Miniaturization

A conventional SAR system is bulky, has high energy consumption, is expensive, and can only be deployed in large platforms, such as unmanned or manned aircraft, satellites, and missiles, which dramatically limits the large-scale implementation of SAR. The shape and mission requirements of modern aircraft are becoming more and more diverse. It is beneficial for SAR to be installed on a smaller and lighter platform, which requires its miniaturization.

Generally, two forms are adopted in the miniature SAR systems: the pulse system and the continuous-wave system. The pulse SAR system is complex, bulky, and heavy. New technology, devices, and materials are needed to achieve the miniaturization, and the system structure need to be optimized. With the development of light antenna and signal processing technology, it is possible to miniaturize SAR. The continuous-wave SAR system is a combination of frequency modulated continuous-wave and SAR technologies. It combines advantages of continuous-wave radar and SAR – small size, low cost, low power consumption, and high resolution – which is an important trend of miniature SAR. These miniature SAR systems are particularly suitable for small UAVs that fly at low heights.

The miniaturization of SAR system has many advantages. In addition to small size, light weight, high technical content, short development cycle, and low manufacturing cost, miniaturization can serve as a test bed for new technology. Therefore, the development and application prospects are very broad. In addition, micro-miniaturization of SAR systems is also one of the important techniques to realize deep-space microwave detection.

At present, the research on miniature SAR (such as the size of kilograms) is still in the stage of verification and is far from being developed and used in large scale. With the development of high-speed digital signal processing technology, motion measurement technology, and various new SAR imaging algorithms, SAR miniaturization is bound to flourish.

## References


1 Zhu, L., Guo, W., and Yu, W. (2009). Analysis of satellite development history and tendency. *Modern Radar* 31 (4): 5–10.

2 R. Horn, A. Nottensteiner, A. Reigber, J. Fischer (2009). F-SAR — DLR's new multifrequency polarimetric airborne SAR. *Proceedings of the 2009 IEEE International Geoscience and Remote Sensing Symposium*, Cape Town, South Africa (July 12–17, 2009). Piscataway, NJ: IEEE.

3 Zhang, K. (2013). *Handbook of Airborne Radar*. Beijing: National Defence Industry Press.

4 Peng, S. (2011). *Key Technology Research of Missile Borne SAR Imaging*, 5–6. Changsha: National University of Defense Technology.

5 Qin, Y. (2008). *Research on Missile Borne SAR Guidance Technology*, 3–4. Changsha: National University of Defense Technology.

6 T. Malenke, T. Oelgart, W. Rieck (2002). W-band-radar system in a dual-mode seeker for autonomous target detection. *Proceedings of the European Conference on*

*Synthetic Aperture Radar*, Cologne, Germany (June 4-6, 2002). Berlin, Germany: VDE Verlag.

7 Nakamura, K., Wakabayashi, H., Naoki, K. et al. (2005). Observation of sea-ice thickness in the sea of Okhotsk by using dual-frequency and fully polarimetric air-borne SAR (pi-SAR) data. *IEEE Transactions on Geoscience and Remote Sensing* 43 (11): 2460–2469.

8 Huang, Q. and Zhang, L. (2011). Micro deformation monitoring system based on GBInSAR technology and its application in dam deformation monitoring. *Water Conservancy and Hydropower Science and Technology Development* 31 (3): 84–87.

9 Zhou, X., Wang, P., and Xing, C. (2012). Research on the measurement of micro deformation of buildings based on GB-SAR. *Surveying and Mapping Geographic Information* 37 (5): 40–43.

10 J. Lu (2015). The technique challenges and realization of space-borne digital array SAR. *Proceedings of the 2015 IEEE 5th Asia-Pacific Conference on Synthetic Aperture Radar*, Singapore (September 1–4, 2015). Piscataway, NJ: IEEE.

# 2

# Radar System Design

## 2.1 Overview

Synthetic aperture radar (SAR) is a kind of radar system loaded and operated on a moving platform. The imaging system is composed of the platform SAR and the ground processing system. Based on the type of loading platform, SAR can be divided into spaceborne SAR (such as on a satellite, a space shuttle, etc.), airborne SAR (such as on an aircraft, an unmanned aerial vehicle [UAV], etc.), and other mobile platforms (such as a missile). The design requirements of a SAR system vary for different platforms. However, the main technology and system composition of the SAR are basically the same. The information system of a SAR is usually composed of three parts, as shown in Figure 2.1a, including the radar system, platform and data link system, and ground data processing and control system.

The basic function of the platform and the data link system is to provide the necessary resources for SAR and to guarantee a normal function of the radar as well as to feed back the processed results of the radar to the ground. The available resources include the space, weight, power consumption, motion parameter measurement, and communication data link required by the radar. Moreover, the platform/data link system provides functions like data processing, recording, and communication, which mainly compresses and formats the SAR raw data or SAR image data after processing. The data can then be recorded or transferred down to the ground receiving station.

The ground data processing and control system mainly performs data decompression, SAR image processing, and data recording, display, and control. The data decompression subsystem decompresses the platform downlink data. If SAR real-time processing is performed on the platform, the ground SAR image data processing only performs the post-SAR image processing, such as radiometric correction, geometric correction, object geometry, physical parameter extraction, etc. Otherwise, the ground SAR image data processing first feeds the raw downlink data for SAR-related processing and then conducts the SAR image postprocessing. The data recording module records the SAR image data, while the display and control module displays the SAR image and also serves as a human interface to control the working state of the SAR system data communications.

The three components of the SAR system scene-matching seeker of the missile loading platform are slightly different. To shorten the time between radar observation and motion control of the missile, generally the radar system performs the signal processing and image processing on the platform. The image information is sent to the missile

*Design Technology of Synthetic Aperture Radar,* First Edition. Jiaguo Lu.

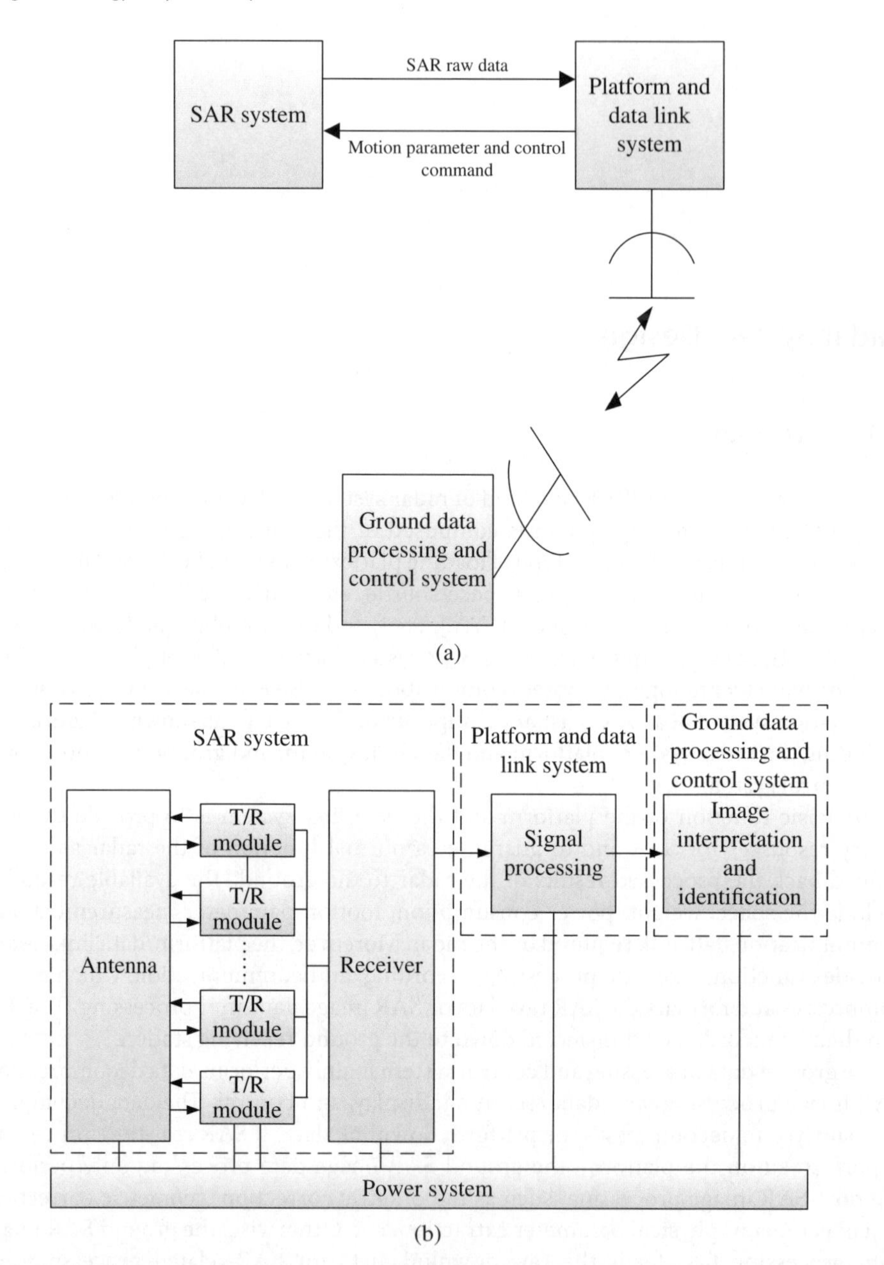


Figure 2.1 Block diagram of SAR system composition. (a) Basic composition of SAR system; (b) basic composition of SAR system (detailed).

control system directly. The telemetry system replaces the data link system, and usually there is no ground data processing system.

A SAR system usually consists of several subsystems, such as the antenna, transmitter, receiver, signal processing, image processing, and image interpretation and intelligence extraction subsystems, as shown in Figure 2.1b.

The antenna system realizes the conversion of the guided wave to a space wave. It radiates energy to the target direction and receives the signals reflected by the target area.

During transmission, the transmitter amplifies the low power signal and outputs fixed power to the antenna system. During receiving, the weak echo is linearly amplified by a low noise amplifier to the required signal level, which is fed to the receiving system.

The major functions of the receiver are to amplify and process the radar echoes, to filter the echoes to remove clutter as much as possible, and to provide a high-performance waveform to the radar as well as to provide one common time reference and phase reference for the radar system.

The signal processing system varies with operation mode. If real-time image processing is performed on the platform, the platform needs to complete not only the raw SAR data acquisition and preprocessing but also the real-time SAR image processing. If image processing is performed on the ground station, the signal processing on the platform is only the raw SAR data acquisition and partial preprocessing.

The subsystem for image interpretation and intelligence extraction transforms the intricate SAR image to available intelligence and quickly interprets, detects, and recognizes the target from the SAR image.

## 2.2 Radar Equations

### 2.2.1 Conventional Radar Equation

For radar, if an isotropic antenna transmits all the power $P_t$ to the ground with a distance of $R$, then the power density is given by

$$\frac{P_t}{4\pi R^2} \tag{2.1}$$

Practically, the radiation of an antenna cannot be isotropic. The level of radiation at its maximum direction is denoted by the antenna gain $G_t$. Then the power density at the maximum gain direction is

$$\frac{P_t G_t}{4\pi R^2} \tag{2.2}$$

For a target that is $R$ away from radar, after illumination by the above power density, the electromagnetic wave will be reflected back, and the obtained backscattering wave density of the target at the radar is

$$\frac{P_t G_t}{4\pi R^2} \cdot \frac{\sigma}{4\pi} \cdot \frac{1}{R^2} \tag{2.3}$$

where $\sigma$ is the radar cross-section area of the target normally called the *backscattering coefficient* and is defined as $4\pi$ times the ratio of the reflected power in unit solid angle in the radar direction to the incident power in unit area. If the target reflects all the incident power, the radar cross section of the target is equal to the area that receives the incident power.

The power of echo intercepted by the receiving antenna, $S$, is equal to the product of the above power density and the effective antenna aperture area $A_r$, that is

$$S = \frac{P_t G_t}{4\pi R^2} \cdot \frac{\sigma}{4\pi} \cdot \frac{1}{R^2} \cdot A_r \tag{2.4}$$

The effective aperture area of the receiving antenna is defined as

$$A_r = \frac{G_r \lambda}{4\pi} \tag{2.5}$$

Substitute Eq. (2.5) into Eq. (2.4), then

$$S = \frac{P_t G_t G_r \lambda^2 \sigma}{(4\pi)^3 R^4} \tag{2.6}$$

Practically, the radar transmitter and receiver share a common antenna and $G_t = G_r = G$. Then the general equation of radar echo power is given by

$$S = \frac{P_t G^2 \lambda^2 \sigma}{(4\pi)^3 R^4} \tag{2.7}$$

Normally the effect of multiple loss factors needs to be considered in the design of a radar system. This effect is usually expressed by system loss factor $L$ ($L > 1$), and Eq. (2.7) is modified to

$$S = \frac{P_t G_t G_r \lambda^2 \sigma}{(4\pi)^3 R^4 L} \tag{2.8}$$

where $G_t$ is the antenna transmitting gain, $G_r$ is the antenna receiving gain, $P_t$ is the average transmitting power of the radar, $\lambda$ is the wavelength of the radar signal, $\sigma$ is the backscattering coefficient of the target, $L$ is overall loss of a radar system, and $R$ is the distance between target and radar.

### 2.2.2 SAR Equation

Compared with conventional radars, SAR needs to consider two issues: one is the expression of $\sigma$, the other is the coherent accumulation effect of the echoes within synthetic aperture time.

According to the radar target scattering theory,

$$\sigma = \sigma^\circ A \tag{2.9}$$

where $\sigma^\circ$ is the normalized backscattering coefficient of ground targets, and $A$ is the effective area of a ground scattering unit, that is

$$A = \rho_a \rho_{rg} \tag{2.10}$$

where $\rho_a$ is the azimuth resolution and $\rho_{rg}$ is the ground range resolution, a projection of the range resolution on the ground.

$$\rho_a = \frac{1}{2}\left(\frac{\lambda}{L_s}\right) R = \frac{1}{2}\left(\frac{\lambda}{T_a V_s}\right) R \tag{2.11}$$

where $L_s$ is length of the synthetic aperture; $L_s = T_a V_s$, $T_a$ is the coherent accumulation time of the echo, $V_s$ is the moving velocity of the SAR, then

$$\sigma = \sigma^\circ \rho_a \rho_{rg} = \sigma^\circ \frac{1}{2}\left(\frac{\lambda}{T_a V_s}\right) R \rho_{rg} \tag{2.12}$$

Equation (2.12) is the signal received from a single echo. For SAR, $n$($n = T_a F_r$) echoes can be coherently accumulated within synthetic aperture time $T_a$, where $F_r$ is the repetition frequency of the transmitted pulse. Therefore, echo intensity is increased by $n$ times.

Considering coherent accumulation, substitute Eq. (2.12) into Eq. (2.8),

$$\begin{aligned} S &= \frac{P_t G_t G_r \lambda^2}{(4\pi)^3 R^4 K_s} \sigma^\circ \frac{1}{2} \left( \frac{\lambda}{T_a v_s} \right) R \rho_{rg} \\ &= \frac{P_t G^2 \lambda^3 \sigma^\circ F_r \rho_{rg}}{2(4\pi)^3 R^3 v_s} \cdot \frac{1}{L} \end{aligned} \tag{2.13}$$

Usually the signal-to-noise ratio (SNR) of the received signal $S/N$ is more essential than the received power level of the radar, where

$$N = kT_s F_n B_n \tag{2.14}$$

where $k$ is the Boltzmann constant, $k = 1.380\ 54 \times 10^{-23}\text{J K}^{-1}, B_n$ is the noise equivalent bandwidth in Hz, $T_s$ is the noise temperature of the receiver, and $F_n$ is the noise factor.

In SAR, the matching acceptance relation of the linear frequency modulation (FM) signal is $\tau = 1/B_n$, which implies that the pulse width $\tau$ after direction compression is approximately equal to the reciprocal of the equivalent noise bandwidth $B_n$. The average power relates to the peak power of the SAR system as follows:

$$P_{av} = \tau F_r P_t = \frac{F_r}{B_n} P_t \tag{2.15}$$

The radar equation can be obtained from Eqs. (2.13)–(2.15).

$$\text{SNR} = \frac{P_{av} G^2 \lambda^3 \sigma^\circ \rho_{rg}}{2(4\pi)^3 R^3 v_s k T_s F_n L} \tag{2.16}$$

where $P_{av}$ is the average transmit power, $G$ is the single path gain of the antenna (both the receiver and the transmitter share the same antenna), $R$ is the slant distance, $\rho_{rg}$ is the ground range resolution, $v_s$ is the speed of the loading platform, $\lambda$ is the wavelength, $k$ is the Boltzmann constant, $T_s$ is the noise temperature, $F_n$ is the noise factor, and $L$ is the system loss.

The noise equivalent scattering coefficient (noise equivalent sigma zero, or $NE\sigma_0$) in a SAR system is commonly used to represent the system sensitivity, which is an important parameter of a radar system. It is defined as average backscattering coefficient at SNR = 0 dB. $NE\sigma_0$ is given by

$$NE\sigma_0 = \frac{2 \times (4\pi)^3 R^3 kTF_n V_{st} L_s}{P_{av} G^2 \lambda^3 \rho_{rg}} \tag{2.17}$$

System sensitivity is one of the key parameters in SAR. The following takes the spaceborne SAR system as an example to discuss the sensitivity of the SAR system.

It can be seen from Eq. (2.17) that the system sensitivity decreases with the increase of the range resolution when the other parameters stay unchanged. To maintain a certain sensitivity, some measures need to be taken, such as increasing the product of the power and the aperture of the system and the aperture size or reducing the height of the platform. Platform height reduction can shorten the slant distance $R$ at a certain viewing angle, which can effectively improve the system sensitivity. More specific analysis is as follows:

(1) The general azimuth resolution is correspondingly improved with the increase of range resolution. These two parameters are correlated. It is well known that the azimuth resolution is theoretically half the aperture size of the antenna in strip-map SAR mode. Therefore, the improvement of the azimuth resolution requires reduction in the azimuth aperture size, causing the antenna aperture area and gain to be reduced. To make the antenna gain not decrease drastically, the aperture size in elevation must be increased. As a result, the beam width in elevation is narrowed and the swath is narrowed.
(2) For a satellite platform, the power supplied by solar panels is limited. The total amount of energy is determined by the size and the conversion efficiency of the solar panel. In the payload of a satellite-borne SAR, due to the limitation of total energy, the system sensitivity is also limited.
(3) The sensitivity can be improved by lowering the orbit height, which shortens the slant distance if the viewing angle is kept constant. However, the swath width is also reduced within the same viewing scope, which will impact the coverage of spaceborne SAR. Therefore, a compromise design is needed.

According to Eq. (2.17), the design of a SAR system requires a comprehensive consideration based on system optimization.

## 2.3 Radar System Parameters

Here, a spaceborne SAR system is employed to illustrate the principal parameters of a SAR system. Mission parameters and system performance parameters are included. The former consists of coverage, revisit period, view scope, and lifetime. The latter comprises frequency, polarization, power, antenna gain, noise factor, loss, and imaging quality-related parameters. The spaceborne SAR system can be designed by properly optimizing the parameters to meet the requirements of the mission.

### 2.3.1 Antenna and Channel Number

In SAR design, the antenna design is related to the signal processing method, which is of the most concern. It is well known that the antenna structure is essential to the mechanical, telecommunications, and thermal performance in a radar system. It also determines the environment adaptability, electromagnetic compatibility, reliability, and cost of the radar system. In this section, the antenna design will be the focus.

The working mode of SAR guides the antenna design. Normally, in strip-map mode, the antenna beam is required to be fixed, without scanning; in scan mode, the antenna beam needs to be steered in the elevation; in spotlight mode, the antenna beam is required to have scanning capability in the azimuth direction; and in terrain observation by progressive scans SAR (TOPSAR) mode, scanning in both elevation and azimuth directions is required for the antenna. The interference mode obtains the elevation map on a single route, requiring the antenna to have dual-channel reception capability; the two antennas corresponding to the two channels have a certain baseline distance in the height direction, and the baseline distance determines the measurement accuracy. In the moving target mode, most cases require the antenna to have two or more channels

**Table 2.1** Relationship between SAR operation mode and antenna form.

| | Strip-map SAR | Scanning SAR | Spotlight SAR | TOPSAR | Moving target | Interference SAR | Remarks |
|---|---|---|---|---|---|---|---|
| Beam requirements | Fixed | Range direction scanning | Azimuth scanning | Range and azimuth scanning | Multi channels in azimuth | Two channels in altitude | |
| Reflector antenna | √[a] | Mechanical scanning | Mechanical scanning | Mechanical scanning | Two antennas | Two antennas | |
| Reflector antenna (switch multifeeding) | √ | √ | Mechanical scanning | Mechanical scanning + switching | Two antennas | Two antennas | Large beam jump |
| Reflector antenna (phase-controlled feeding) | √ | √ | Mechanical scanning | Mechanical scanning + electric scanning | Two antennas | Two antennas | Small scanning range |
| Concentrated feed array antenna | √ | Mechanical scanning | Mechanical scanning | Mechanical scanning | Multiple channels | Two antennas | |
| Phased array antenna | √ | √ | √ | √ | Multiple channels | Two antennas | |

[a] √ denotes the capability of antenna in different operation modes.

in the azimuth direction. The number of channels and the channel spacing are related to the performance of the moving target.

An antenna beam can be scanned mechanically and electronically. SAR antennas can be divided into two types: reflector antennas and planar array antennas. Table 2.1 lists the relationships between radar operation modes and antenna types. As shown, for the basic mode, i.e. strip-map mode, due to the invariant beam pointing, all types of antennas meet the requirement, whereas additional requirements are imposed on antenna scanning or channel number for the other modes.

Reflector antennas have been developed from the initial single-fed single-beam reflector to multifed multibeam, switch-fed, phase-controlled reflector, etc. To get enough power aperture products, the transmitter is usually required to have high-output power for the reflector.

Planar array antennas can be divided into two categories: concentrated feed planar array and planar phased array. The former is usually applied in the fields of low-frequency and low-efficiency antennas due to the loss of the feed line. The latter has been widely used on different SAR loading platforms due to the advantages of flexible beam control and high reliability.

For airborne SAR, compared with satellite-borne, the antenna selection is wider due to slightly fewer limitations of the platform on the load size and weight.

In small platforms, such as UAVs and missiles, concentrated feed reflector antennas or planar arrays are widely used, due to their relatively simple task. The concentrated feed reflector antenna completes the elevation and azimuth scanning, achieves the function of strip map, scanning, and spotlight of the SAR, and realizes ground moving target indication (GMTI), through a three-dimensional rotating platform.

In spaceborne SAR, most antennas are required to scan in two dimensions. From the perspective of cost, for small-scale spaceborne SAR, a reflector antenna is the first choice, because spotlight and scanning mode can be achieved by the overall rotation of the satellite. However, with the development of phased array technology, the phased array antenna has become the dominant choice for spaceborne SAR antennas. The main radiation units are a planar waveguide slot antenna and a microstrip antenna.

#### 2.3.1.1 Reflector Antenna

A typical block diagram of a reflector antenna is shown in Figure 2.2a. The antenna system consists of reflector, feed source, circulator, transmitting/receiving protection switch, and transmission line of the feed. During transmitting, the radio-frequency (RF) signal of the transmitter goes through a circulator and enters the primary feed source of the reflector antenna via a transmission line, then a highly directive beam is formed at the radiation reflector and is radiated into space. During receiving, echoes are collected by a high-gain reflector antenna and are fed into the primary feed source. The free-space electromagnetic wave is converted into a guided wave signal, which finally enters the receiver after going through the circulator and protection switch.

The antenna pattern is shaped by the design of the reflecting surface and the primary feeding source. The antenna in azimuth and elevation directions can be scanned by mechanical rotation or platform rotation. In SAR applications, this type of antenna

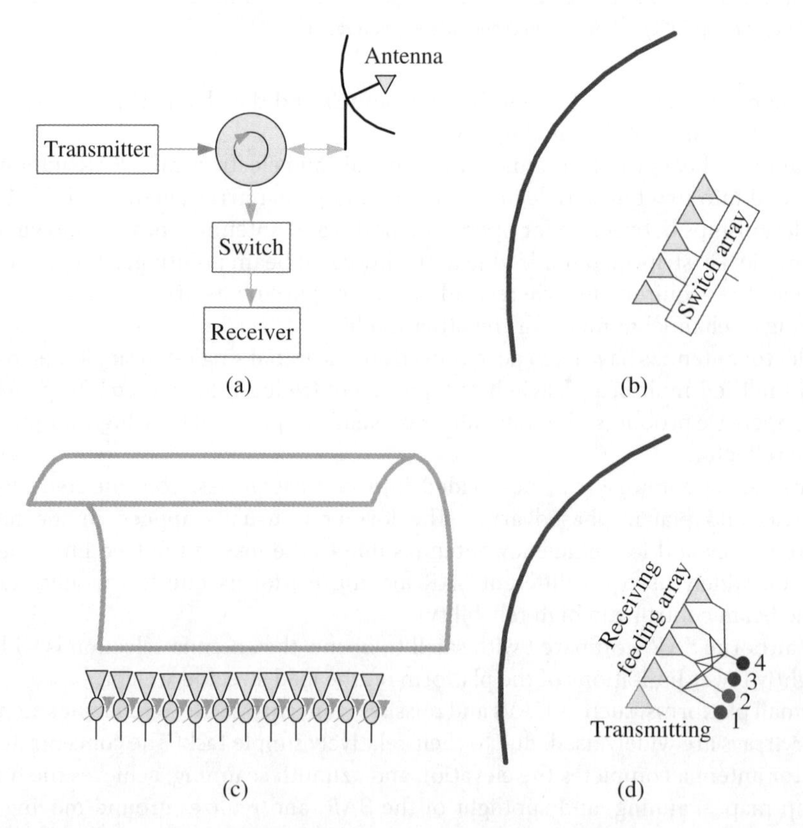


**Figure 2.2** Reflector antennas. (a) Basic composition of reflector antenna; (b) switch-fed reflector; (c) 1D phase scanning reflector; (d) multibeam wide-swath reflector.

is suitable for small airborne and missile platforms as well as spaceborne platforms. It offers the advantages of small size, light weight, and high efficiency and can also meet the beam-scanning requirements via mechanical scanning. In addition, dual polarization and single/dual circular polarization can be easily realized by simply changing the feed source; in the meantime, double/multiband shared aperture can be designed in the feed source.

However, the disadvantages of reflector antenna are obvious. In case of concentrated feed, there is no redundancy in the system design, and a single failure of the unit in the RF link will lead to the failure of the whole system. The power of concentrated feed is high, which can introduce such risks as microwave spark breakdown, spaceborne environment micro-discharge, and high-voltage discharge.

A combination of mechanical and electric scanning can be used for reflector antennas. One-dimensional (1D) scanning is realized by mechanical rotation of the platform. The feed source switch array is used to get different beam coverage, as shown in Figure 2.2b. Because of the limited number and space of the feed sources, this type of antenna array has the defects of a large beam-scanning step angle and small scanning angle. Compared with two-dimensional (2D) mechanical scanning reflector antennas, the key for this type of antenna is the switch. On one hand, the insertion loss and switching time are minimized. On the other hand, the reliability of the high-power switch needs to be considered.

A parabolic shape can be used for 1D electric scanning. The 1D electric scanning using phased array feed is shown in Figure 2.2c. The antenna points to the elevation direction, while the vertical radiation pattern is obtained by antenna beam forming, and the parabolic cylinder only reflects horizontally. Multiple high-power transmit/receive (T/R) modules are generally used in this type of antenna to realize the required power aperture product, which avoids the reliability risk of a single high-power transmitter. In the horizontal direction, the beam scanning and beam forming are obtained by controlling the amplitude and phase of the T/R modules.

One-dimensional mechanical scanning with one-dimensional wide swath can be realized by a reflector antenna, as shown in Figure 2.2d. For this type of antenna, multiple feed sources and transmitting shaped wide beams are commonly used. The receiver uses multiple beams simultaneously. Different receiving feed source combinations are used to achieve different beam swath width. Multibeam acceptance can be achieved through multibeam-forming network, or can be fulfilled in digital domain.

#### 2.3.1.2 Planar Array Antenna

A planar array antenna comprises a concentrated feed planar array antenna and a phased array planar antenna. The composition of a concentrated feed 2D mechanical scanning planar antenna is similar to that of a reflector antenna, but they differ in the antenna radiation plane, which is changed from a reflector to a planar antenna. A planar array antenna can be a waveguide slot antenna array, a microstrip patch antenna array, or a printed dipole antenna array. The power synthesis/distribution of the radiation unit is done by the feeding network. The required radiation pattern is obtained by different weighting.

Phased array planar array antennas can be divided into 1D phased arrays and 2D phased arrays. A 1D phased array antenna scans a beam in one dimension and fixes the beam in other directions. The fixed beam can be scanned mechanically.

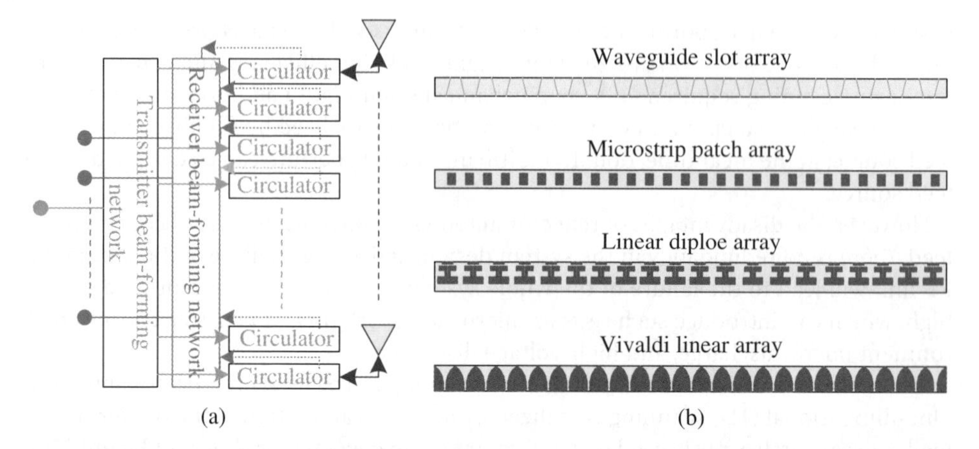


**Figure 2.3** Elevation wide-coverage planar antenna array. (a) Basic composition; (b) optional 1D linear array.

A 1D mechanical scanning combined with 1D electric scanning planar antenna array usually constitute a 1D linear array in horizontal via the feeding network. A passive phased array is constructed in the elevation direction by high-power and low-loss phase shifters. A high-power transmitter with concentrated feed or each linear array connecting to a high-power T/R module is used to construct a 1D active phased array. In this way, the phased scanning in the elevation direction is realized by controlling the phase shifters, and in the azimuth direction, the beam scanning is realized by a mechanical rotation platform.

A power distribution/synthetic network is used in a 1D mechanical scanning plus 1D wide-swath planar antenna array to achieve transmitting wide beam and receiving multiple beams as a 1D electronic scanning planar array. The other dimensions use a power divider/combiner for beam coverage, as shown in Figure 2.3a. During transmitting, the elevation beam forming is realized by a beam-forming network. During receiving, stacked beams are obtained by a multibeam-forming network, and the coverage area corresponds to the transmitted beam. The multibeam can be formed both in RF and in the digital domain. In the same way, if beam scanning is needed in another dimension, it is generally realized by mechanical means. Figure 2.3b gives the options of the 1D fixed beam antenna array form, including waveguide slot array, microstrip patch array, linear dipole array, and Vivaldi linear array. The antenna efficiency is dependent upon the synthetic feeding line loss.

A 2D phased array planar antenna is achieved by controlling a phase shifter to obtain the phase shift of the beam in 2D space and to realize 2D beam scanning. A 2D phased array antenna has a flexible beam and high reliability, so it tends to be more widely used in different platforms.

In spite of flexibility in beam scanning, a phased array antenna is limited by fundamental physical principles, and the beam can only be scanned within 60°.

Traditional spaceborne SAR usually adopts single channel. In a single-channel system, the azimuth resolution is inversely proportional to the range direction swath width. The number of independent samples collected by the system is usually increased to break this limitation. If multichannel is used, the dimension of the signal collection system

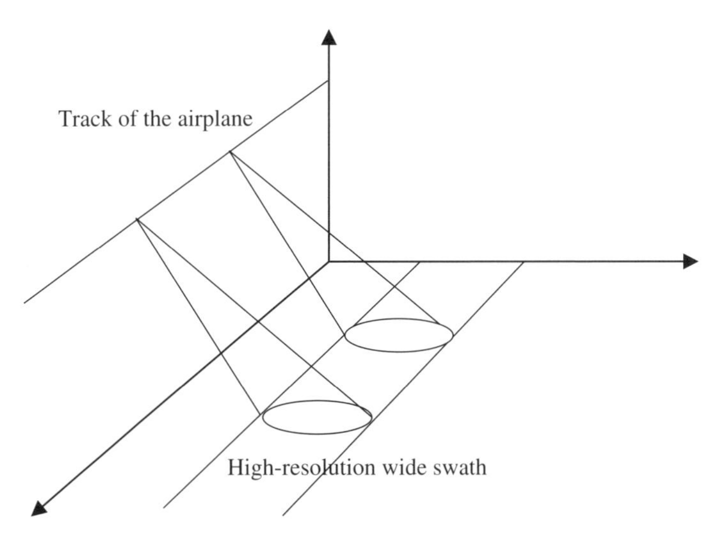


**Figure 2.4** Imaging geometry of strip-map SAR.

will increase. Furthermore, the swath width can be extended, or the azimuth resolution can be improved. To some extent, this method alleviates the contradiction between high resolution and wide swath.

#### 2.3.1.3 Single-Input, Single-Output

The strip-map mode is a typical model of a conventional single-input, single-output (SISO) system, which directly uses the fixed pointing beam to illuminate and realizes continuous imaging of the illuminated area, as shown in Figure 2.4. To meet the requirements of high-imaging resolution, spotlight mode is usually adopted, where the center of the azimuth beam always points to the center of the imaging area. A larger Doppler bandwidth can be obtained, and a high-azimuth-resolution imaging can be achieved, as shown in Figure 2.5. Another typical imaging mode is ScanSAR. The wide swath is obtained by range scanning. However, the decreased illumination time of a specific target will lead to resolution reduction, as shown in Figure 2.6. To obtain high resolution while achieving wide-swath imaging, such modes as sliding spotlight, mosaic, and TOPS are proposed for a conventional SAR system. However, due to the inherent space and energy distribution limitations of a SISO system, the conflict between high resolution and wide swath cannot be tackled.

#### 2.3.1.4 Single-Input, Multiple-Output

Single-input, multiple-output (SIMO) means that the radar forms multiple receiving beams, covering adjacent areas. A SIMO system includes single-phase center multi-beams (SPCMB) and displaced-phase center multiple beams (DPCMB). SPCMB can be divided into single phase center multiple elevation beams (SPCMEB), single-phase center multiple azimuth beams, and some extend modes [1], according to the alignment direction of the receiving sub-beams.

Multiple elevation beam mode is shown in Figure 2.7. A wide-beam transmitting pulse in elevation is used to realize a wide swath in azimuth. High pulse repetition frequency (PRF) is used to ensure azimuth resolution and ambiguity-to-signal ratio, which will lead

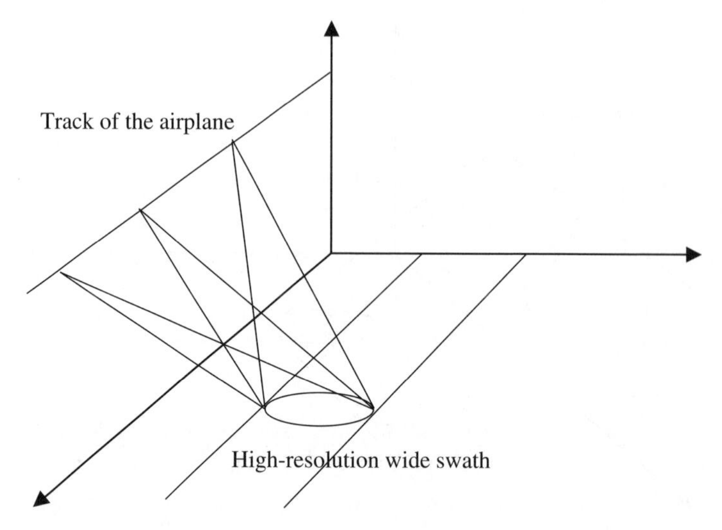


**Figure 2.5** Imaging geometry of spotlight SAR.

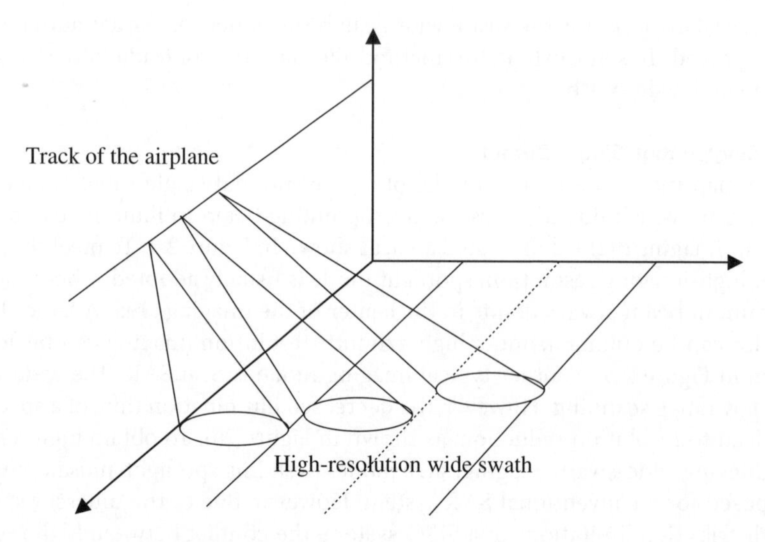


**Figure 2.6** Imaging geometry of ScanSAR.

to blind regions in range direction [2]. These blind regions separate a complete width swath into several subswaths. When receiving echoes, SPCMEB receives the echoes of every subswath by elevation sub-beams formed in a network. Due to the very small interval between the subswaths and the receiving sub-beams, defects of range ambiguity for SPC-MEB mode occur.

Azimuth multibeam mode (shown in Figure 2.8) is a type of wide transmitting and narrow receiving mode that uses a small aperture transmitting pulse to realize a wide azimuth swath. Multiple adjacent narrow beams with the same phase center are used to receive echoes. In this mode, each sub-beam receives only partial echoes, therefore, it is acceptable if the PRF is larger than the Doppler bandwidth of the receiving sub-beam.

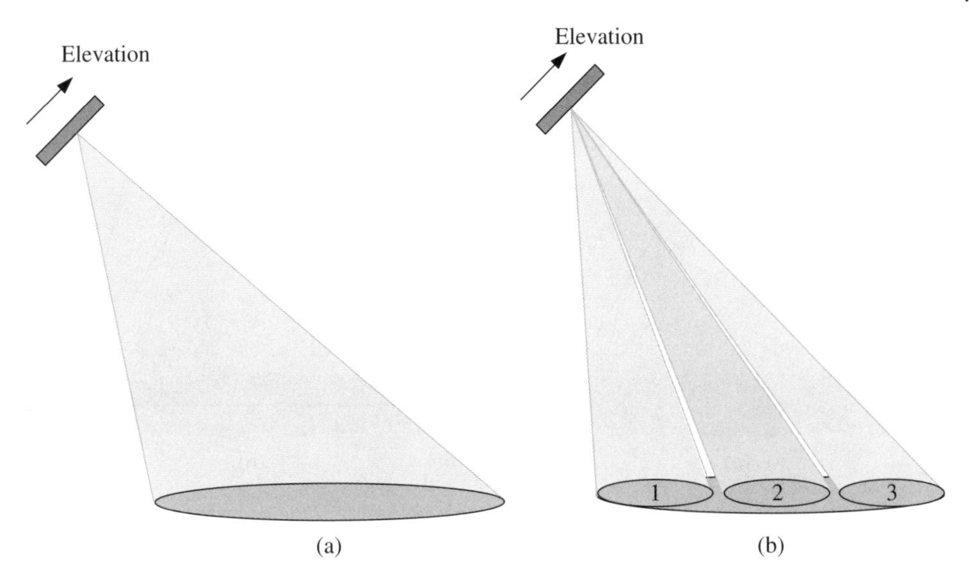


**Figure 2.7** Schematic of elevation multibeam with single-phase center. (a) Wide-beam transmitting; (b) narrow-beam receiving.

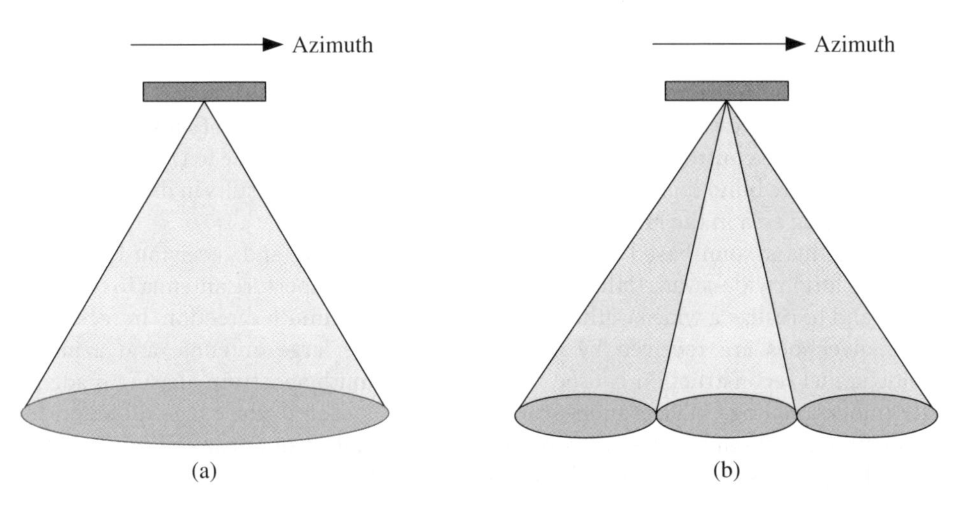


**Figure 2.8** Schematic of azimuth multibeam with single-phase center. (a) Wide-beam transmitting; (b) narrow-beam receiving.

Decreasing PRF can improve the range direction swath width. The disadvantage is that crosstalk exists between beams, which will increase the azimuth ambiguity of the system.

Small antenna aperture transmitting pulse is also used to achieve wide swath under DPCMB mode. In receiving, it uses multiple azimuth wide beams with displaced-phase centers to receive echoes. Many azimuth sampling points can be obtained within one pulse repetition interval by this means. Therefore, a wide swath can be realized

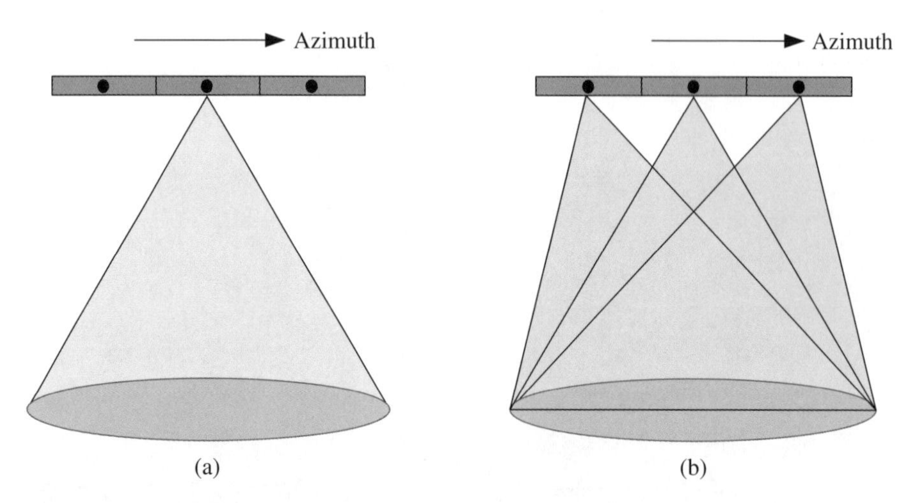


**Figure 2.9** Schematic of multibeam with displaced-phase center. (a) Wide-beam transmitting; (b) narrow-beam receiving.

by reducing the system PRF. Figure 2.9 shows a schematic of multibeam with displaced-phase center.

In 2D (range and azimuth) SIMO mode, the radar antenna is generally composed of a number of subarrays, among which one is transmitter/receiver subarray. The rest of the arrays are receivers. Multiple azimuth channels are used to reduce PRF, thus improving the swath width. In the meantime, they can be used to make the elevation transmitting beam cover multiple subswaths under the same PRF. And the echoes of subswaths can be acquired through control of null-steering in elevation. However, due to the gap between beams, there are blind regions between two subswaths, which results in discontinuities of the imaging area in the range direction.

The antenna system based on separation of transmitting and receiving is used in high-resolution wide-swath (HRWS) mode. It uses a small aperture antenna to transmit pulses and to realize a wide swath in the range and the azimuth direction. In receiving, terminal echoes are received by a 2D multiaperture large antenna, and azimuth multichannel reconstruction is used to eliminate azimuth spectrum aliasing made by PRF undersampling. Multichannel scan-on-receive is used in the range direction [3] to generate a high-gain pen-shaped beam-scanning swath from near end to far end to offset the gain loss due to the small transmitting aperture size.

#### 2.3.1.5 Multiple-Input, Multiple-Output

Multiple-input, multiple-output (MIMO) has the structure of multiple transmitting antennas and multiple receiving antennas. Spaceborne MIMO-SAR is generally divided into two categories: distributed spaceborne MIMO-SAR and compact spaceborne MIMO-SAR. As the name suggests, each satellite transmits orthogonal coding waveform in distributed spaceborne MIMO-SAR, and the relationship with the multisatellite can be used to improve the radar performance. The overwhelming advantage of this system is the reduction of payload size, weight, and power consumption requirements in a single spaceborne SAR. The larger baseline distance improves interference measurement precision and the moving target detection accuracy and expands the degree of the system freedom, which helps to realize HRWS. But the disadvantage is that strict time, space, and frequency synchronization are required between satellites, which makes the design more complicated. Compact spaceborne MIMO-SAR uses a number of subaperture antennas to transmit different orthogonal coding signals (such as space-time block coding signal, chaotic signal, and multicarrier frequency signal). The relationship of the apertures is used to increase the degree of the system freedom and to improve the radar performance.

### 2.3.2 Antenna Size

Antenna size is one of the most important parameters in the design of spaceborne SAR systems. It is directly related to the design of the power aperture product of the payload system. Antenna size defines the sensitivity and image quality of a SAR system. It also defines the size, weight, and power consumption requirements of the payload system on the satellite platform. Many constraining factors need to be considered in the design, especially ambiguity, resolution, swath width, and system sensitivity ($NE\sigma_0$).

#### 2.3.2.1 Ambiguity Limit

The echoes of the target must be received and collected within one pulse so as to avoid transmitting interference and to satisfy the ambiguity requirement in the range direction. Therefore, the slant-range-projected swath width of SAR imaging $W_s$ must satisfy

$$W_s < \frac{c}{2f_p} < \frac{cL_a}{4V_{st}} \tag{2.18}$$

where $c$ is the electromagnetic wave propagation velocity, $f_p$ is the PRF, and $L_a$ is the antenna azimuth dimension.

$$W_s = \theta_r R \tan\theta = \frac{\lambda R}{L_r} tan\theta \tag{2.19}$$

where $\theta_r$ and $L_r$ are the range direction antenna beam width and size, respectively, $\theta$ is the incident angle of the antenna beam, and $R$ is the distance between the radar and the target. By combining Eq. (2.18) with (2.19), the antenna aperture area should meet

$$A = \frac{4V_{st}\lambda R}{C} \tan\theta \tag{2.20}$$

Equation (2.20) shows that the antenna aperture area has a lower limit to meet the ambiguity requirements in the range and the azimuth direction. As the incident angle of the antenna beam and the slant distance of the target are expanding, the minimum area needs to be increased. Taking into account the affecting factors of transmitting, the wavelength of the radar signal, nadir point interference, pulse width, antenna alignment errors (including satellite attitude and deformation error), and terrain elevation difference, the minimum antenna area that meets the ambiguity limit is shown in Eq. (2.21), where $k = 4–8$ is dependent on the viewing error.

$$A_{min} = \frac{kV_{st}\lambda R}{C} \tan\theta \tag{2.21}$$

In practice, the minimum antenna area serves as a guideline, which does not need to be strictly followed.

#### 2.3.2.2 Swath Width and Resolution Limit

SAR antenna azimuth size $L_a$ depends on the system requirements of azimuth resolution. While the range size of an antenna is designed, the antenna dimension in range direction $L_r$ needs to consider the minimum antenna area, antenna gain, swath width, and the output SNR as a whole. Sometimes, the antenna beam width in the range direction is extended to obtain a wide swath. Although extending the beam width in the range direction decreases the antenna gain to some extent, it has little impact on the overall system performance.

In strip-map mode, the antenna azimuth size $L_a$ should conform to the following relationship:

$$L_a \leq \frac{2\delta_x}{k_r} \tag{2.22}$$

where $\delta_x$ is the azimuth resolution and $k_r$ is the extending factor. The range direction antenna size $L_r$ is dependent on the swath width in the range direction, and the relationship is

$$L_r \leq \frac{0.886\lambda R}{W_g \cos\theta} \tag{2.23}$$

where $W_g$ is the swath width in the range direction, $\theta$ is the incident angle of the antenna beam, and $R$ is the distance between the radar and the target.

#### 2.3.2.3 $NE\sigma_0$ Restriction

As shown in Eq. (2.17), when the resolution is determined, to improve system sensitivity and reduce $NE\sigma_0$, average transmitting power, the antenna aperture area, and the antenna efficiency need to be boosted. Due to the fact that $NE\sigma_0$ is inversely proportional to the square of the antenna aperture area and radiation efficiency, increasing the effective aperture area or the antenna efficiency is the most obvious way to improve the sensitivity of the system.

### 2.3.3 Resolution and Swath Width

#### 2.3.3.1 Resolution

The design of instantaneous signal bandwidth depends on the range resolution, which is the ground distance resolution of the image. They have the following relationship:

$$\rho_{gr} = \frac{k_r k_1 c}{2B\sin\theta} = \frac{\rho_r}{\sin\theta} \tag{2.24}$$

where $B$ is the instantaneous signal bandwidth, $k_r$ is the imaging weighted extending coefficient in the range direction, $k_1$ is the extending factor induced by the system magnitude-phase characteristic, $\theta$ is the incident angle, and $\rho_r$ is the range resolution.

The values of $k_r$ and $k_1$ are very important in the design. Generally, their empirical value can be $k_r k_1 = 1.1$.

For azimuth resolution, according to the basic principles of SAR

$$\rho_a = \frac{\lambda}{2\theta_B} \tag{2.25}$$

where $\lambda$ is the wavelength and $\theta_B$ is the cone angle of the antenna beam that the ground target is covered during aperture synthesis. Obviously, azimuth resolution of SAR is independent from the SAR beam angle. However, the synthetic aperture time at different beam positions varies with the beam angle. Within one strip, usually the same period is used during aperture synthesis. Therefore, the azimuth resolution of the image in the near range is better than that of the image in the far range.

For range resolution, the range resolution of spaceborne SAR $\rho_r$ is determined by instantaneous bandwidth of the radar system.

$$\rho_r = \frac{c}{2B} \tag{2.26}$$

Images acquired by spaceborne SAR systems with the same bandwidth have different ground distance resolutions. The smaller the viewing angle, the lower the ground distance resolution. That is, in the same strip, the ground resolution in the short range is low, and the ground resolution in the long range is high.

#### 2.3.3.2 Swath Width

The ratio of swath width to azimuth resolution is one of the important parameters taken into account in the design of spaceborne SAR system. It shows the relationship between geometric relations and system parameters of spaceborne SAR. In contrast, airborne SAR is rarely restricted by this condition. Typically, the swath width/azimuth resolution is analyzed from the perspective of the PRF. In practice, this ratio is around 1 : 10,000. Namely, under 1 m resolution, the swath width is around 10 km. However, in the case of high resolution, the ratio of swath width to azimuth resolution also needs to be considered from the perspective of sensitivity design. It is not enough to consider only the PRF.

To meet the sensitivity conditions, strict requirements are imposed on the power aperture product of the system. The azimuth aperture of the antenna should be shortened to improve the azimuth resolution. Moreover, it is required to increase the range direction aperture size of the antenna to ensure the system power aperture product. Then the beam width in the range direction, the imaging strip width, and the ratio of swath width to azimuth resolution will all be decreased accordingly. Therefore, given a certain resolution, there is an upper threshold of the ratio of swath width to azimuth resolution to ensure the sensitivity of the system.

It is necessary to optimize parameters under these constraints, so as to achieve high resolution and wide swath. Multiphase center and multiazimuth beam are relatively preferred. For example, there are three phase centers in the azimuth direction by using the single-input, double-output mode, as shown in Figure 2.10. The transmitting phase center is located in the middle, which generates a single transmitting beam. At the edge are two receiving phase centers, which generate the two receiving beams. Figure 2.10 shows the principle of how to reduce PRF in single-input, double-output mode. The actual PRF can be reduced to a half.

Another advantage of SIMO is that the PRF is reduced when the duty cycle is fixed, which implies that the pulse width can be increased. Therefore, the transmitting power of each pulse is also increased, which brings the benefits of improved SNR, which is 3 dB for a single-input, double-output system.

### 2.3.4 Pulse Repetition Frequency

The Doppler bandwidth in the coverage area of the actual aperture beam is $\pm v/L_a$ (as shown in Figure 2.11). From the sampling theory, PRF must be higher than the Doppler

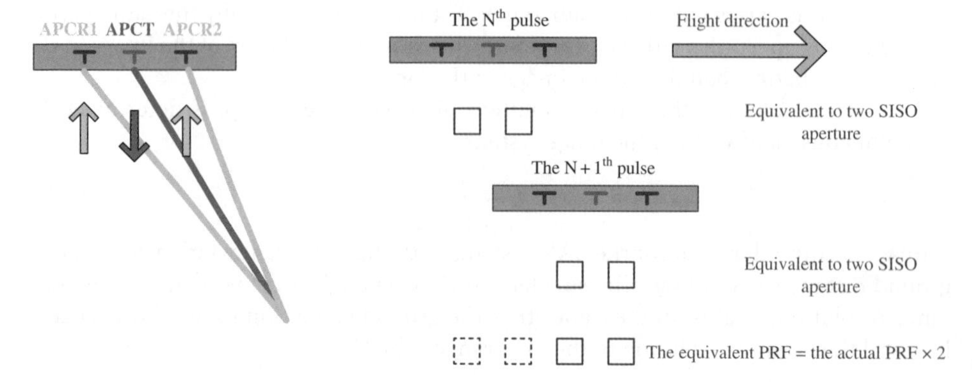


**Figure 2.10** The principle of single-input, double-output mode. APCR: actual phase center of receiving; APCT: actual phase center of transmitting.

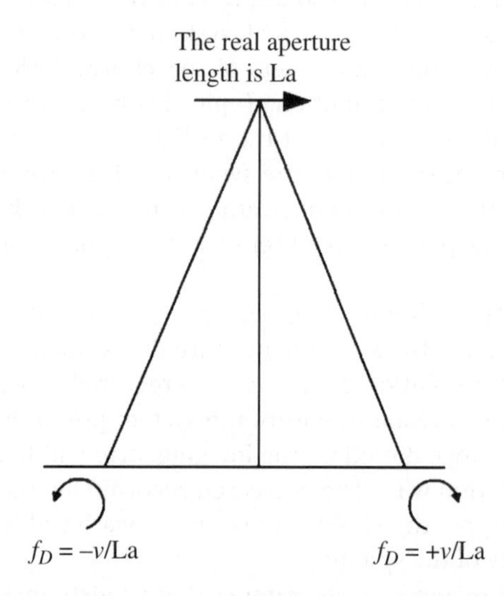


**Figure 2.11** The Doppler bandwidth in the coverage area of actual aperture beam.

bandwidth

$$\mathrm{PRF} \geq 2v/L_a \tag{2.27}$$

where $v$ is the moving speed of the platform, and $L_a$ is the antenna azimuth size.

There is another restriction on PRF (shown in Figure 2.12) to avoid range ambiguity. For side-view radar, it is

$$\mathrm{PRF} < \frac{c}{2W_{max}} \tag{2.28}$$

It is slightly more complicated for a spaceborne system. As radar usually receives more than one pulse echo transmitted from the same pulse, all echoes of one swath width are allowed if they arrive within one repetition period.

$$\mathrm{PRF} < \frac{c}{2w_g \sin\theta} \tag{2.29}$$

where $\theta$ is the antenna incident angle, and $w_g$ is swath width.

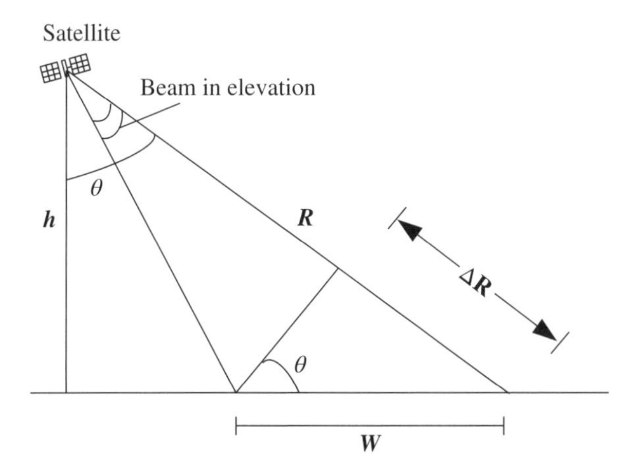


**Figure 2.12** The restriction of swath with width due to range ambiguity in spaceborne SAR.

Thus, once the azimuth resolution is defined, the value of PRF can be obtained. In the meantime, the maximum swath width is also determined. Here, it is assumed that the antenna beam is ideal in both the azimuth and range directions, without any side lobe. In fact it is impossible, which will lead to an ambiguity problem in the azimuth and range directions. It is usually required to select a suitable PRF for spaceborne SARs so that the receiving and transmitting system is always kept at the lowest ambiguity.

In an array antenna, $v \cdot PRT$ can be regarded as the interval of the antenna element. Similar results [Eq. (2.27)] can be obtained to suppress grating lobes.

For spaceborne SAR, the combination of Eqs. (2.27) and (2.29) will show the limitation of the minimum antenna area. Thus, the product of antenna size in range and azimuth, $L_r$ and $L_a$, is given by

$$L_r L_a > \frac{4vr\tan\theta}{c} \tag{2.30}$$

The above requirements may not need to be considered all the time for airborne SAR, but for spaceborne SAR, the swath width is limited.

PRF is one of the important parameters of SAR. It affects the choice of the other radar parameters as well as the radar performance. It is closely related to other parameters. Among them, parameters that affect the selection of PRF are the moving velocity, length of the antenna, swath width, incident angle, pulse width, height of the platform, and so on. The PRF value also affects such parameters as peak transmitting power and data rate, and it affects the radar ambiguity in the range and azimuth directions. It is critical to choose a suitable PRF in radar design. The factors in the following sections need to be considered in PRF design.

#### 2.3.4.1 Doppler Bandwidth

For a SAR system, PRF is equivalent to the sampling of the Doppler echo of the target in the azimuth direction, as shown in Figure 2.13, where the radar explores target $T$ in a speed of $v$ along a straight line.

Assuming that the target enters the radar main lobe, when the radar reaches point $B$, and the target leaves the main lobe when the radar reaches point $E$, then the difference of the Doppler frequency shift of the target $T$ point obtained by the radar between points

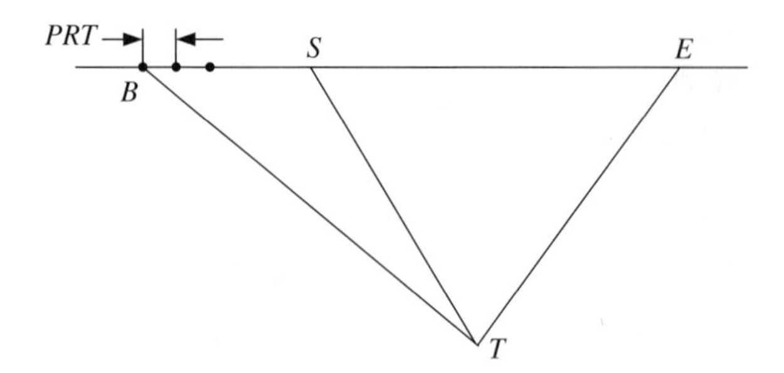


**Figure 2.13** PRF and azimuth sampling.

$B$ and $E$ is the Doppler bandwidth of point $T$ in the radar echo. To meet the Nyquist sampling theorem,

$$\mathrm{PRF} \geq B_a = \frac{v}{\rho_a} \tag{2.31}$$

#### 2.3.4.2 Data Reception Interspersed with Transmitting Event

The transmitted and received pulses of spaceborne SAR are frequently separated by several pulse repetition periods, and the echoes returned to the antenna are likely masked by the subsequent transmitted pulses, making the echoes lost by the receiver. Therefore, the echoes within the swath must completely fall into the receiving window, as shown in Figure 2.14.

Suppose that the echo of the transmitted radar pulse can only be received after the $i$th pulse repetition cycle, as in Figure 2.14. The condition that the echo receiving window is not covered up by the transmitted pulse is

$$\begin{aligned} i \cdot PRT + \tau_p + \tau_{pt} &\leq T_{near} = \frac{2R_n}{c} \\ (i+1) \cdot PRT + \tau_p + \tau_{pt} &\leq T_{far} = \frac{2R_f}{c} \end{aligned} \tag{2.32}$$

where $\tau_p$ is the pulse width, $PRT$ is the pulse repetition time, $\tau_{pt}$ is the pulse protection time, and $T_{near}$ and $T_{far}$ are the echo delay times in near-end $R_n$ and far-end $R_f$, respectively. From the above equations, the range of $PRF$ is chosen by

$$\frac{i}{2R_n/c - \tau_p - \tau_{pt}} \leq PRF \leq \frac{i+1}{2R_f/c + \tau_p + \tau_{pt}} \tag{2.33}$$

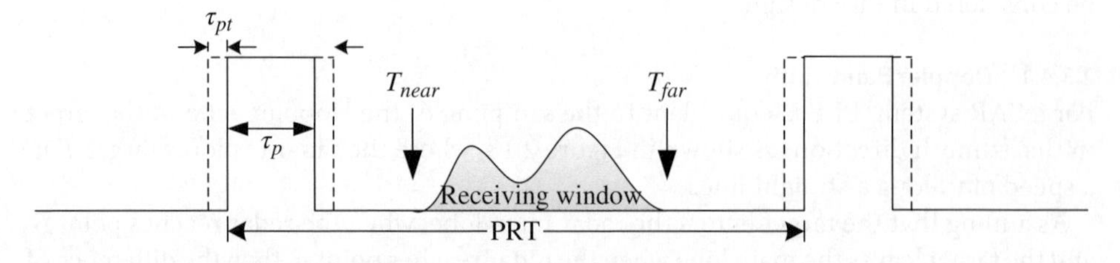


**Figure 2.14** Data reception interspersed with transmitting event.

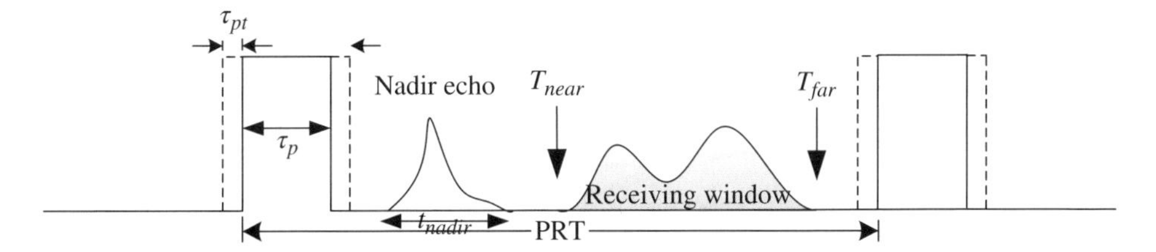


**Figure 2.15** Schematic of avoiding nadir echo.

#### 2.3.4.3 Avoiding Nadir Echo

The spatial path of the nadir echo is the shortest while the backscattering coefficient is large when the electromagnetic wave is incoming vertically and the radar echo is intensified. Therefore, even illuminated by side lobe, the intensity of the echo is relatively strong, which can cause serious ambiguity of the useful signal in the swath. It is required to avoid interference of the nadir echo by PRF selection, as shown in Figure 2.15.

Suppose $t_{nadir}$ is the delay time of the nadir echo, and the gap between the near-end echo delay time $T_{near}$ and the far-end echo delay $T_{far}$ is larger than $j$ times the pulse repetition time. Then the condition of avoiding the nadir echo is

$$\begin{aligned} t_{nadir} + j \cdot PRT + \tau_p + \tau_{pt} \leq T_{near} = \frac{2R_n}{c} \\ t_{nadir} + (j+1) \cdot PRT - \tau_p - \tau_{pt} \geq T_{far} = \frac{2R_f}{c} \end{aligned} \tag{2.34}$$

where $t_{nadir} = \frac{2H}{c}$, and $H$ is the height relative to the ground. From Eq. (2.34), the range of PRF is given by

$$\frac{j}{2R_n/c - \tau_p - \tau_{pt} - 2H_r/c} \leq PRF \leq \frac{j+1}{2R_f/c + \tau_p + \tau_{pt} - 2H_r/c} \tag{2.35}$$

From the above three requirements, the near end and far end of the swath can be obtained after the radar viewing angle is defined, and the PRF range at different viewing angles, i.e. the *zebra map*, can be drawn. Figure 2.16 shows an example of the zebra map of a spaceborne SAR. The dark strips represent the interference of the signal transmitted, and the strip width represents the time of the pulse width plus the protection time $\tau_{RP}$, which happens before and after the pulse. The light strips represent the interference of the nadir signal, and the strip width represents the width of the nadir echo (normally assumed to be twice the pulse width). The blank areas represent the selectable range, and the vertical lines are the selected PRF, whose length represents swath width (Figure 2.16 shows the scope of the viewing angle). As the viewing angle increases, the length becomes shorter. This is because variation scope of the viewing angle for a fixed swath width is smaller as the viewing angle becomes larger.

### 2.3.5 Ambiguity

The ambiguity feature of SAR means that the nonartificial interference echoes from outside the swath and the useful echoes inside the swath come into the radar receiver simultaneously. After imaging signal processing, on one hand, SNR of the radar image

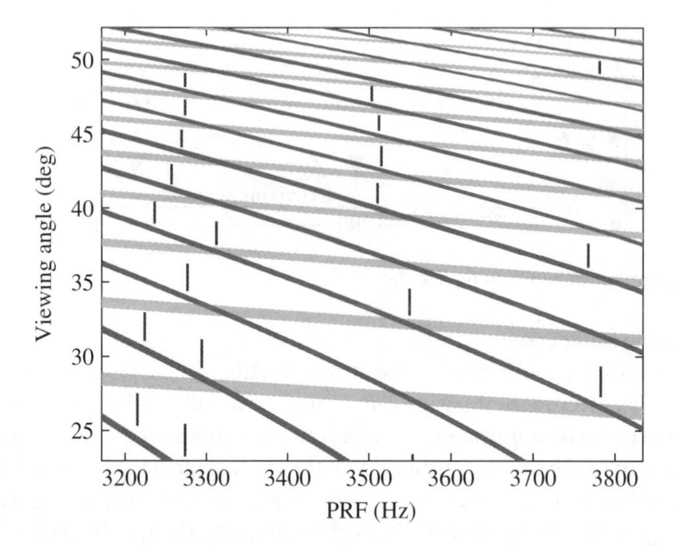


**Figure 2.16** Zebra map of a spaceborne SAR.

decreases; on the other hand, it may cause a false image (ghost) in the low-reflection region of the radar image. The ambiguity characteristic of SAR is not only related to swath width and the azimuth resolution but also to the antenna area, the shape of the pattern, and the selection of the PRF. For an airborne SAR system, the echoes are generally received within one pulse repetition time due to the close operation distance. Moreover, the PRF is generally several times larger than the azimuth Doppler bandwidth. Therefore, the ambiguity issue of airborne SAR is not prominent. However, the ambiguity issue is very prominent in spaceborne SAR, due to the long operation distance and the fast moving speed. The ambiguity analysis plays an important role in the design of spaceborne SAR. It is the key factor that decides the success or failure of the spaceborne SAR design.

The ambiguity is measured by the ambiguity-to-signal ratio of SAR, that is, the ratio of the ambiguity power to the power of the useful signals. Usually, the coupling signal between the range direction and the azimuth ambiguity can be neglected in analyzing system ambiguity or computing ambiguity-to-signal ratio.

#### 2.3.5.1 Range Ambiguity

*Range ambiguity* refers to echoes from outside the swath, and it enters the radar receiver together with the useful echoes inside the swath, leading the radar image quality to decrease after signal processing. In spaceborne SAR, the echoes of one transmitted pulse can only be received after one or several pulses, which causes a serious ambiguity problem. Figure 2.17 shows the range ambiguity, where $T_P$ is the sampling period, $R_1$ is the distance between radar and target, $T_w$ is the target recording window, and $T_w{}'$ is the width of the main lobe.

The range ambiguity-to-signal ratio (RASR) is calculated by

$$\mathrm{RASR} = \frac{\sum_{i=1}^{N} S_{ai}}{\sum_{i=1}^{N} S_i} \tag{2.36}$$

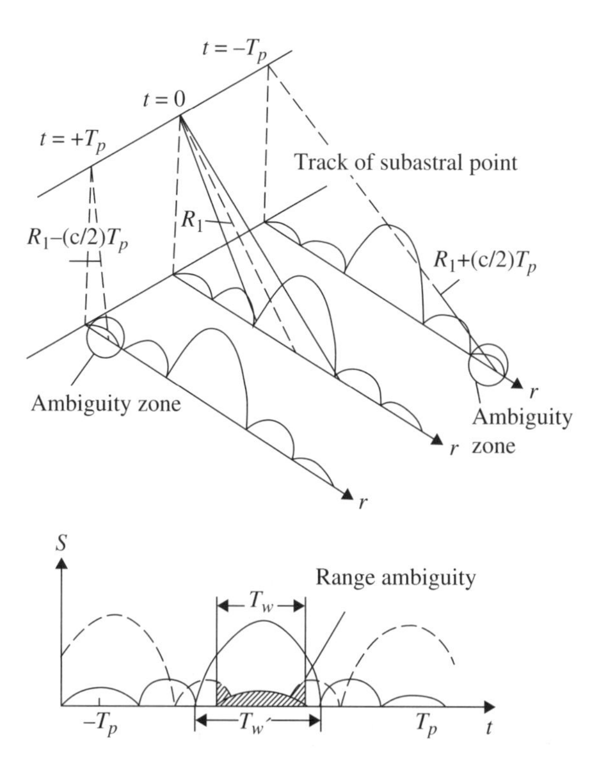


**Figure 2.17** Schematic of range ambiguity.

where $S_{ai}$ is the ambiguity power of the *i*th sampling point inside the echo recording window, and $S_i$ is signal power of the *i*th sampling point inside the echo recording window.

#### 2.3.5.2 Azimuth Ambiguity

The Doppler frequency difference between the target echoes in some angles and the main beam is an integral multiple of the PRF azimuth ambiguity caused by Doppler frequency foldback. Azimuth ambiguity is mainly caused by finite samplings of the Doppler spectrum at an interval of PRF. Because the echo spectrum of the target is repeated in PRF, echo signals outside the main spectrum will fold back into the main spectral region, as shown in Figure 2.18, where $B_D$ is the Doppler bandwidth, $B_P$ is the signal processing bandwidth, and $f_p$ is the PRF.

The ratio of ambiguity to the expected signal is defined as the azimuth ambiguity-to-signal ratio (AASR), which is calculated by

$$\mathrm{AASR} = \frac{\sum\limits_{\substack{m=-\infty \\ m \neq 0}}^{\infty} \int_{-Bp/2}^{Bp/2} G^2(f_d + mf_p)df_d}{\int_{-Bp/2}^{Bp/2} G^2(f_d)df_d} \tag{2.37}$$

where $G$ is the azimuth antenna pattern, and $m$ is normally −10 to 10. The influence of $m$ on AASR is very small outside this range. Of course, the range can be enlarged.

The key of ambiguity-to-signal ratio calculation is to compute the relationship between azimuth round-trip far-field antenna power pattern $G^2$ and $f_d$. For a uniform

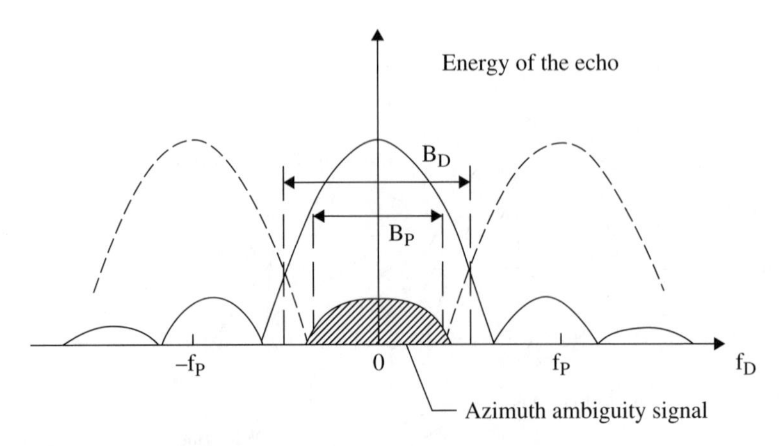


**Figure 2.18** Schematic of azimuth ambiguity.

weighted antenna aperture, the antenna power pattern is

$$G = \left[ \frac{\sin\left[\pi\left(^{L_a}/_{\lambda}\right)\sin\theta\right]}{\pi\left(^{L_a}/_{\lambda}\right)\sin\theta} \right]^2 \tag{2.38}$$

Assuming $f_d = \frac{2v_s}{\lambda}\sin\theta \approx \frac{2v_s}{\lambda}\theta$, then $\sin\theta = \frac{\lambda}{2v_s}f_d$, substituting it to $G$, then

$$G = \left[ \frac{\sin\left[\pi\left(^{L_a}/_{2Vs}\right)f_d\right]}{\pi\left(^{L_a}/_{2Vs}\right)f_d} \right]^2 \tag{2.39}$$

and

$$G^2(f_d + mf_p) = \left[ \frac{\sin[\pi(L_a/(2V_s))(f_d + mf_p)]}{\pi(L_a/(2V_s))(f_d + mf_p)} \right]^4 \tag{2.40}$$

where $L_a$ is antenna aperture size, $V_s$ is the moving velocity of the satellite, $f_d$ is the Doppler frequency of the pulse, and $f_p$ is the PRF.

### 2.3.6 Beam Position Design

The beam position includes such parameters as beam width in the range direction, instantaneous signal bandwidth, and position of the swath. Spaceborne SAR has the geometrical relationship shown in Figure 2.19. It is shown that

$$W_g \approx \frac{\theta_r R_m}{\cos\theta}$$

$$R_m^2 = (R_e + H)^2 + R_e^2 - 2R_e(R_e + H)\cos\gamma$$

$$R_m = \frac{\sin(\theta - \alpha)R_e}{\sin\alpha}$$

$$\theta = \arcsin\frac{(R_e + \mathrm{H})\sin\alpha}{R_e}$$

$$\alpha = \arcsin\frac{R_e\sin\theta}{(R_e + \mathrm{H})}$$

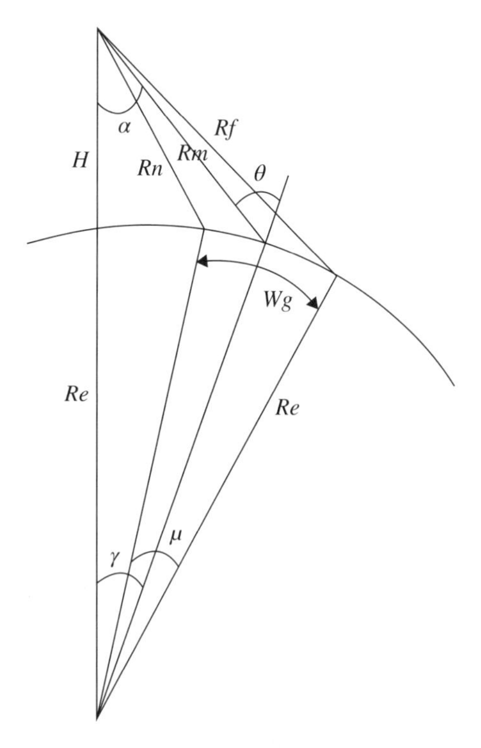


**Figure 2.19** Geometrical relationship of spaceborne SAR.

$$\gamma = \theta - \alpha$$

$$\alpha_n = \alpha - \frac{\theta_r}{2}$$

$$\alpha_f = \alpha + \frac{\theta_r}{2}$$

$$T_W = \frac{2W_s}{C} + \tau = \frac{2W_g \sin\theta}{C} + \tau \tag{2.41}$$

where $R_m$ is the center of the slant distance, $R_n$ is the short-range distance, $R_f$ is the long-range distance, $H$ is the height of the satellite, $R_e$ is the earth's radius (6378.14 km), $\alpha$ is the central viewing angle, $\theta$ is the central incident angle, $\gamma$ is the geocentric angle, $\alpha_n$ and $\alpha_f$ are the near-end and the far-end viewing angles, respectively, $\theta_r$ is the range direction beam width at a certain beam position, $T_W$ is the echo window width, $W_g$ is the swath width, $W_s$ is the corresponding width in slant distance of the ground scanning width, $C$ is the speed of light ($3 \times 10^8$ m/s$^{-1}$), and $\tau$ is the transmitted pulse width.

#### 2.3.6.1 Range Direction Beam Width

Range direction beam width mainly considers ground swath width $W_g$. It is generally required that the instantaneous ground swath width $W_g$ is fixed to maintain the received echo power and the data rate. According to Eq. (2.41), the required swath width $W_g$ can be ensured by changing the range direction beam width $\theta_r$. In fact, even if $W_g$ and $\rho_{rg}$ are constant, the data rate is not constant. Data rate is related to echo window width, sampling rate, PRF, and bit number. If the swath width $W_g$ is constant, $\theta_r$ is inversely proportional to $R_m/\cos\theta$. In fact, $R_m/\cos\theta$ increases and $\theta_r$ decreases with the increase of incident angle.

#### 2.3.6.2 Instantaneous Signal Bandwidth

Instantaneous signal bandwidth is mainly related to the ground range resolution, and they have the following relationship:

$$\rho_{gr} = \frac{k_r \times c}{2B\sin\theta} \tag{2.42}$$

where $\rho_{gr}$ is the ground range resolution, $B$ is the instantaneous signal bandwidth, $k_r$ is the weighted broadening factor, and $c$ is the speed of light.

It is shown that the higher the ground range resolution, the wider the instantaneous signal bandwidth. To ensure a constant $\rho_{gr}$, the signal bandwidth requires frequent change, which will greatly increase complexity of the system hardware design. The signal bandwidth is designed in sections to reduce the system complexity. Then the actual range resolution is not always kept constant.

#### 2.3.6.3 Swath Position Selection

During parameter selection for a spaceborne SAR system, the approximate range of PRF must be defined first in accordance with the system azimuth resolution and the range resolution as well as the ambiguity-to-signal ratio. Then an appropriate beam position in the diamond shape of a zebra map is selected according to the system swath width and the viewing angle. The beam position is generally selected according to the following principles:

(1) If azimuth ambiguity is ensured, try to choose the swath position at low PRF to decrease the data rate.
(2) Sufficient distance shall be left between two swath ends and the interference band of the transmitted signal or nadir signal.
(3) When multiple swaths are required, try to make their length the same. Reserve sufficient overlap between all swaths to enable connections between the image blocks. Ten percent overlap of the beam position is normally suggested.

In addition, it is usually required to go through several iterations during beam position design to obtain the optimal performance parameters.

## 2.4 Imaging Mode

SAR can be operated in different modes so as to provide radar images with a variety of resolutions, swath widths, or polarization in different applications. The operation modes of SAR include strip-map mode, scanning mode, spotlight mode, sliding spotlight mode, mosaic mode, and TOPS mode. The features of common frequently operated modes are listed in Table 2.2.

**Table 2.2** Features of SAR operation modes.

| No. | Operation mode | Antenna requirements | Continuous imaging capability | Features |
|---|---|---|---|---|
| 1 | Strip-map | No scanning | Yes | Traditional imaging mode, mature in technology |
| 2 | Scanning | Range scanning | Yes | Wide and large imaging can be generated; generally used in large-scale survey but with low-azimuth resolution, strong scalloping effect, and difficult radiation correction |
| 3 | Spotlight | Azimuth scanning | No | Fine and high-resolution imaging can be generated but imaging area is small; generally used in fixed-point surveillance |
| 4 | Sliding spotlight | Azimuth scanning | No | High-resolution imaging can be generated, and imaging area is between strip-map and spotlight |
| 5 | Mosaic mode | Range-azimuth 2D scanning | No | High-resolution and wide-swath imaging can be generated, at the cost of increasing the antenna scanning scope |
| 6 | TOPS mode | Range-azimuth 2D scanning | Yes | Wide-swath imaging can be generated at the expense of azimuth resolution, and the scalloping effect is eliminated by azimuth antenna scanning |

### 2.4.1 Strip-Map Mode

Strip map is the most fundamental SAR operation mode [4–6]. In strip-map mode, the antenna beam maintains a constant vertical side looking as the radar platform moves on, and the beam sweeps the ground at a constant speed. Figure 2.20 shows a schematic diagram of strip-map mode SAR operation.

As shown in Figure 2.20, when the platform is between position $A$ and position $C$, target $P$ is inside the radar beam. When the platform is at position $B$, target $P$ is covered by the beam, and the synthetic aperture length is approximated as

$$L_s = R_0\theta_b \tag{2.43}$$

where $R_0$ is the slant range between point $P$ and the center of the synthetic aperture, $\theta_b$ is the azimuth beam width, which is approximated by $\lambda/L_a$, then the synthetic aperture time is

$$T_S = L_s/v \tag{2.44}$$

where $v$ is the speed of the platform.

#### 2.4.1.1 Signal Model

The chirp signal, transmitted by the SAR to the ground during flight, is expressed as

$$s(\tau) = rect\left(\frac{\tau}{T_p}\right) exp(j2\pi f_0\tau + j\pi k_\gamma \tau) \tag{2.45}$$

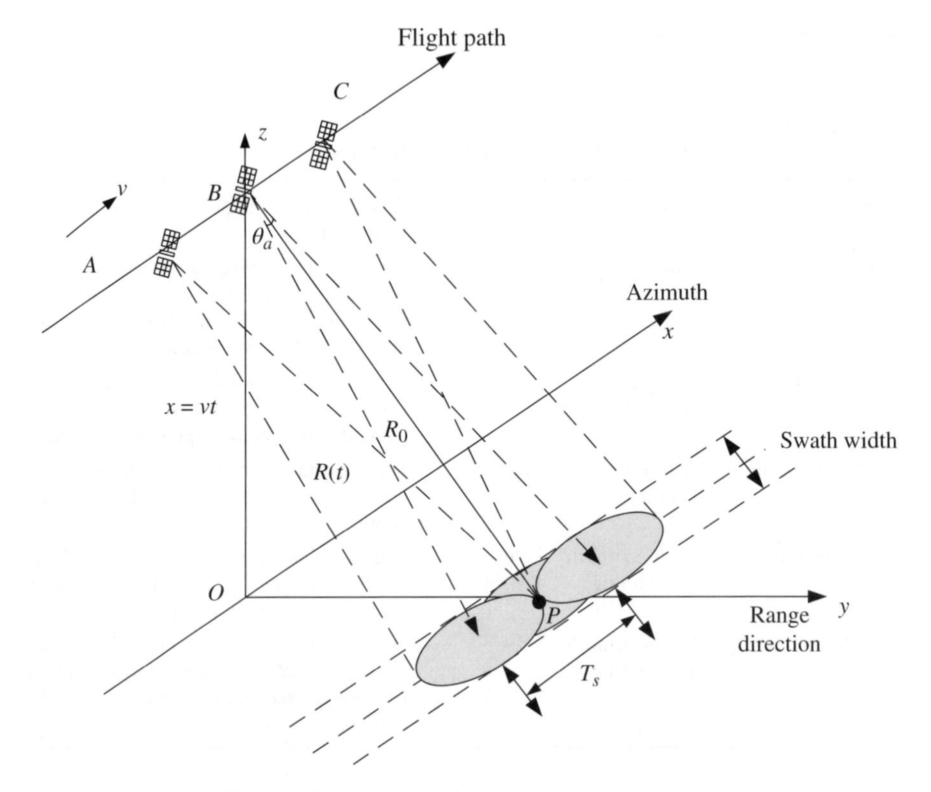


**Figure 2.20** Schematic diagram of strip-map mode SAR operation.

where $T_p$ is the transmitted pulse width, $f_0$ is the transmitted carrier frequency, and $k_\gamma$ is the chirp rate of the chirp signal. Then the echo of target $P$ received by radar is

$$s(t,\tau) = \sigma \cdot rect\left(\frac{\tau - 2R(t)/c}{T_p}\right) exp\left(j2\pi f_0\left(\tau - \frac{2R(t)}{c}\right) + j\pi k_\gamma\left(\tau - \frac{2R(t)}{c}\right)\right) \quad (2.46)$$

where $t$ and $\tau$ are the azimuth and the range direction time, respectively, often referred to as *slow time* and *fast time*, respectively. According to the geometrical relationship shown in Figure 2.20, $R(t)$ can be approximated as

$$R(t) = \sqrt{R_0^2 + x^2} = \sqrt{R_0^2 + (vt)^2} \approx R_0 + \frac{v^2t^2}{2R_0} \quad (2.47)$$

where $x = vt$ is the azimuth radar position.

From Eqs. (2.45)–(2.47), phase difference between the transmitted signal and the echo is given by

$$\phi = -\frac{4\pi R(t)}{\lambda} = -\frac{4\pi}{\lambda}R_0 - \frac{2\pi v^2t^2}{\lambda R_0} \quad (2.48)$$

where the first term is the fixed phase caused by the slant distance, and the second term is the phase that varies with the position $x$. The spatial Doppler frequency caused by

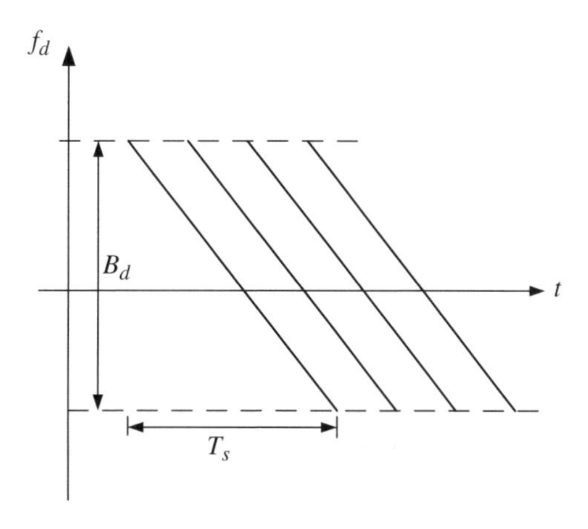


**Figure 2.21** Azimuth frequency process of a target in strip-map mode.

phase change is

$$f_d = \frac{1}{2\pi}\frac{d\phi}{dt} = -\frac{2v^2 t}{\lambda R_0} \tag{2.49}$$

The chirp slope is

$$r = \frac{df_{\mathrm{d}}}{d\mathrm{t}} = -\frac{2v^2}{\lambda R_0} \tag{2.50}$$

Within the synthetic aperture length, the Doppler frequency bandwidth is given by

$$B_d = f_r T_s = \frac{2vL_s}{\lambda R_0} \tag{2.51}$$

Based on the above analysis, the azimuth frequency process of a target in strip-map mode can be obtained as shown in Figure 2.21.

#### 2.4.1.2 Resolution and Swath Width

Based on the analysis of azimuth frequency characteristic of the target, azimuth resolution of strip-map SAR ideally is

$$\rho_a = \frac{v}{B_d} = \frac{L_a}{2} \tag{2.52}$$

The SAR swath width in strip-map mode is related to range direction beam width, antenna viewing angle, platform height, size of the the radar receiving echo window, and so on. As shown in Figure 2.22, the spaceborne SAR geometric relation is used as an example to analyze the range direction swath width in strip-map mode.

In Figure 2.22, both the near-end slant-range $R_n$ and the far-end slant-range $R_f$ are given by

$$R_n = \sqrt{(R_e + H)^2 + R_e^2 - 2R_e(R_e + H)\cos(\eta_n - \gamma_n)} = R_e\frac{\sin\psi_n}{\sin\gamma_n} \tag{2.53}$$

$$R_f = \sqrt{(R_f + H)^2 + R_f^2 - 2R_f(R_f + H)\cos(\eta_f - \gamma_f)} = R_e\frac{\sin\psi_f}{\sin\gamma_f} \tag{2.54}$$

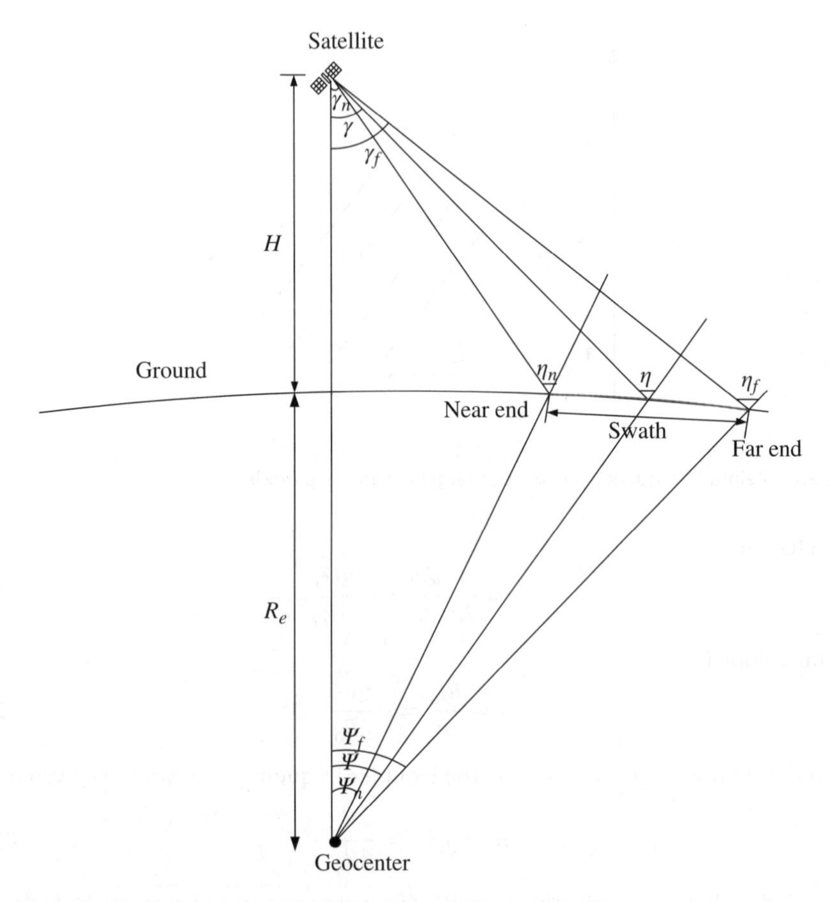


**Figure 2.22** Schematic diagram of strip-map SAR swath.

where $\eta_n$ and $\eta_f$ are the incident angle at the near end and the far end, respectively, $\gamma_n$ and $\gamma_f$ are the antenna viewing angle at the near end and far end, respectively, and $\psi_n$ and $\psi_f$ are the geocentric angle at the near end and far end, respectively.

Therefore, the maximum slant range within the swath is

$$W_r = R_f - R_n \leq \frac{c}{2} T_w \tag{2.55}$$

where $T_w$ is the length of the radar receiver window. Then the ground swath width is given by

$$W_g = R_e \cdot (\psi_f - \psi_n) \tag{2.56}$$

### 2.4.2 Scanning Mode

ScanSAR is proposed to meet the demands for wide swath [7, 8]. In ScanSAR, the antenna first transmits a pulse and receives the corresponding echoes in one beam-pointing direction, producing a sub-band echo data block. Then the antenna changes the beam-pointing direction, transmitting and receiving again, acquiring echo data of another subswath. Repeating in this way, the echo data of several adjacent and

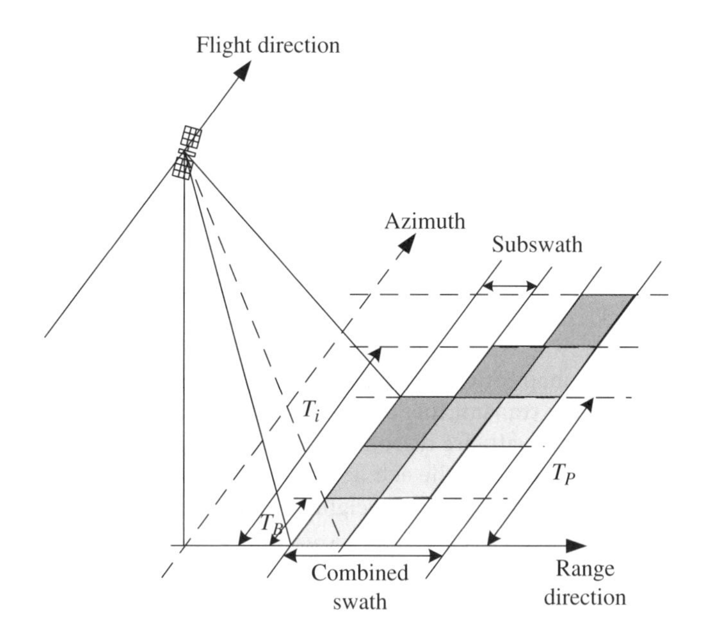


**Figure 2.23** Schematic diagram of ScanSAR.

parallel subswaths can be obtained. Then a wider swath can be obtained after image processing and range direction splicing. However, the scanning mode has a decreased azimuth resolution, to some extent. Figure 2.23 shows the schematic diagram of ScanSAR. It is shown that the overlay length of the beam along the azimuth is much more than the beam-moving distance in dwell time, at every position. Therefore, a gapless continuous SAR image of all combined swaths can be obtained as long as dwell time of the beam at every position is allocated properly.

ScanSAR requires that the radar system be capable of scanning fast in one dimension, so that the antenna beam can work alternately at multiple wave positions along the range direction. When a SAR operates in ScanSAR mode, the radar antenna first transmits a series of pulses in one beam-pointing direction and receives the corresponding echoes. Then the antenna changes the range direction pointing. The antenna beam skips to another pointing direction and continues to illuminate. The corresponding illumination strip of every beam pointing is called a *subswath*, or *sub-band* for short. After $N_s$ times of skipping, $N_s$ sub-bands will be generated in the range direction in the scene. The width of each sub-band is approximately equal to the swath width in strip-map mode under the same conditions. Thus, swath width of the ScanSAR in the range direction is expanded to about $N_s$ times the strip-map mode SAR.

#### 2.4.2.1 Sequential Relationship

In ScanSAR, the continuous illumination time of the antenna in one subswath is called *dwell time* and denoted as $T_B$. The data sampled within dwell time is called *dwell data*. The dwell data, obtained continuously within one subswath, is called *burst* [9]. The one-time scan by the antenna in every subswath is called a *scan cycle*. The associated time is called the *regression time* and denoted as $T_r$. $T_i$ denotes the synthetic aperture time. For a ScanSAR system with single-looking azimuth, the relationship between dwell time, regression time, and synthetic aperture time is given by

$$T_i = T_r + T_B \tag{2.57}$$

Assuming switch time of the antenna is $T_c$ and dwell time of each sub-band is $T_{B,i}$, then for a ScanSAR system with $N_s$ subswaths, the signal period is expressed as

$$T_r = \sum_{i=1}^{N_s} (T_{B,i} + T_c) \tag{2.58}$$

#### 2.4.2.2 Signal Model

The signal transmitting and receiving means in ScanSAR and strip-map SAR are identical. However, in strip-map SAR, the signal transmitting and receiving are continuous, whereas in ScanSAR, the transmitting and receiving alternate with beam switch, and echoes between each subswath are discontinuous. Echoes in the same subswaths also show block discontinuity. Usually the echoes of ScanSAR refer to the ones in the same subswaths. If we ignore antenna pattern weighting, it is given by

$$s(t,\tau) = \sigma \cdot rect\left(\frac{\tau - nT_r}{T_B}\right) rect\left(\frac{\tau - 2R(t)/c}{T_p}\right) \cdot exp\left(j2\pi f_0\left(\tau - \frac{2R(t)}{c}\right)\right.$$
$$\left. + j\pi k_\gamma\left(\tau - \frac{2R(t)}{c}\right)\right) \tag{2.59}$$

where $k_\gamma$ is the chirp rate of the chirp signal, $T_B$ is the beam dwell time, $T_r$ is the beam regression time, $f_0$ is the center frequency of the transmitted carrier, and $R$ is the distance between radar and target.

The ScanSAR echo data can be regarded as truncated strip-map SAR echo data. Figure 2.24 shows the Doppler process of ScanSAR echoes.

#### 2.4.2.3 Resolution and Swath Width

The swath width of ScanSAR is denoted as SW, given by

$$SW = R_e \cdot (\psi_{f,f} - \psi_{n,n}) \tag{2.60}$$

where $R_e$ is the earth's radius, and $\psi_{f,f}$ and $\psi_{n,n}$ are the distal geocentric angle of the farthest subswath and the proximal geocentric angle of the nearest subswaths, respectively.

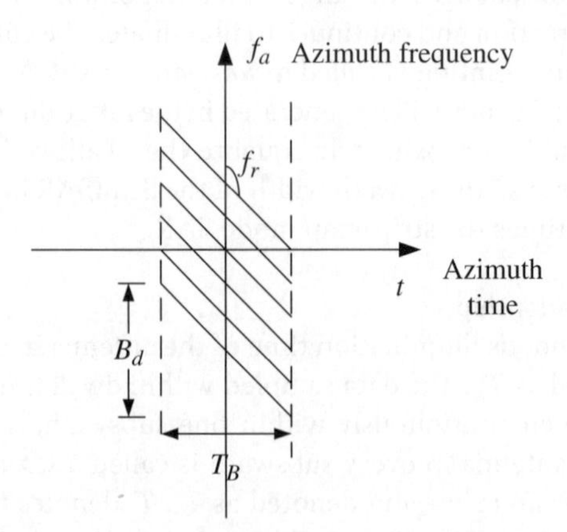


**Figure 2.24** Azimuth frequency course of ScanSAR target.

ScanSAR improves swath width by sacrificing the azimuth resolution. Within each subswath, the range direction data characteristics and the imaging methods are basically the same as those in strip-map mode. The azimuth resolution is related to the Doppler processing bandwidth. ScanSAR Doppler processing bandwidth is

$$B_d = f_r \cdot T_B \tag{2.61}$$

where $f_r$ is the Doppler chirp rate, and $T_B$ is the dwell time of the sub-band. The azimuth resolution of ScanSAR is given by

$$\rho_a = \frac{V_g}{B_d} = \frac{V_g}{f_r \cdot T_B} \tag{2.62}$$

where $V_g$ is the velocity of the beam on the ground. If $T_B \gg T_c$, then $T_B$ is approximated as

$$T_B \approx T_r/N_s \approx T_i/(N_s + 1) \tag{2.63}$$

Thus, the relationship between azimuth resolution and number of subswaths is

$$\rho_a = \rho_{a,strip}(N_S + 1) \tag{2.64}$$

where $\rho_{a,strip}$ is the azimuth resolution of the strip-map SAR under the same condition, which means the azimuth resolution of ScanSAR is at least $N_S + 1$ times less than that of the strip-map SAR.

#### 2.4.2.4 Scalloping Effect

During ScanSAR imaging, only partial synthetic aperture data in the azimuth direction is used for imaging. This kind of "cutoff" operation makes point targets in different azimuth positions correspond to different parts of the antenna pattern, as shown in Figure 2.25. Therefore, the image amplitude in ScanSAR mode is modulated by amplitude of the antenna pattern. The phenomena that the amplitude of the ScanSAR image is periodically nonuniform in the azimuth is the so-called *scalloping effect* [10, 11].

The scalloping effect has the following characteristics: it emerges along the azimuth and extends to the whole or part of subswaths in the range direction. In addition, it is

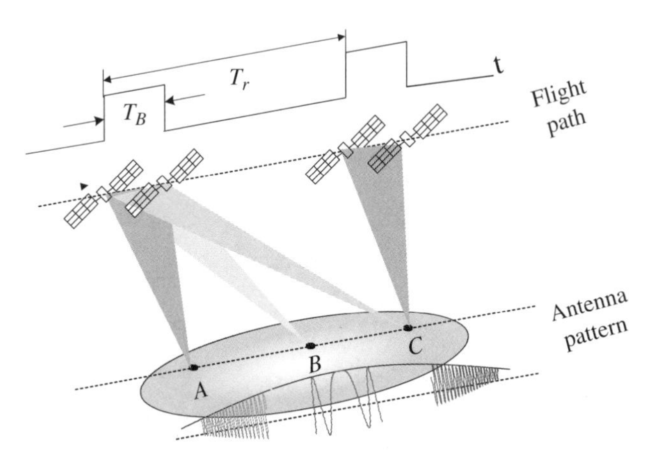


**Figure 2.25** Mechanism of scalloping effect in ScanSAR.

periodic. The space interval is equal to the moving distance of the beam in the azimuth direction during dwell time. The scalloping effect deviation is maximal where two adjacent bursts convert. The amount of scalloping effect deviation depends on the Doppler deviation and the processing method of the imaging processor.

To minimize the influence of the scalloping effect, such methods as increasing the scanning number (as in azimuth multiviewing) within one imaging cycle and making the main lobe of the azimuth antenna pattern flat can be used.

### 2.4.3 Spotlight Mode

Spotlight mode, a kind of fine resolution imaging mode [12, 13] is the main way to realize high-resolution imaging in a small area. It achieves a relatively long synthetic aperture time by continuously illuminating the target region via controlling azimuth beam pointing of the radar antenna. Thus, the high-resolution image in a small area, difficult to achieve in conventional strip-map mode, can be obtained. Figure 2.26 shows the schematic diagram of a spotlight SAR. In the whole synthetic aperture time, the radar antenna is continuously adjusted to focus on the same ground area. In this way, the limitation of azimuth resolution, determined by azimuth aperture length of the antenna in strip-map mode, can be broken. A longer synthetic aperture size can be obtained, leading to a higher azimuth resolution.

Compared with strip-map mode, spotlight mode produces a smaller imaging area if the antenna length stays the same. It can realize not only high-azimuth-resolution

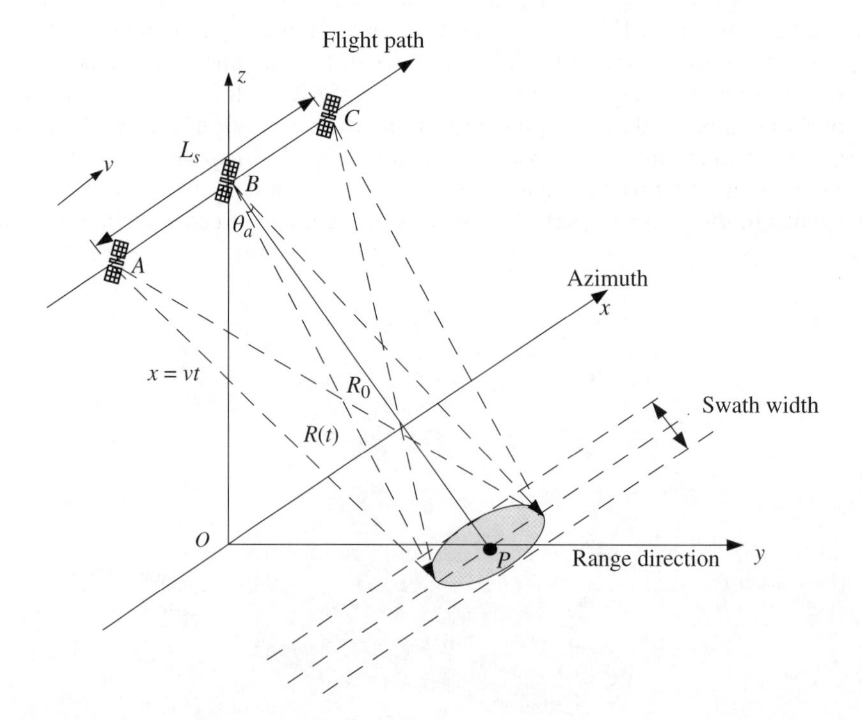


**Figure 2.26** Schematic diagram of spotlight SAR.

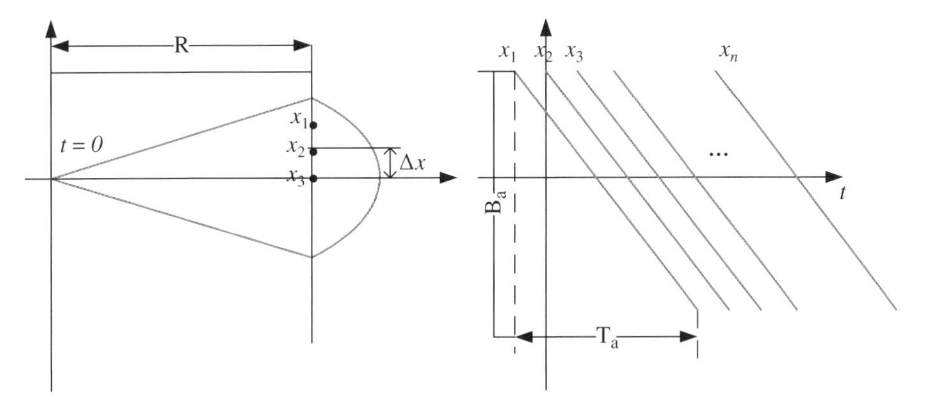


**Figure 2.27** Echo process of strip-map SAR in azimuth.

imaging in multiple small scenes but also multiviewing imaging in the same area during a single imaging process to improve recognition ability.

### 2.4.3.1 Signal Characteristics

In spotlight mode, the frequency process of the radar echoes in the range direction is identical to that in the strip-map mode. However, because the target in the imaging area is always illuminated by the radar beam during the imaging process, its frequency process in the azimuth direction is different from that of the strip-map SAR.

The geometric relation of side-looking imaging is exemplified to describe the azimuth frequency process of strip-map SAR illustrated in Figure 2.27, where $x_1$, $x_2$, and $x_3$ are the azimuth positions of point targets. Points $x_1$, $x_2$, and $x_3$ arrive at and leave the radar beam illumination area successively as platform moving. Their azimuth frequency processes all offer the same synthetic aperture time. The time difference of frequency process between two adjacent points is $\Delta t = \frac{\Delta x}{v}$. At the same time, the frequency difference is $\Delta f_a = \frac{2v\Delta x}{\lambda R}$. The azimuth bandwidth is only related to the synthetic aperture time $T_s$, which is $B = B_a = \frac{2v^2}{\lambda R} T_s$.

The point target within the illumination area is always illuminated by the radar beam during spotlight mode imaging. The azimuth frequency process is shown in Figure 2.28. The duration of the azimuth frequency process at $x_1$, $x_2$, and $x_3$ is the synthetic aperture time of the platform flying through $T_s$. The frequency process of each point has the same start time and end time, which is different from strip-map mode. The azimuth bandwidth is not only related to the synthetic aperture time but also the azimuth range of imaging area $W_a$, which is given by

$$B = B_a + B_d = \frac{2vW_a}{\lambda R} + \frac{2v^2}{\lambda R} T_a \tag{2.65}$$

It can be seen from Eq. (2.65) that the azimuth Doppler bandwidth of the spotlight SAR echo consists of two parts. One is the instantaneous Doppler bandwidth of the echo $B_a$; the other is the Doppler bandwidth of the echo from the point target $B_d$.

### 2.4.3.2 Resolution

The spotlight SAR and the strip-map SAR are quite different in achieving the azimuth resolution. In spotlight SAR, the azimuth beam is controllable, which can be used to

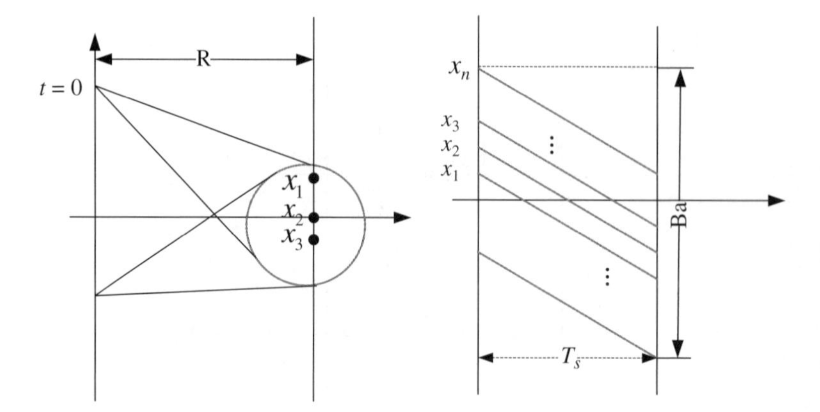


**Figure 2.28** Echo process of spotlight SAR in azimuth.

extend the synthetic aperture time. The azimuth resolution of spotlight SAR is

$$\rho_a = \frac{1}{4}\frac{k_a\lambda}{\sin(\Delta\theta_a/2)} \tag{2.66}$$

where $k_a$ is the azimuth broadening factor, $\lambda$ is the radar operating wavelength, and $\Delta\theta_a$ is the beam rotation angle, namely, the rotation angle of the radar beam center during the synthetic aperture time.

### 2.4.4 Sliding Spotlight Mode

The sliding spotlight mode is a SAR working mode between the strip-map mode and the spotlight mode [14–16]. In the strip-map mode, the beam of the radar antenna is perpendicular to the track of the platform and points to a fixed direction. With the movement of the platform, the footprint of the antenna beam is moving continuously on the ground, producing a strip-map imaging. Theoretically, there is no restriction on the extension of the imaging along the azimuth. However, the antenna length limits the azimuth resolution.

To improve the azimuth resolution, the antenna beam in spotlight mode is controlled to point to the same ground point during the whole data acquisition process, increasing the effective azimuth length. However, this is at the expense of extending the azimuth beam illumination time.

The major feature of sliding spotlight mode is that it controls the antenna beam to point to a place far away from the center of the imaging scene. This may make azimuth imaging wider than that in the spotlight mode and the azimuth resolution better than that in the strip-map mode. If the antenna beam in sliding spotlight points to the illuminated ground center, it is called the *pure spotlight mode.* If the antenna beam is directed to an infinite point, it is the strip-map mode. The strip-map mode and the spotlight mode can be regarded as two special cases in the sliding spotlight mode.

Figure 2.29 shows the schematic diagram of the sliding spotlight mode, where $X_l$ is flight length of the platform, $X$ is the footprint length of the antenna beam, and $X_g$ is the swath length of the illuminated ground.

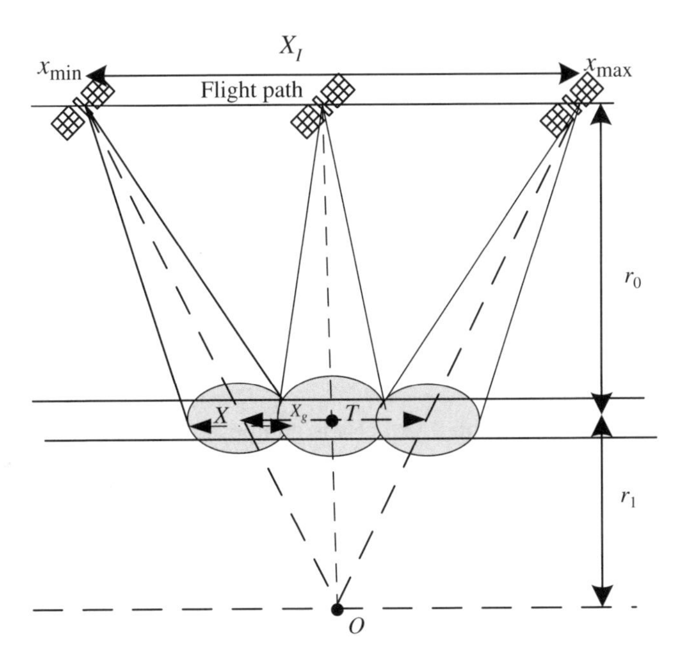


**Figure 2.29** Schematic diagram of sliding spotlight SAR.

In the sliding spotlight mode, the sliding spotlight factor $A$ is defined as

$$A = \frac{r_1}{r_1 + r_0} = \frac{v_f}{v_s} \tag{2.67}$$

where $v_s$ is the platform speed, $v_f$ is the moving speed of the antenna beam on the ground, $r_1$ is the shortest distance between the scene and the focal point, and $r_1 + r_0$ is the shortest distance between the sensor and the focal point.

The sliding spotlight mode must have a tradeoff between the azimuth resolution and the azimuth imaging width.

#### 2.4.4.1 Signal Model

Assuming the chirp rate of the transmitted chirp signal is $K_r$, wavelength of the RF signal is $\lambda$, pulse width is $T_p$, azimuth time is $t_a$, range time is $\tau$, light speed is $c$, and platform speed is $v_s$, then the echo of the point target $(t_0, r_c)$ is given by

$$s(t_a, \tau, t_0, r_c) = exp\left[jK_r\pi\left(\tau - \frac{2R(t_a, r_c)}{c}\right)^2\right] exp\left(-j4\pi\frac{R(t_a, r_c)}{\lambda}\right) \cdot rect\left[\frac{\tau - \frac{2R(t_a, r_c)}{c}}{T_p}\right] rect\left[\frac{t_a - t_0/A}{L_b/v_s}\right] \tag{2.68}$$

where $R(t_a, t_0, r_c) = \sqrt{r_c^2 + v_s^2(t_a - t_0)} \approx r_c + \Delta R$, $L_b$ is the footprint of the beam at $R$.According to the geometric relation, we know that the relationship between the effective synthetic aperture length $L_s$ at the center of the sliding spotlight scene $T$ and the beam footprint $L_b$ is

$$L_s = L_b \cdot \frac{v_s}{v_f} = \frac{L_b}{A} \tag{2.69}$$

As $0 < A < 1$ in sliding spotlight mode, we get $L_b < L_s$.

Due to the continuous change of the beam pointing in the process of sliding spotlight imaging, the instantaneous Doppler center of the echo in the sliding spotlight mode is a function of azimuth time $t_a$

$$f_{dc} = \frac{2v_s}{\lambda}\sin\theta_s = \frac{2v_s}{\lambda}\frac{v_s t_a}{\sqrt{v_s^2 t_a^2 + r_{rot}^2}} \tag{2.70}$$

where $\theta_s$ is the squint angle.

The Doppler bandwidth of the SAR echo is composed of beam width and Doppler center frequency offset

$$B = B_a + B_{shift} = T_a k_{a,r} + T k_{rot} \tag{2.71}$$

where $T_a$ is the aperture synthetic time, $k_{a,r} = -\frac{2v_s^2}{\lambda r}$ is the Doppler frequency of slant-range vector $r$, $T$ is the total imaging time, $k_{rot} = -\frac{2v_s^2}{\lambda r_{rot}}$ represents the variation rate of the Doppler center, which is related to selection of the slant range in the rotation center $r_{rot}$, and $v_s$ is the platform speed. According to Eq. (2.71), Figure 2.30 shows the Doppler process of the echo in the sliding spotlight SAR.

It can be seen that the linear variation of the Doppler center will lead to expansion of azimuth spectrum, sometimes even much greater than PRF.

### 2.4.4.2 Resolution and Swath Width

The sliding spotlight mode controls the moving speed of the beam on the ground by regulating antenna scanning speed. The scanning speed is not only related to operation range of the swath center and the moving speed of the platform but also to speed of the beam footprint on the ground. In sliding spotlight mode, if the shortest distance between the center of the scene and the track is $R$, then the antenna scanning speed is

$$\omega = \frac{\Delta\theta}{\Delta t} = \frac{v_s - v_f}{R} \tag{2.72}$$

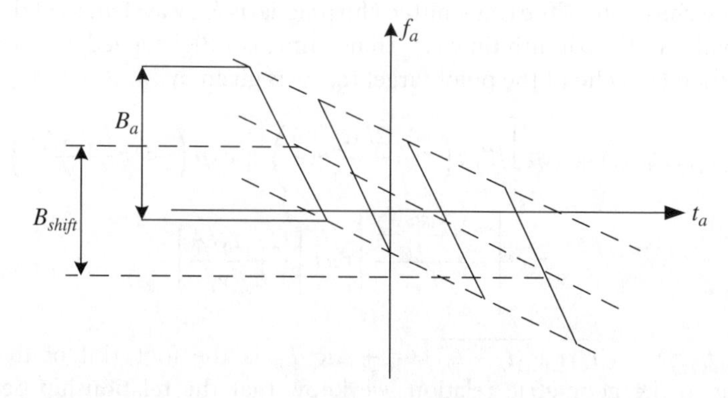


**Figure 2.30** Doppler process of the echo in the sliding spotlight SAR.

In terms of antenna scanning speed, the strip-map mode is a special case in the sliding spotlight mode where $v_s = v_f$, whereas the spotlight mode is a special case of the sliding spotlight mode where $v_f = 0$.

If the antenna scanning range is $(-\theta_{max}, \theta_{max})$, the azimuth swath width in the sliding spotlight SAR is

$$W_a = \frac{2\theta_{max}}{|\omega|} + L_b \tag{2.73}$$

For the strip-map mode ($v_s = v_f$), the azimuth swath width is unlimited. In the spotlight mode ($v_f = 0$), the azimuth swath width is $L_b$. In the sliding spotlight mode, the azimuth swath width is $\frac{2\theta_{max}}{|v_s - v_f|} v_f$, wider than that in the spotlight mode.

The azimuth resolution of SAR is related to the azimuth bandwidth of the radar echo signal. The azimuth resolution of sliding spotlight SAR is

$$\rho_a = \frac{L_a}{2} \cdot \frac{v_f}{v_s} \tag{2.74}$$

where $L_a$ is the antenna azimuth aperture size, $v_s$ is the moving speed of the platform, and $v_f$ is the moving speed of the antenna beam on the ground.

The azimuth resolution for sliding spotlight SAR is $v_f/v_s$ times that of the strip-map mode. It can be seen that the azimuth resolution for sliding spotlight SAR is related to not only the antenna azimuth size but also the platform moving speed and the moving speed of the beam footprints on the ground. The azimuth resolution for sliding spotlight SAR can be adjusted by controlling the speed of the beam footprints.

### 2.4.5 Mosaic Mode

The mosaic SAR is proposed to provide a high-resolution and wide-swath image [17, 18]. In fact, it is a spotlight (sliding spotlight) scanning mode. It achieves wide-swath imaging by scanning multiple sub-bands in the range direction. A high-azimuth resolution is acquired by applying spotlight or sliding spotlight within every sub-band.

Figure 2.31 shows the schematic diagram in the mosaic mode. In this mode, the antenna scans in the range direction to obtain wide-swath SAR images. First the antenna beam points to the near end of the wide swath and dwells for a time long enough to synthesize a SAR image in a single beam illuminated area. Then the beam points to the next position, a SAR image of the illuminated area is synthesized, and so on. In azimuth, to obtain high resolution, and to overcome the disadvantage of low-azimuth resolution in the scanning mode, spotlight imaging or sliding spotlight imaging is used. A relatively high-azimuth resolution can be achieved by controlling the movement of the illuminated ground area of the radar antenna.

#### 2.4.5.1 Sequential Relationship

In the mosaic mode, the signal cycle is shown in Figure 2.32, where $T$ is the signal period, $T_{d_i}$ is the dwell time of the beam in the sub-band, $T_c$ is the switching time of the antenna beam, and $N_B$ is the number of sub-bands. $T$ can be expressed as

$$T = \sum_{i=1}^{N_B} T_{d_i} + T_c \tag{2.75}$$

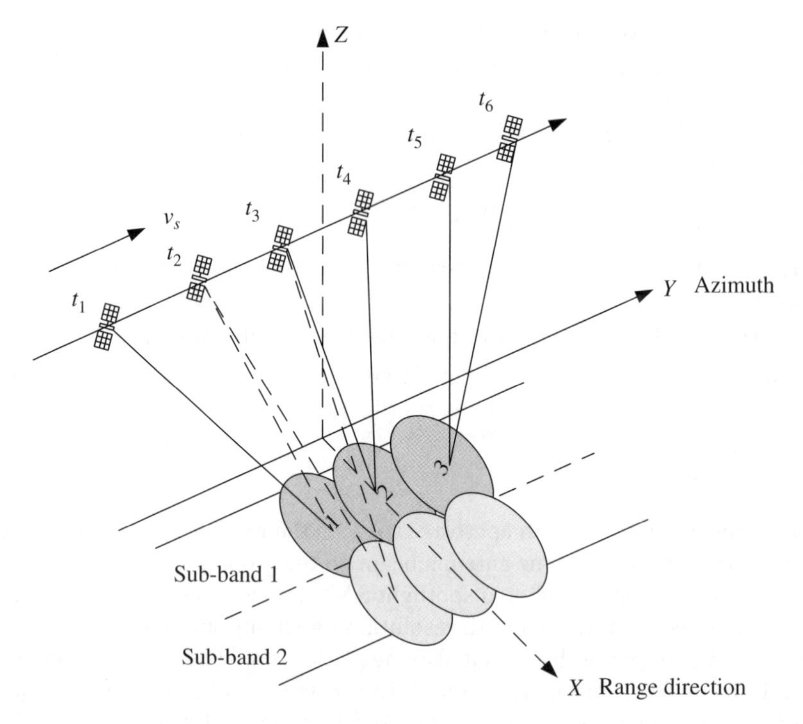


**Figure 2.31** Schematic diagram of the mosaic SAR.

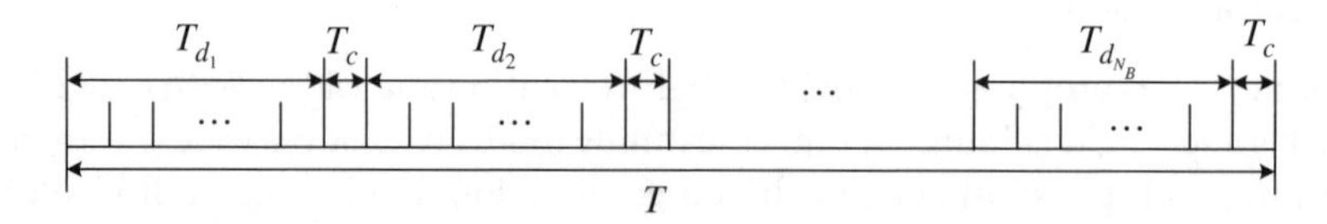


**Figure 2.32** Signal cycle in mosaic mode.

To ensure connection between images in the azimuth direction, there must be a small overlap between two scanning periods. Therefore, the signal period $T$ cannot exceed one synthetic aperture time $T_s$.

In addition, in the mosaic mode, if the required azimuth resolution is $\rho_a$, azimuth swath width is $W_a$, and the number of sub-bands is $N_B$, then the required azimuth resolution is $\rho_{a\_req}$ of the azimuth beam in each subswath as given by

$$\rho_{a\_req} = \frac{\rho_a}{N_B + 1} \tag{2.76}$$

If the platform speed is $v_s$, within one subswath, the sliding speed of the beam on the ground is

$$v_f = v_s \cdot \frac{\rho_{a\_req}}{\rho_{strip}} \tag{2.77}$$

where $\rho_{strip}$ is the resolution in the strip-map mode.

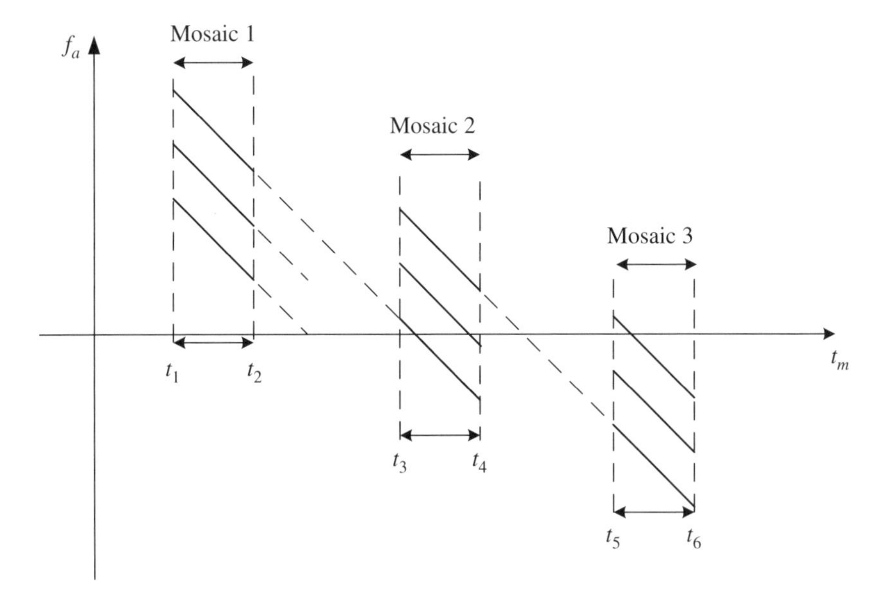


**Figure 2.33** Doppler process of echoes in mosaic mode.

The dwell time of the beam in one subswath is given by

$$T_d = \frac{W_a}{v_f} \tag{2.78}$$

#### 2.4.5.2 Signal Characteristics

We may know from the working principle of the mosaic mode that every mosaic unit works in spotlight mode. The Doppler center frequency of every unit is different. To achieve a seamless illuminated scene connection, there must be an overlap between each mosaic unit. Figure 2.33 shows the Doppler process of a single subswath echo signal.

#### 2.4.5.3 Resolution

The ground resolution is similar to that of a traditional scanning SAR, which is related to the incident angle and the transmitted signal bandwidth in the mosaic mode. In the azimuth direction, although the resolution can be effectively improved by the spotlight mode or the sliding spotlight mode, the synthetic aperture time is divided into $N_B$ subswaths. Therefore, compared to full aperture spotlight or slight spotlight SAR, the azimuth resolution is at least $N_B + 1$ times worse.

### 2.4.6 TOPS Mode

The scanning mode SAR is widely used because of its wide swath. However, because the ScanSAR uses discontinuous bursts in each subswath, the amplitude of the ScanSAR image is periodically nonuniform in azimuth, which is the so-called scalloping effect. In addition, the incomplete illumination of the azimuth antenna beam makes the AASR and $NE\sigma^0$ vary obviously with the change of the azimuth position, which is not feasible for subsequent data processing.

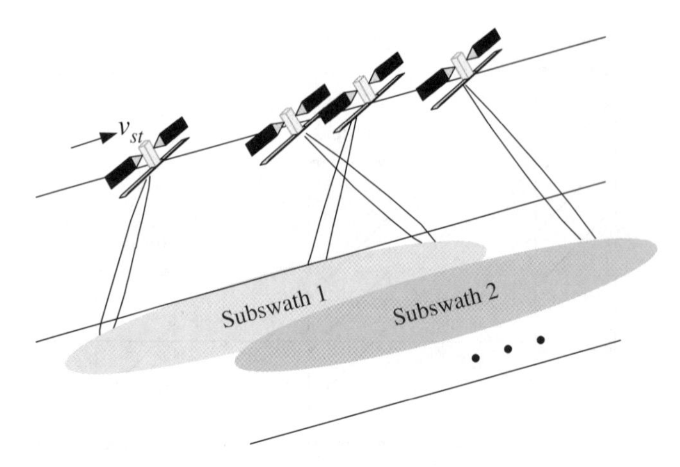


**Figure 2.34** Schematic diagram of TOPS mode.

TOPS mode is then proposed to overcome the drawbacks in the ScanSAR mode [19–21]. The radar beam not only regulates the range direction pointing periodically but also scans from back to front in azimuth to speed up the acquisition of the ground information. With this unique working method, TOPSAR not only can obtain a swath width exactly the same as ScanSAR but also overcomes the obvious scalloping effect of ScanSAR well. Furthermore, the problem of the nonuniform AASR in azimuth direction is solved [22, 23]. Figure 2.34 shows the schematic diagram of TOPS mode.

In ScanSAR mode, to enable the radar to operate in every subswath periodically, the synthetic aperture time is properly divided into several time slots, which are then assigned to each subswath. Every subswath produces an image. In the TOPSAR mode, the radar also needs to operate periodically in every subswath. Compared with ScanSAR, it speeds up the acquisition of the ground information through active azimuth scanning, which allows the radar to operate periodically among multiple subswaths.

To speed up ground information acquisition by the radar, the azimuth beam-scanning direction is made consistent with movement direction of the radar loading platform, namely, from back to front (opposite to beam-scanning direction in spotlight mode). Consequently, the effect is also contrary to that in the spotlight mode, resulting in a decrease in azimuth resolution. However, in the same time interval, TOPSAR can obtain longer strip data than standard strip-map SAR.

#### 2.4.6.1 Sequential Relationship

Similar to ScanSAR, spaceborne TOPSAR also operates in discrete burst mode. The main purpose of the TOPSAR time sequence is to ensure continuity of the SAR image in each subswath by properly allocating time of the radar in each subswath. The timing setting of TOPSAR is related to the system azimuth resolution requirement, the number of scanning sub-bands, azimuth beam width, beam-scanning range, and scanning speed.

Before defining the TOPSAR signal time sequence, the azimuth beam width $\theta_b$ and the beam-scanning speed $\omega_{r,i}$ in every subswath are required to be defined as per the azimuth resolution requirement of the system. To ensure continuous imaging of each subswath in the azimuth, the burst length of each subswath is required to satisfy the following:

$$(\omega_{r,i}T_{b,i} - \theta_0)r_{0,i} + v_g T_{b,i} = (1+\varepsilon)v_g \sum_{i=1}^{N_s} T_{b,i} \tag{2.79}$$

where $\omega_{r,i}$ is the angular velocity of azimuth beam scanning in the $i$ th subswath, $T_{b,i}$ is burst length of the $i$ th subswath, $r_{0,i}$ is the shortest slant distance between the radar and the $i$ th subswath, $\varepsilon$ is overlap ratio between effective imaging regions output by two adjacent bursts in one subswath, and $N_s$ is the number of subswaths.

#### 2.4.6.2 Signal Characteristics

The echo signal of TOPSAR is different from that in strip-map mode and ScanSAR mode in the azimuth direction. In ScanSAR mode, the azimuth antenna pattern is fixed, whereas in TOPSAR mode, the antenna pointing varies with time.

Assuming the rotating velocity of the antenna beam is $k_\psi$, then at time $\tau$, the beam center points to

$$\psi_{dc} = k_\psi \tau \tag{2.80}$$

The variation of antenna pointing introduces a Doppler center change rate:

$$k_a = \frac{\partial\left(-\frac{2v_s}{\lambda}\sin(\psi_{dc}(\tau))\right)}{\partial \tau} \approx -\frac{2v_s}{\lambda}k_\psi \tag{2.81}$$

Therefore, the Doppler process of TOPSAR echo is different from that in ScanSAR. Figure 2.35 shows the Doppler process in TOPSAR mode, where $B_f$, $B_s$, and $B_d$ are the azimuth instantaneous bandwidth (i.e. the azimuth beam width), the total bandwidth of the azimuth burst signal, and the effective Doppler bandwidth of the point target, respectively; $T_d$ and $k_R$ are the dwell time of the corresponding azimuth beam on the target and the chirp rate in azimuth, respectively; and $t_c$ and $t_z$ are the time when the point target is at the beam center and the time when the Doppler is zero, respectively.

#### 2.4.6.3 Azimuth Resolution

Assuming the moving speed of the SAR platform is $v_{st}$ and the angular velocity of the azimuth beam scanning is $\omega_r$, then the time-varying azimuth two-way antenna gain can

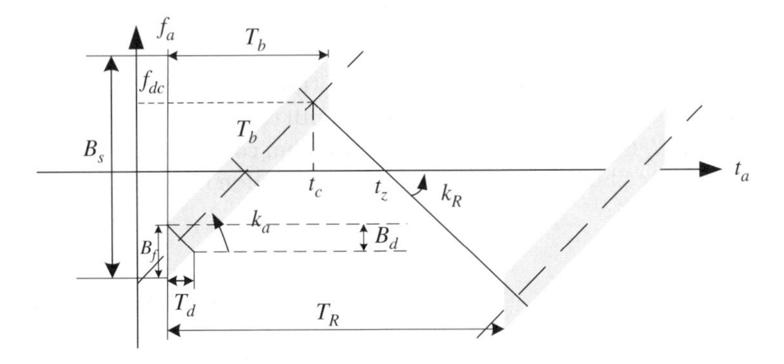


**Figure 2.35** Doppler process of TOPSAR.

be expressed as

$$G_{a\_TOPS}(t) \approx G_0 \sin c^2 \left[\frac{L_a}{\lambda}\left(\frac{v_g t}{R_0} + \omega_r t\right)\right] \approx G_0 \sin c^2 \left[\frac{L_a}{\lambda}\frac{v_g t}{R_0}\left(1 + \frac{\omega_r R}{v_g}\right)\right] \tag{2.82}$$

where $G_0$ is the antenna gain, $L_a$ is the azimuth antenna aperture length, $R$ is the shortest slant distance between radar and target, and $v_g$ is the beam velocity on the ground regardless of azimuth beam scanning. In contrast with traditional modes where azimuth beam pointing is constant, an azimuth beam-scanning factor (shrinkage factor) is introduced in the antenna pattern in the TOPSAR mode:

$$\alpha(R) = 1 + \frac{R\omega_r}{v_g} \approx \alpha(R_0) \gg 1 \tag{2.83}$$

where $R_0$ is the shortest slant distance between radar and the center of the swath. Here, Eq. (2.83) can be equivalent to the antenna pattern in the strip-map mode with constant beam pointing. The equivalent antenna length is $L_e = \alpha L_a$, and the corresponding azimuth resolution is

$$\rho_{az} \approx \alpha L_a/2 \tag{2.84}$$

In the TOPSAR mode, the angular velocity of beam-scanning $\omega_r$ must be limited, so that the beam can cover the scene continuously. The total signal bandwidth of the target in the subswath should be at least equal to the signal bandwidth of the target between the subswaths. For a single target, the total dwell time can be approximated as

$$B_T = k_a(T_B - T_D) \geq |k_R|(T_R - T_B + T_D) \tag{2.85}$$

where $T_R$ is the return time required by the antenna to go through every subswath once.

Assuming each subswath has the same $T_B$, $N_s$ is the number of subswaths, $T_R = T_B N_s$, and $T_B \gg T_D$, Eq. (2.85) can be approximated as

$$k_a T_B \geq |k_R|(N_s - 1)T_B \tag{2.86}$$

Therefore, the lower limit of $k_a$ is

$$k_a = |k_R|(N_s - 1) \tag{2.87}$$

and

$$\alpha = 1 + \frac{k_a}{|k_R|} = N_s \tag{2.88}$$

The above analysis shows that both spaceborne TOPSAR and ScanSAR obtain a wide swath by sacrificing azimuth resolution. The azimuth resolution decreases as the number of subswaths $N_s$ increases. For the ScanSAR system composed of $N_s$ subswaths and $N_l$ multiviewing, the obtainable optimal azimuth resolution is

$$\rho_{scan} \approx (N_l N_s + 1)\frac{L_a}{2} \tag{2.89}$$

In the TOPSAR mode, it can be seen from Eqs. (2.84) and (2.88) that the azimuth resolution is given by

$$\rho_{TOPS} \approx N_l \cdot \alpha \cdot \frac{L_a}{2} \approx N_l N_s \cdot \frac{L_a}{2} \tag{2.90}$$

From Eqs. (2.89) and (2.90), it is known that a slightly better azimuth resolution can be obtained in the TOPSAR mode than ScanSAR under the same conditions.

## 2.5 Moving Target Working Mode

According to the characteristics of the detected target and its background, the moving target working mode is generally divided into the ground moving target mode, the marine moving target mode, and the airborne moving target mode. The moving target detection in different modes is closely related to the target background clutter characteristics, the target characteristics, the radar system parameters, and the signal processing methods.

### 2.5.1 GMTI

GMTI mainly completes the ground movement target detection, the positioning, and the tracking and obtains such information as location, speed, and direction of the moving target.

SAR/GMTI systems with the GMTI function generally have two working mechanisms. One is the single-channel GMTI system, where the antenna scans for moving target detection. Assuming the target is always located outside the main lobe clutter, the moving target detection is realized by filtering out this clutter. The advantage of the single-channel system is the small amount of equipment and signal processing required; its disadvantages are its bad performance at low-speed moving target detection and low positioning accuracy. Therefore, this kind of GMTI system is usually small and slow. The other structure is the multichannel system with single input and multiple output. The whole antenna aperture is used to transmit, and multiple subantenna apertures are used to receive. It adopts full antenna aperture transmission and multiple subantenna aperture receiving and uses the characteristics of static clutter between multiple channels to perform clutter suppression and effectively detect moving targets. Compared with a single-channel system, it has minimum detectable velocity (MDV) in a slowly moving target. In addition, through channel grouping, two kinds of clutter suppression channels are produced and used for high-precision moving target positioning. The disadvantage of the multichannel system is the large amount of equipment and signal processing required. Therefore, the SAR/GMTI radar with this kind of system is usually used in large-scale airborne or spaceborne platforms.

The SAR/GMTI system is usually in a side-looking mode, detecting ground moving targets in the side of the track. According to the application, there are three operation modes for GMTI. The first one is wide-area scanning. The radar beam takes the normal direction of the antenna as the center to perform wide range scanning in the azimuth direction, so as to search for the moving target within the scanned area. The track of the moving target can be obtained by multiple continuous scanning. Figure 2.36 shows the schematic diagram of a multichannel system for wide-area scanning.

The second mode is section tracking. This method is based on the wide-area target situation obtained by the wide-area scanning mode. To achieve monitoring of specific moving targets or key areas, the small-sector tracking mode is adopted. The method

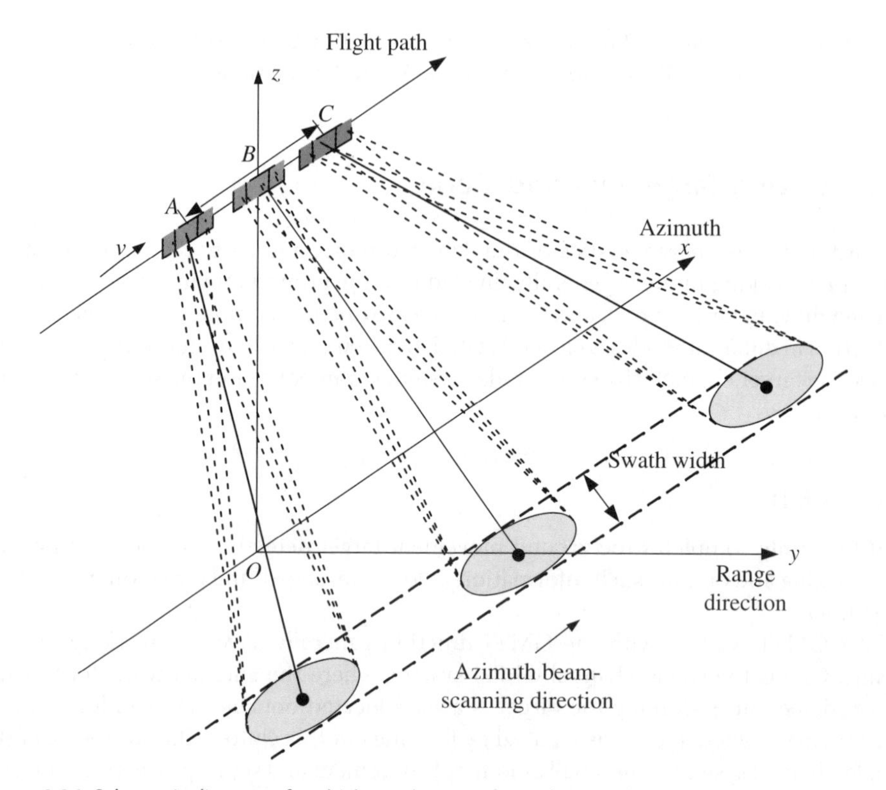


**Figure 2.36** Schematic diagram of multichannel system for wide-area scanning.

features a high tracking data rate. It can track moving targets with high speed. Figure 2.37 shows the schematic diagram of a multichannel system scanning the ground sections.

The third mode is simultaneous SAR/GMTI. In this mode, the radar antenna beam control method is consistent with that in the SAR imaging method, either in the strip-map imaging mode or in the spotlight imaging mode. When moving targets in the swath are simultaneously detected, imaging of the region can be completed as well as getting the terrain information and the moving target information. The moving target image superimposed on the SAR image can be obtained by interferometric positioning, visually displaying position information of the moving target on the SAR image. Furthermore, we can also extract moving target information to obtain images of the moving target, which can be superimposed on the SAR image. Figure 2.38 shows the schematic diagram of a multichannel system scanning the ground in simultaneous SAR/GMTI mode.

#### 2.5.1.1 Signal Characteristics

The geometric relation of the radar platform and the ground moving target in a slant plane is shown in Figure 2.39. When $t = 0$, the radar platform is at (0, 0), and the radar platform moves in the azimuth direction with the velocity $v_a$. At that time, the moving target $T$ is located at $(x_0, R_c)$. The moving target is moved in a uniform linear motion, and radial velocity and azimuth velocity of the moving target are $v_r$ and $v_x$, respectively.

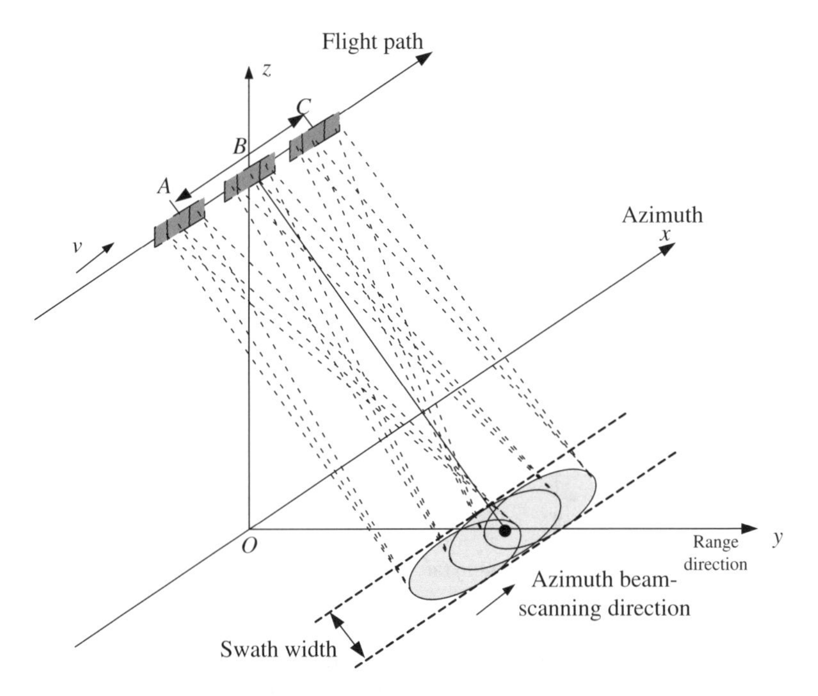


**Figure 2.37** Schematic diagram of multichannel system scanning the ground sections.

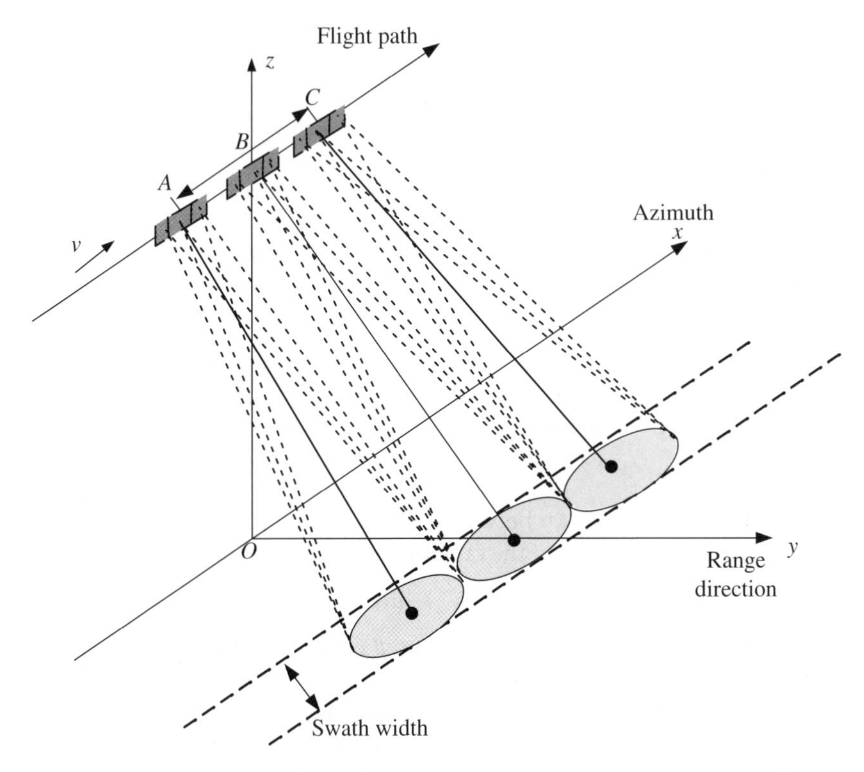


**Figure 2.38** Schematic diagram of the multichannel system scanning the ground in simultaneous SAR/GMTI mode.

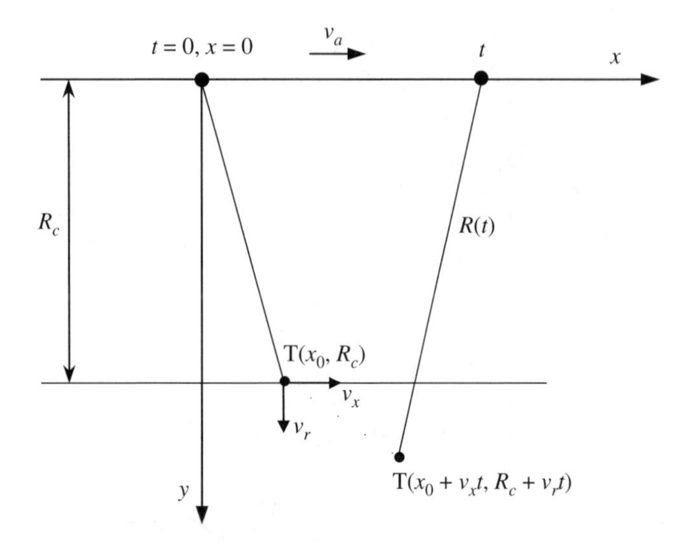


**Figure 2.39** Geometric relation of the moving target in a slant plane.

$R_c$ is the vertical distance from the moving target to the flight path of the radar platform at that moment, and the position $x_0$ of the target when $t = 0$ is its true azimuth position.

Assuming the coordinates of the radar platform are $(v_a t, 0)$ at time $t$ and the coordinates of the moving target in the slant plane are $(x_0 + v_x t, R_c + v_r t)$, the relative instantaneous distance between the platform and the target at time $t$ is

$$R(t) = \sqrt{(x_0 + v_x t - v_a t)^2 + (R_c + v_r t)^2} \tag{2.91}$$

The Doppler center frequency of the moving target echo is

$$f_{dc} = -\frac{2}{\lambda}\frac{dR(t)}{dt}\bigg|_{t=0} = \frac{2x_0 v_a}{\lambda R_c} - \frac{2v_r}{\lambda} - \frac{2x_0 v_x}{\lambda R_c} \tag{2.92}$$

Due to additional radial movement of the target, the Doppler effect is the additional Doppler frequency offset compared to a static target, which is given by

$$\Delta F = \frac{2v_r}{\lambda} \tag{2.93}$$

where $v_r$ is the radial (forward) velocity of the moving target relative to the platform and $\lambda$ is the radar operating wavelength.

#### 2.5.1.2 MDV

Due to the unknown relationship between the moving direction of the ground target and the radar, the radial velocity of the target relative to the radar is given by

$$v_r = v_m \cdot \sin\beta \tag{2.94}$$

where $v_m$ is the speed of the moving target, and $\beta$ is the angle between the target moving direction and the radar moving direction. To detect as many targets with different velocities and directions as possible, a conventional radar system uses a relatively small MDV. The MDV in a GMTI system is related to the radar wavelength, platform speed, antenna aperture, and number of channels.

For a single-channel SAR/GMTI system, the moving target is detected mainly in a "clean area," that is, outside the azimuth clutter spectrum. The condition of a moving target in a clean area is that the Doppler frequency shift $f_d$ due to radial movement is larger than the Doppler width of the radar main lobe clutter $F_d$. The frequency shift of a moving target is

$$f_d = \frac{2v_r}{\lambda} \tag{2.95}$$

The Doppler clutter width is

$$F_d = \frac{2v_a}{d} \tag{2.96}$$

The MDV of the target is

$$\text{MDV} = \frac{\lambda \cdot v_a}{d} \tag{2.97}$$

where $\lambda$ is the wavelength of the radar, $v_a$ is the velocity of the platform, and $d$ is the azimuth antenna aperture.

For multichannel SAR/GMTI systems, the clutter suppression is usually performed before the moving target detection. The SNR after clutter suppression decides the MDV. The SNR after space-time signal processing is expressed as

$$SCNR_{out} = SCNR_{in} - L_p + G_{st} \tag{2.98}$$

where $SCNR_{in}$ is the signal-to-clutter noise ratio defined by system performance and scenarios, $L_p$ is the processing loss due to pulse compression and other signal processing, and $G_{st}$ is the gain of space-time processing (improvement factor). Generally, a moving target with high radial velocity can be easily detected due to relatively high processing gain. The processing gain decreases with the decrease of radial velocity, and the processing gain can even be negative. The target cannot be detected when the radial velocity of the target is lower than a threshold. Therefore, the MDV of the moving target is calculated according to the system parameters, the improvement factor, and the constant false alarm rate (CFAR) threshold.

#### 2.5.1.3 Azimuth Angle Measurement Accuracy

There are two ways to measure the azimuth angle of a moving target. The amplitude modulation characteristics of the azimuth antenna are used to locate the mass center. The interferometric measurement is used to locate the position of the target. The former is usually used in a single-channel SAR/GMTI system, and the latter is applied to multichannel SAR/GMTI system.

For a single-channel SAR/GMTI system, the antenna is usually in the scanning mode. After beam sweeping over a moving target, assuming the antenna beam width is $\theta$ in degrees and the angular scanning velocity is $\omega$, the time needed to scan the moving target is $T = \theta/\omega$. When dividing the full time $T$ into $N$ segments, the corresponding azimuth angle of each section of the radar antenna beam is $\phi(n), n = 0 \sim N-1$, and the moving target is detected in every segment. If the moving target is not in the blind area, $N$ continuous detection points and their corresponding amplitude $U(n)$ can be obtained by theory. The amplitude envelope of the amplitude $U(n)$ of the $N$ detected points is consistent with the antenna pattern. Therefore, by interpolating the beam azimuth $\phi(n)$ and amplitude $U(n)$ of the $N$ point, a more accurate azimuth $\phi(n_0)$ corresponding to the maximum amplitude $U(n_0)$ point can be obtained, that is, the azimuth of the moving target. The moving target positioning error of single-channel SAR/GMTI is mainly defined by the amplitude characteristics of the $N$ detection points and value of $N$. Normally, measurement accuracy of the azimuth angle is given by

$$\sigma_\alpha \geq \max(B, V) \tag{2.99}$$

where $B = \frac{k_1\theta}{\sqrt{N^x}}$, $V = \frac{k_2\theta}{x\sqrt{N}}$, $k_1, k_2$ are constant and are close to 1, and $x$ is the SNR of the moving target.

For the GMTI system with $M$ channels, the receiving channel is usually divided into two groups to achieve multichannel clutter suppression and high-precision interferometric positioning. To ensure the signal-to-clutter noise ratio (SCNR) of the moving target in priority, channel 1 to $M-1$ are grouped together, and channel 2 to $M$ are grouped together. The distance of the equivalent centers of the clutter suppression channel, coming from space-time processing in two receiving groups, is $D_{ax}$. The expression of the moving target positioning using interferometry is

$$X_\mathrm{a} \approx \frac{\lambda R_c \phi}{2\pi D_{ax}} \tag{2.100}$$

where $\phi$ is the interferometric phase, $R_c$ is the target slant distance, and $D_{ax}$ is the interferometry baseline length. The azimuth angle of the moving target is given by

$$\alpha_a = \arcsin\left(\frac{X_a}{R_c}\right) = \arcsin\left(\frac{\lambda\phi}{2\pi D_{ax}}\right) \tag{2.101}$$

Therefore, for a multichannel GMTI system, measurement of the azimuth angle of the moving target is mainly related to the interferometric phase. There are two main factors affecting the accuracy of the interferometric phase of the moving target. One is the impact of residual background clutter and the noise of target pixels on the target signal. Another is the residual phase error after interchannel calibration.

#### 2.5.1.4 Detection Capability

The capability of moving target detection is a combination of discovery probability, detection probability, and false alarm probability. The probability of the moving target detection is based on the percentage statistics of a number of moving targets that can be detected, from the application point of view. In the case of a specific moving target SNR, for a single-channel GMTI system, the moving target cannot be separated from the main clutter region because the radial velocity is too low, or the azimuth spectrum of the moving target signal is folded into the main clutter region due to the low PRF of the system; either of them will cause the moving target to be undetectable. For a multichannel GMTI system, if target radial velocity is too low, the space-time processing gain will decease to negative. In addition, there is a PRF blind speed in space-time processing. The probability of detection can be defined as a ratio of the radial velocity range that can be detected to a radial velocity period of the corresponding PRF. The radial velocity period of the corresponding PRF is $\frac{\mathrm{prf}\cdot\lambda}{2}$. Combined with the analysis of MDV of the moving target, the probability of detecting a moving target can be expressed as

$$\eta = 1 - \frac{2 \cdot \mathrm{MDV}}{0.5\mathrm{PRF} \cdot \lambda} = 1 - \frac{4 \cdot \mathrm{MDV}}{\mathrm{PRF} \cdot \lambda} \tag{2.102}$$

where $\lambda$ is the system wavelength.

Normally, the target detection probability and false alarm probability are of concern because they will affect the design of the GMTI system, such as the SNR and detection threshold.

### 2.5.2 Marine Moving Target Indication

Marine moving target indication (MMTI) achieves detection, tracking, and positioning of marine targets by obtaining position, velocity, and moving direction of the target on the sea surface (such as ships, boats, fishing boats, etc.) through detection.

There are two kinds of operation modes for the marine moving target, the wide-area marine search mode and the marine surveillance mode. Figure 2.40 shows the wide-area marine search mode, where a wide range on the sea surface is detected to obtain the ship target information, the track of the moving ship target is established, and static landmarks, such as coasts and islands, are reserved.

Figure 2.41 shows the marine surveillance mode based on the detection of a key target by wide-area surveillance or access to a critical spot. One or several small key sea areas are selected. The antenna beam center is pointed to the center of the monitored area, and a sector angle is selected to monitor the prominent targets or sea areas. The radar antenna is scanned under standard working conditions in the azimuth direction to continuously monitor the critical sea areas within a certain angle and generate a high-resolution imaging of the suspected target, obtain inverse synthetic aperture radar image of the target, and produce a 2D profile and the length and width of the target.

#### 2.5.2.1 Signal Characteristics

In MMTI, the characteristics of the target signal are similar to those of the ground moving target, and the Doppler characteristic has an additional Doppler frequency offset relative to static targets. The offset is proportional to the relative radial velocity of the target.

**Figure 2.40** Working principle diagram of wide-area marine search mode. MWAS: marine wide-area search

**Figure 2.41** Working principle diagram of maritime surveillance mode.

The difference between marine moving target detection and ground moving target detection is particularity of sea clutter. Because the sea surface is subject to wind, there are abundant Doppler shift components in sea clutter. There is a certain distribution of the radial velocity of the wave scatters that makes Doppler frequency with the sea clutter somewhat distributed. Doppler spectrum width with the sea clutter is directly related to marine conditions. The wider the sea is, the wider the frequency spectrum of the sea clutter. Typically, the Doppler spectral width of the sea clutter can be expressed as

$$\sigma_f = \frac{2\sigma_v}{\lambda} \tag{2.103}$$

where $\lambda$ is the radar wavelength, and $\sigma_v$ and $\sigma_f$ are the standard deviation of speed and frequency, respectively.

According to the Doppler spectrum width of sea clutter, it can be concluded that decorrelation time of sea clutter is $\frac{\lambda}{2\sqrt{2\pi\sigma_f}}$. In general, the decorrelation time of ocean clutter is far longer than the radar pulse repetition period. Therefore, the sea clutter is related in adjacent pulse echoes.

To suppress the sea clutter, frequency agile technology is frequently used. It is based on the statistical characteristics difference of the target echo and sea clutter under frequency agility to decrease the correlation between target and clutter and to improve the accumulated gain of the target. In addition, frequency agility affects fluctuation of the target. It increases the echo fluctuation rate and enables the slow fluctuation change (no correlation during antenna scan) to fast fluctuation (no correlation between pulses).

By using appropriate frequency agility rules, the marine target detection capability can be effectively improved. Because frequency dependence of the target is related to the radial size of the target, the frequency difference of the general frequency agility should be at least larger than one wave path difference. Thus, the decorrelation frequency interval of the target is

$$\Delta f > \frac{c}{2d} \tag{2.104}$$

where $c$ is the light speed, and $d$ is the radial size of the target.

For frequency agility, the accumulated gain is mainly due to fluctuation loss after the fixed frequency is removed. The frequency agility gain $G_{FAT}$ is related to the number of accumulated pulses $N_e$ and detection rate. The agility gain is a function of not only the number of accumulated pulses but also the required detection rate. The higher the detection rate is, the greater the gain. In general, when accumulated pulses exceed a certain number, the gain tends to saturate. This relation can be expressed accurately by an empirical formula:

$$G_{FAT} = [L_f(1)]^{1-\frac{1}{N_e}} \tag{2.105}$$

where $L_f(1)$ is the single pulse fluctuation loss, which is related to the rate of detection.

#### 2.5.2.2 Operating Range

In MMTI, the main targets on the sea surface are naval ships and boats. Sea clutter and visual range are the main factors affecting the detection distance. In general, the radar system is in pulse mode. The main factor affecting the detection is the SCNR.

Similar to SNR calculation, the ratio of sea clutter echo to noise power (signal-to-clutter ratio) inside the beam illuminated area within one pulse is also a function of the action range $R$:

$$C/N = \frac{P_{av}G_t(\theta,\phi)G_r(\theta,\phi)\lambda^2\sigma_c}{(4\pi)^3KT_0F_nC_BL_sR^4} \tag{2.106}$$

where $P_{av}$ is the average transmitter power, $G_t(\theta,\phi)$ is the transmitter antenna gain, $G_r(\theta,\phi)$ is the receiver antenna gain, $\lambda$ is the wavelength of the system, $K$ is the Boltzmann constant, $T_0$ is the system noise temperature, $C_B$ is the bandwidth mismatch factor, $L_s$ is the system loss, $R$ is the action range, and $\sigma_c$ is the radar cross section of sea clutter, given by

$$\sigma_c = R\sigma_{c0}\theta_a\left(\frac{c\tau}{2}\right)\sec\psi \tag{2.107}$$

where $\theta_a$ is the horizontal beam width, $c$ is the light speed, $\psi$ is the grazing angle, and $\sigma_{c0}$ is the surface reflection coefficient corresponding to different grazing angles, whose value is related to the specific marine situation.

SCNR at different distances can be obtained by the above clutter-to-noise ratio $C/N$ and signal to noise ratio $S/N$

$$SCNR = \frac{S/N}{C/N+1} \tag{2.108}$$

The radar action range can be calculated by the ratio of SCNR to the required detection factor.

#### 2.5.2.3 Detection Rate and False Alarm Rate

The sea clutter characteristics are not only subject to wind force, wind direction, rain, surge, ocean currents, and many other natural factors but also depend on the radar system operating system, such as working frequency, sampling rate, bandwidth, transmitting pulse width, polarization mode, incident angle, etc. In general, the marine target detection method is based on the study of statistical signal theory. A large amount of measured data is used to fit the statistical distribution model of sea clutter in different environments, then, according to the characteristics of the statistical distribution model, the optimal (or suboptimal) CFAR detection algorithm is derived to achieve the target detection in the sea clutter background.

The sea clutter amplitude is generally subject to Rayleigh distribution, Weibull distribution, lognormal distribution, and $K$ distribution.

### 2.5.3 Airborne Moving Target Indication

Airborne moving target indication (AMTI) mainly detects airborne moving targets, such as aircraft, missiles, etc., and obtains flight target motion information, as shown in Figure 2.42. For detection background, both ground clutter and sea clutter are different. The intensity of ground clutter is very large. In time domain, the target echo is completely submerged by ground clutter in the received signal. Ground clutter is a kind of surface scattering clutter. Its intensity is related to the radar frequency, the radar main beam illumination region, the radar transmitter power, the antenna main lobe to side lobe ratio, the antenna beam incident angle, the unit area backscattering coefficient of the clutter, and so on.

There are three kinds of airborne moving target detection modes of airborne SAR/AMTI systems: tracking while scan (TWS), tracking and searching (TAS), and target tracking (TT). In TWS, while the target is tracked continuously, the full airspace is scanned. And a tracking filter is used on the detected target, so as to realize simultaneous multiple-target tracking.

In TAS, searching and tracking are performed simultaneously and independently, depending on the demands of the radar. Unlike TWS, searching and tracking rates of TAS are different and may be different for different targets. The searching and tracking

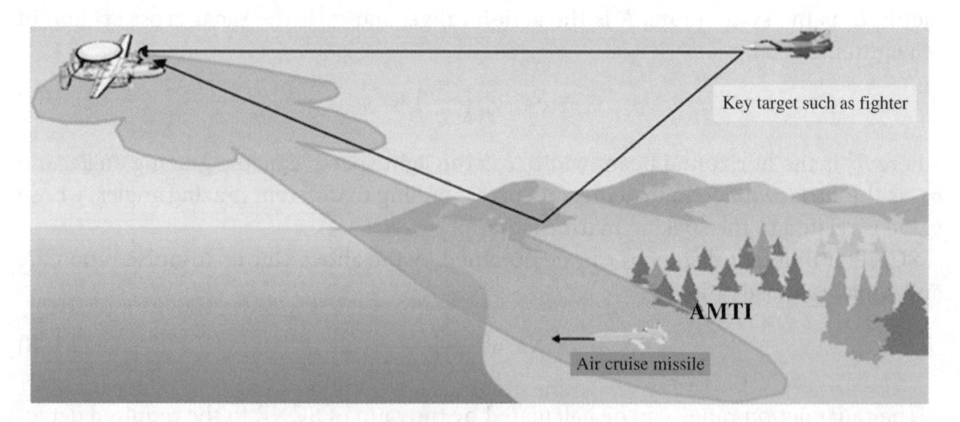


Figure 2.42 Diagram of airborne moving target indication.

mode generally requires the system has agility control capability of arbitrary beam pointing.

TT can be subdivided into single-target tracking (STT) and multiple-target tracking (MTT). STT tracks only a certain target or a key observation target. Its tracking data rate is very high. All radar resources are assigned to this mission to ensure that the tracking mission is performed at sufficient data rates. MTT tracks multiple targets. In general, different targets may be divided into different priorities, and the allocated tracking time is different as well.

#### 2.5.3.1 Signal Characteristics

When the radar is detecting airborne targets, the main beam points to the ground, and the main lobe clutter intensity is strong. The ground footprint of the beam is large, and the range is fuzzy inside the main lobe, making the target submerged by clutter in the range direction. The target cannot be detected in time domain. Because of the differences in the target and the clutter in the velocity domain (frequency domain), a pulsed-Doppler (PD) radar system must be used for target detection.

The relative position and velocity between the target and the radar platform define the relative relationship between target echo and clutter in the range-Doppler 2D spectrum as well as directly affecting the radar detection performance. Due to high speed of a spaceborne radar platform, which is much greater than that of the airborne platform, the signal characteristics of airborne AMTI and spaceborne AMTI are quite different. Taking satellite AMTI as an example, clutter characteristics in the AMTI mode are described below.

Figure 2.43 shows a Doppler diagram of the clutter in spaceborne AMTI mode. It can be seen that, without considering the rotation of the earth, Doppler frequency shift of the clutter reflected at any point on the ground is

$$f_{dC_{\max}} = \frac{2V_R}{\lambda}\cos\theta_{cone} \tag{2.109}$$

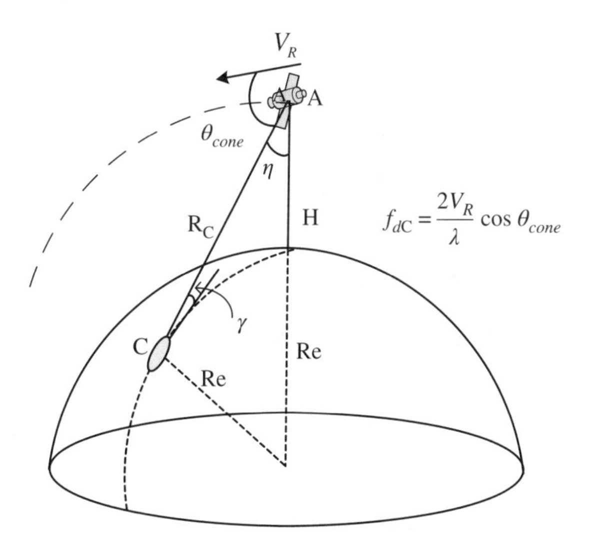


**Figure 2.43** Doppler diagram of the clutter in spaceborne AMTI mode.

where $V_R$ is the moving velocity of the radar platform, $\lambda$ is the wavelength, and $\theta_{cone}$ is the angle between the line that connects the clutter scattering point and the satellite and the satellite velocity, i.e. the cone angle.

It can be seen from Eq. (2.109) that when the clutter scattering point is located on the ground projection of the satellite track, that is, in Figure 2.43, $AC$ and $V_R$ are in the same plane, and the clutter range $R_c$ is equal to the visual range, which has a maximum clutter Doppler frequency, i.e.

$$f_{dC_{\max}} = \frac{2V_R}{\lambda}\cos\theta_{cone(\min)} = \frac{2V_R}{\lambda}\sin\eta_{\max} = \frac{2V_R}{\lambda}\frac{R_e}{R_e+H} \tag{2.110}$$

then the side lobe clutter range is $\left(-\frac{2V_R}{\lambda}\frac{R_e}{R_e+H}, \frac{2V_R}{\lambda}\frac{R_e}{R_e+H}\right)$.

In Figure 2.44, at a given cone angle $\theta_{cone}$ between the illumination direction of the antenna main lobe and the velocity direction of the satellite platform, the geometric relations of the target in a given plane (parallel to the ground plane) at different flying directions are given. Here, the Doppler frequency of the target is

$$f_{dT} = \frac{2V_T}{\lambda}\cos\alpha\cos\gamma + \frac{2V_R}{\lambda}\cos\theta_{cone} \tag{2.111}$$

where $V_T$ is the target moving velocity, $\alpha$ is the angle between $V_T$ and $TD$, $\gamma$ is the angle between $AT$ and $TD$, $T$ is the target position, and $TD$ is the projection of the sight line between the radar and the target $AT$ in the target plane. At a given main lobe illumination direction, $\theta_{cone}$ has a fixed value, therefore, $\frac{2V_R}{\lambda}\cos\theta_{cone}$ is a constant. The analysis of target Doppler needs to be focused on $\frac{2V_T}{\lambda}\cos\alpha\cos\gamma$.

It is known from the above analysis that in the moving target detection of the satellite platform, the very wide Doppler frequency coverage of the ground clutter makes most target detection areas under heavy ground/sea clutter. To detect the target, a clutter suppression algorithm (such as space-time adaptive processing [STAP]) must be used first to suppress clutter, such that the target can be detected.

In the process of clutter suppression, it is required to choose an appropriate PRF to improve the clutter suppression ability of the radar system. Optimizing frequency

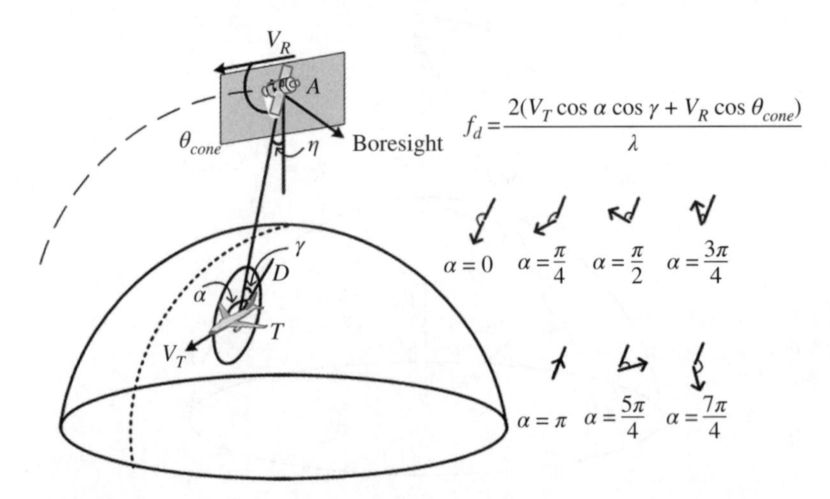


**Figure 2.44** Schematic diagram of Doppler structure of the target.

selection, on one hand, can reduce the blind region of the distance and velocity detection. On the other hand, it can reduce the clutter intensity and improve the clutter suppression effect. At a low-repetition frequency, the large velocity blind region is usually not capable of measuring speed. At medium-repetition frequency, there are different degrees of ambiguities in both range and velocity. The number of the range ambiguity is lower than that of the high-repetition frequency, and it is more affected by the side lobe clutter. There is an obvious bend clutter region in the range-Doppler spectrum. Although there is no ambiguity in speed at high-repetition frequency, the number of ambiguities in the range direction is relatively increasing. Although there is no clutter bending, clutter of the main lobe broadens, which will affect target detection at low speed. Therefore, high-repetition frequency is usually used in detection of long-distance, high-speed moving targets.

#### 2.5.3.2 Operating Range of Airborne Target

In airborne target detection, due to the heavy ground/sea clutter, it is required to suppress clutter in the range-Doppler 2D spectrum before target detection. Therefore, when detecting an airborne target, a PD system is usually used in the spaceborne AMTI radar. When the radar detects the target in the side lobe clutter zone, the operating range of the radar mainly depends on the ratio of the signal to the sum of the clutter and noise, and the radar action range equation is given by

$$R_{max}^4 = \frac{P_{av} G^2 \lambda^2 \sigma_t}{(4\pi)^3 L_\Sigma k T_0 B_d F_n \left( \frac{S}{C+N} \right)_{min}} \tag{2.112}$$

where $P_{av}$ is the average transmitter power, $L_\Sigma$ is the total loss of the radar system, $B_d$ is the Doppler filter bandwidth, $\left( \frac{S}{N} \right)_{min}$ is the minimum detectable SNR output by the Doppler filter, and $C$ is the clutter power. The action range in clutter background is calculated through multiplying free space action range by an attenuation factor $\alpha$ caused by clutter to noise ratio $C/N$. The attenuation factor $\alpha = 1/\sqrt[4]{1 + C/N}$.

#### 2.5.3.3 Minimum Detectable Velocity of the Moving Target

An AMTI radar system can measure the Doppler velocity of the target directly. This speed is the projection of the target velocity on the line of sight between target and radar, which is normally called the *radial velocity* of the target. In this mode, the clutter competes with the target. To detect the target, a clutter suppression algorithm (such as PD/STAP) is used to suppress the clutter before target detection. As the main lobe clutter has a certain width and clutter energy is large, the SNR even after the clutter suppression algorithm still cannot meet the detection requirements and is usually within a certain range of the main lobe clutter spectrum. In this case, the clutter spectrum distribution and the specific algorithm design determine the MDV of the radar system. The schematic diagram of the velocity nondetectable region corresponding to the radial velocity of the target and the MDV is shown in Figure 2.45, where the radial velocity is

$$V_r = V_T \cos\alpha \cos\gamma \tag{2.113}$$

Assuming $V_r^{detect\_min}$ is the minimum detectable velocity of the system, then for an airborne target at a speed of $V_T$, and at a given grazing angle $\gamma$, the nonmeasurable flight

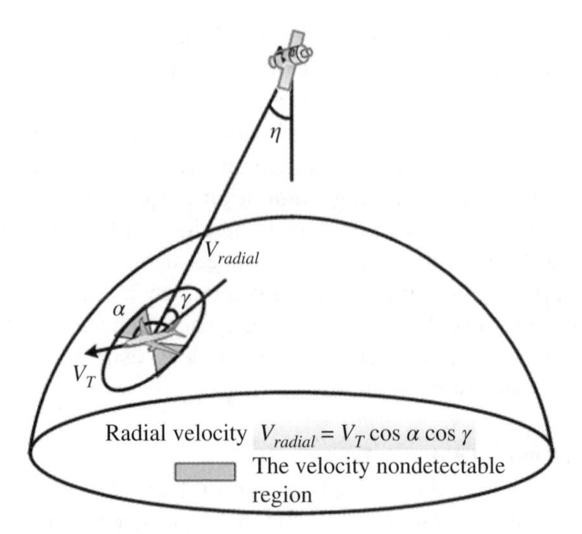


**Figure 2.45** Schematic diagram of the velocity nondetectable region corresponding to the radial velocity.

angle range of the target is calculated by

$$\alpha_1 = \arccos \frac{V_r^{detect\_min}}{V_T \cos\gamma}$$
$$\alpha_2 = \arccos \frac{-V_r^{detect\_min}}{V_T \cos\gamma} \tag{2.114}$$

## References


1 Gebert, N. (2009). *Multi-Channel Azimuth Processing for High-Resolution Wide-Swath SAR Imaging*. Karlsruhe, Germany: University of Karlsruhe.
2 Currie, A. and Brown, M.A. (1992). Wide-swath SAR. *IEE Proceedings F-Radar and Signal Processing* 139 (2): 122–135.
3 C. Heer, F. Soualle, R. Zahn (2003). Investigations on a high resolution wide swath SAR concept. *Proceedings of the 2003 IEEE International Geoscience and Remote Sensing Symposium*, Toulouse, France (July 21–25, 2003). Piscataway, NJ: IEEE.
4 Liu, Y. (1999). *Radar Imaging Technology*. Harbin: Harbin Institute of Technology Press.
5 Bao, Z., Xing, M., and Wang, T. (2005). *Radar Imaging Technology*. Beijing: Publishing House of Electronics Industry.
6 Brown, W.M. and Fredricks, R.J. (1969). Range-Doppler imaging with motion through resolution cells. *IEEE Transactions on Aerospace and Electronics Systems* 5 (1): 98–102.
7 Moore, R.K., Claassen, J.P., and Lin, Y.H. (1981). Scanning spaceborne synthetic aperture radar with integrated radiometer. *IEEE Transactions on Aerospace and Electronic Systems* 17 (3): 410–421.

8 A. Moreira, J. Mittermayer, R. Scheiber (2000). Extended chirp scaling SAR data processing in stripmap, ScanSAR and spotlight imaging mode. *Proceedings of the European Conference on Synthetic Aperture Radar*, Munich, Germany (May 23–25, 2000). Piscataway, NJ: IEEE.

9 Bamler, R. (1995). Optimum look weigting for burst-mode and ScanSAR processing. *IEEE Transactions on Geoscience and Remote Sensing* 33 (3): 722–725.

10 Eldhuset, K. and Weydahl, D.J. (2011). Geolocation and stereo height estimation using TerraSAR-X spotlight image data. *IEEE Transactions on Geoscience and Remote Sensing* 49 (10): 3574–3581.

11 A. Meta, P. Prats, U. Steinbrecher, J. Mittermayer, R. Scheiber (2008). TerraSAR-X TOPSAR and ScanSAR comparison. *Proceedings of the European Conference on Synthetic Aperture Radar*, Friedrichshafen, Germany (June 2–5, 2008). Berlin, Germany: VDE Verlag.

12 Mittermayer, J., Moreira, A., and Loffeld, O. (1999). Spotlight SAR data processing using the frequency scaling algorithm. *IEEE Transactions on Geoscience and Remote Sensing* 37 (5): 2198–2124.

13 Lanari, R. et al. (2000). Spotlight SAR data focusing based on a two-step processing approach. *IEEE Transactions on Geoscience and Remote Sensing* 39 (9): 1993–2004.

14 Alberto, M., Josef, M., and Rolf, S. (1996). Extended chirp scaling algorithm for air – and spaceborne SAR data processing in stripmap and ScanSAR imaging modes. *IEEE Transactions on Geoscience and Remote Sensing* 34 (5): 1132–1136.

15 J. Mittermayer, R. Lord, E. Borner (2003). Processing for TerraSAR – X using a new formulation of the extended chirp scaling algorithm. *Proceedings of the 2003 IEEE International Geoscience and Remote Sensing Symposium*, Toulouse, France (July 21–25, 2003). Piscataway, NJ: IEEE.

16 Ossowska, A. and Rainer, S. (2009). Processing of sliding spotlight mode data with consideration of orbit geometry. *Proceedings of SPIE* 7502: 750200.

17 Naftaly, U. and Nathansohn, R.L. (2008). Overview of the TECSAR satellitehardware and mosaic mode. *IEEE Geoscience and Remote Sensing Letters* 5 (3): 423–426.

18 Shum, H.-Y. and Szeliski, R. (2000). Systems and experiment paper: construction of panoramic image mosaics with global and local alignment. *International Journal of Computer Vision* (2): 101–130.

19 Pau, P., Rolf, S., Josef, M. et al. (2010). Processing of sliding spotlight and TOPS SAR data using baseband azimuth scaling. *IEEE Transactions on Geoscience and Remote Sensing* 48 (2): 770–780.

20 Zen, F.D., Guarnieri, A.M. et al. (2006). *IEEE Transactions on Geoscience and Remote Sensing* 44 (9): 2352–2360.

21 Meta, A., Mittermayer, J., Prats, P. et al. (2010). TOPS imaging with TerraSAR-X: mode design and performance analysis. *IEEE Transactions on Geoscience and Remote Sensing* 48 (2): 759–769.

22 Longmei, X., Lei, S., and Jialong, G. (2011). Analysis and simulation of ambiguities on spaceborne sliding spotlight SAR. *Aerospace Shanghai* (1): 1–6.

23 Nicolas, G., Gerhard, K., and Alberto, M. (2010). Multichannel azimuth processing in ScanSAR and TOPS mode operation. *IEEE Transactions on Geoscience and Remote Sensing* 48 (7): 2994–3008.

# 3

# Antenna System

## 3.1 Overview

An antenna realizes the conversion of a guided wave to a space electromagnetic wave. It converges the radiation energy toward the target and improves the gain. Receiving means that it converges the reflected signal from the target area. With the rapid increase in the demand for synthetic aperture radar (SAR) applications, it is a big challenge for radar system engineers to continue to increase the clarity of the image. A high-efficiency antenna is one of the most effective ways to improve the image quality of radar. Broadband frequency is also a prerequisite of high resolution.

Most antenna radiation has directivity, which means the radiation power is high in some directions and low in the others. The radiation pattern is used to represent the distribution of the antenna radiation energy in space. It shows relative magnitudes of the radiated power in different directions. The radiation pattern is the most important and basic parameter of the antenna. The pattern is normally represented by two orthogonal main planes. Antenna beamwidth represents the concentration of the antenna radiation in a certain plane. The beamwidth or 3 dB beam width is defined as the angle between two half-power levels. The smaller the beamwidth is, more power is radiated. There are many lobes in the antenna pattern, such as the main lobe, the first side lobe, and the second side lobe. The ratio of the maximum value of the side lobe to the maximum value of the main lobe is defined as side lobe level, which is generally expressed in dB, as shown in Figure 3.1.

Gain is another important antenna parameter. It is defined as the ratio of the radiation intensity of the antenna in the maximum radiation direction to radiation intensity of an ideal isotropic antenna at the same direction under the same input power. It represents concentration of the antenna radiation energy. The radiation energy reduction due to the loss of antenna itself is also taken into account. There are also many other antenna parameters, such as polarization, input impedance, frequency bandwidth, effective area, and so on. To realize excellent antenna radiation performance requires systematic consideration, which involves the design, manufacturing, testing, and so on.

Reflector antennas, centralized/distributed array antennas, and phased array antennas have been successfully used in SAR. The planar array antenna, both conventional and phased controlled, plays an important role in SAR systems. The design of planar array antennas is discussed in this chapter.

*Design Technology of Synthetic Aperture Radar,* First Edition. Jiaguo Lu.

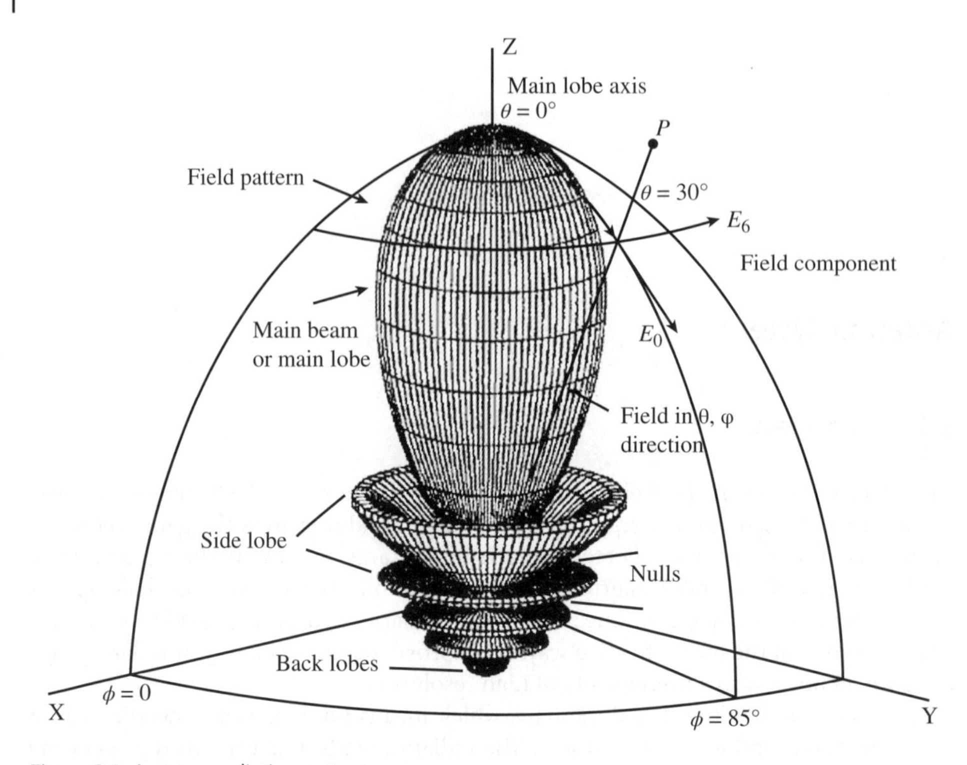


**Figure 3.1** Antenna radiation pattern.

## 3.2 Antenna Design and Analysis

In the design and analysis of planar array antennas, many factors should be considered, such as technical requirements of a SAR system, realization of the planar array antenna, project cost, realization of the antenna system, antenna array structure and antenna elements, performance deviation caused by processing, assembly, and system integration. Compared with other platforms, a spaceborne SAR system features a long operation distance, large size, light weight, and high reliability in the harsh space environment. The satellite antenna is analyzed as an example below.

The spaceborne SAR system experiences the ground, launch process, on-orbit operation, etc. During antenna design and simulation, the influence of these conditions is taken as a key consideration.

As we all know, a certain antenna power aperture product is a prerequisite of operating a spaceborne SAR system. In principle, the antenna size is as large as possible, and power is as low as possible. However, a variety of factors should be considered in the actual design. Antenna beam-forming and beam-scanning capability are the foundation for realizing the multimode operation and improving the observation ability of a spaceborne SAR system. A high-precision calibration system can effectively guarantee the quantitative ground observation of a spaceborne SAR system. Therefore, it is required to comprehensively analyze the antenna aperture size, beam forming, scanning capability, and internal calibration parameters from the spaceborne SAR system and planar array antenna point of view. These are common issues for spaceborne SAR systems and spaceborne SAR antennas.

### 3.2.1 Basic Parameters

The basic parameters of a SAR antenna are described in the following sections.

#### 3.2.1.1 Bandwidth

The working bandwidth of a planar array antenna generally refers to instantaneous working bandwidth. The resolution of SAR is closely related to the working bandwidth – the wider the bandwidth, the higher the resolution of the radar. The planar array antenna changes the phase of the antenna element, so that the electromagnetic wave energy radiated by each antenna is in the same phase at a certain direction in space, forming the beam of the array antenna. Wavefront is defined as the plane perpendicular to the incident direction of an electromagnetic wave. When an electromagnetic wave is traveling perpendicular to the array, the wavefront is parallel to the antenna array. After the phase difference of each transceiver channel unit is offset, the electromagnetic wave of each unit in the array can be superimposed in phase, and the array beam is made. When the electromagnetic wave incident angle is not perpendicular to the array because of the phase difference between the wavefront of each unit, the performance of the conventional phased array antenna is related to the frequency, which limits the working bandwidth. Therefore, delay lines are usually required to eliminate or weaken the beam dispersion effect in broad bandwidth in high-resolution SAR.

#### 3.2.1.2 Scanning Range

The antenna beam-scanning range should be related to its work mode for the SAR system. In general, antenna beam scanning is not required in the conventional strip-map mode (without considering yaw drag), spotlight mode requires the antenna beam be capable of scanning in the azimuth, and scanning mode requires the antenna be capable of scanning in the range direction. The size of the scanning angle is defined by the height of the platform, viewing angle, swath width, and resolution. Under usual circumstances, SAR with wide-area moving target monitoring mode requires a larger beam-scanning angle. The antenna beam-scanning range is mainly related to the choice of antenna element, element spacing, element arrangement, and element number.

#### 3.2.1.3 Beam Width

The amplitude and phase of a single radiating element in a planar phased array antenna can be independently controlled. To effectively utilize the power and improve the performance of the radar system, the amplitude weighting is not performed on the antenna array in common transmitting. If necessary, only phase weighting is carried out, and the receiving state is amplitude or phase weighted according to the requirement. In the case of uniform distribution, the empirical formula for antenna beam width is

$$\theta_B = \frac{51\lambda}{L} \tag{3.1}$$

where $\theta_B$ is the 3dB beam width in degrees, and $L$ is the antenna size. When the beam is scanned to the angle $\theta_0$, the empirical formula of beam width is

$$\theta_B(\theta_0) = \frac{\theta_B}{\cos\theta_0} \tag{3.2}$$

In the case of amplitude weighting, the antenna beam width will be further broadened.

#### 3.2.1.4 Antenna Gain

Antenna gain is defined as energy gain of the antenna in the maximum radiation direction, whose empirical formula is given by

$$G(\theta_0) = 10\log_{10}\frac{A}{\theta_B\phi_B} \tag{3.3}$$

where $\theta_B$ and $\varphi_B$ are the 3dB beamwidth in two main planes in degrees, and $A$ is a constant whose value is 2600~3200. The value of $A$ is related to depth of weighting. Normally, $A$ is 3200 for equal amplitude weighting and 2800 for amplitude weighting of −30 dB side lobe.

#### 3.2.1.5 Side Lobe Level

A lower side lobe level is always desired. The desired amplitude and phase distribution during design are obtained by optimization. By controlling amplitude and phase distribution of the antenna aperture, low antenna side lobe is achieved. The realization of low side lobe of the antenna is limited because various errors exist objectively and are difficult to eliminate – for instance, reduction of the antenna gain, beam widening, cost increase, and improving the accuracy of machining.

### 3.2.2 Antenna Aperture Size

Similar to conventional ground/airborne phased array radar antennas, spaceborne SAR phased array antennas pursue maximum power aperture product, low side lobe level, and low cost under multiple constraint conditions.

Antenna size in the spaceborne SAR system is the most important design parameter. It is related not only to the ambiguity, azimuth resolution, and swath width but also the SAR resolution, sensitivity, and image quality.

#### 3.2.2.1 From the Ambiguity Point of View

The minimum unambiguity area of the SAR antenna is shown in Eq. (3.4). If the height of the satellite orbit is 632 km, then the minimum antenna area of an X-band phased array antenna that meets the requirements of the ambiguity will increase as the antenna viewing angle goes up, as shown in Figure 3.2.

$$A_{min} = \frac{kV_s\lambda R}{c}\tan\theta \tag{3.4}$$

where $\theta$ is the incident angle of the antenna beam, $V_s$ is platform moving speed, $R$ is the distance between the radar and the target, $c$ is the velocity of the electromagnetic wave, and $\lambda$ is the operation wavelength of the radar.

For a specific power aperture product, large aperture size will be chosen to obtain the maximum antenna round-trip gain and try to reduce the transmitter power to save satellite energy. The power aperture product is usually calculated when the viewing angle is maximized (the operation distance is the longest).

In the spaceborne case, as the moving speed of the beam in the illuminated area (ground speed $V_g$) is slower than that of the platform (space velocity $V_s$), in calculating azimuth resolution, a coefficient of $k_v = V_g/V_s \approx 0.9$ needs to be multiplied. Even so,

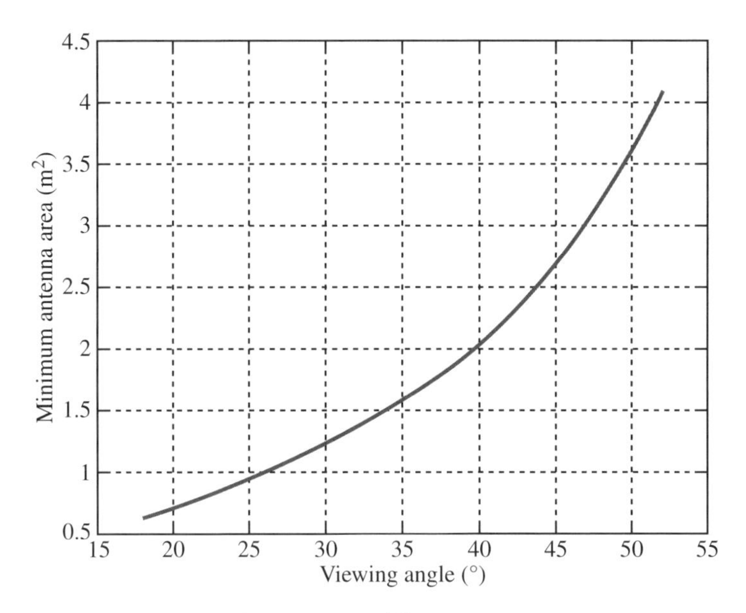


**Figure 3.2** Antenna minimum unambiguity area at different perspectives.

the azimuth resolution of spaceborne SAR is very close to half of the antenna aperture in the azimuth. For instance, the antenna azimuth dimension should be less than 6 m for a 3 m azimuth resolution, and the antenna azimuth dimension should be less than 2 m to achieve a 1 m azimuth resolution. However, the antenna aperture size during design should not be undersized, and the effective aperture size of the antenna aperture after phase weighting shall be taken into account.

The size of the antenna aperture in the range direction also needs to meet the requirements of the swath width. The swath width is defined by the antenna beam width in the range direction, echo window width, sample number in the range direction, analog-to-digital (A/D) sampling frequency, etc. When the time-sequence relationship of the radar system is defined, according to the time-sequence relationship between transmitting and receiving the radar signal as well as the corresponding electromagnetic wave propagation, scattering, and geometrical relationship of the echo, the swath width can be calculated. Specifically, the actual needed beam width in the range direction, echo window time width, pulse repetition frequency, etc. can be calculated with different viewing angles and incident angles, short range and long range.

The azimuth and range ambiguities in the spaceborne SAR system are important parameters related to the image quality. Restricted by the ambiguity condition, the minimum unambiguity antenna area under different $k$ factors can be calculated from Eq. (3.4). Figure 3.3 shows the curves of the minimum antenna area vary with the incident angle at $k = 4, 5, 6, 8$, at an orbit altitude of 632 km. It can be seen from Figure 3.3, that at an incident angle of 62° (i.e. viewing angle of 53°), the minimum antenna area is $A \geq 10.5\,\text{m}^2$ for $k = 6$ and $A \geq 14\,\text{m}^2$ for $k = 8$. If the ambiguity $\leq -20$ dB (in the azimuth and range directions) and $k = 8$, the antenna area should be $A \geq 14\,\text{m}^2$ according to Figure 3.3.

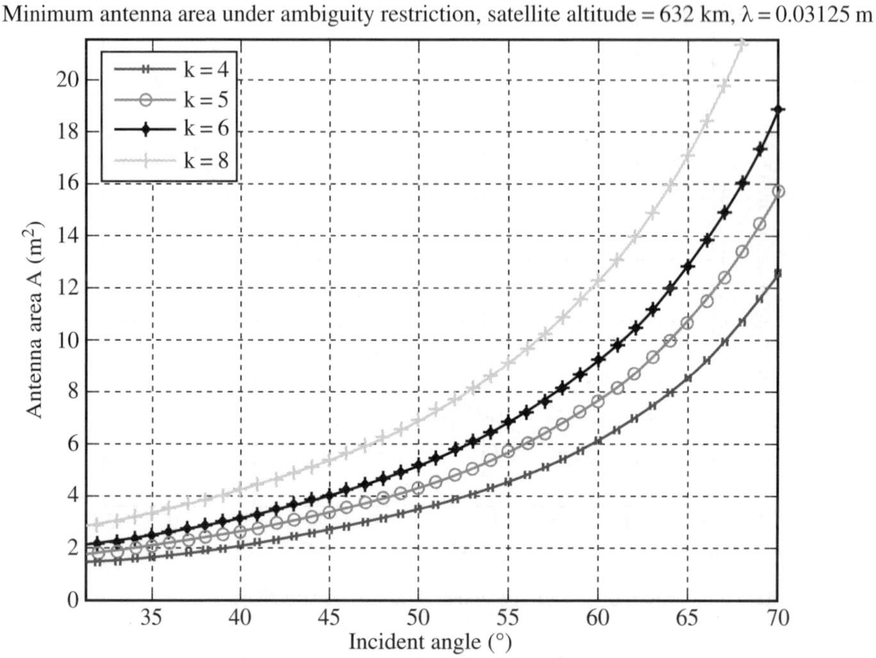

Minimum antenna area under ambiguity restriction, satellite altitude = 632 km, λ = 0.03125 m

**Figure 3.3** Calculated minimum antenna area under ambiguity restriction.

#### 3.2.2.2 Resolution Limitation

In strip-map mode, if resolution is $\delta_x$, the antenna size in azimuth $L_a$ is given by

$$L_a \leq \frac{2\delta_x}{k_r} \tag{3.5}$$

For example, if $\delta_x = 0.6$ m and $k_r = 1.2$, then $L_a = 1$ m is obtained by Eq. (3.5).

#### 3.2.2.3 Swath Width Limitation

The swath width in the range direction of SAR imaging is defined by the beam width in the range direction, as shown in Eq. (3.6).

$$W_g = \frac{\lambda R}{L_r \cos\theta} \tag{3.6}$$

For active phased array antennas, the beam width in the range direction can be broadened as required by the swath width. Under strip-map mode, with a resolution of 0.6 m and a swath width of 15~20 km, simulation results are shown in Figure 3.4, and the antenna size in the range direction meets $L_r > 3.2$ m when the incident angle is 60°.

#### 3.2.2.4 System Sensitivity Limitation

The signal-to-noise ratio (SNR) of SAR images is related to system sensitivity, as shown in Eq. (2.17). The system sensitivity of SAR is represented by the noise equivalent sigma zero ($NE\sigma_0$); the higher the SNR for a SAR image is required, the lower the $NE\sigma_0$ required.

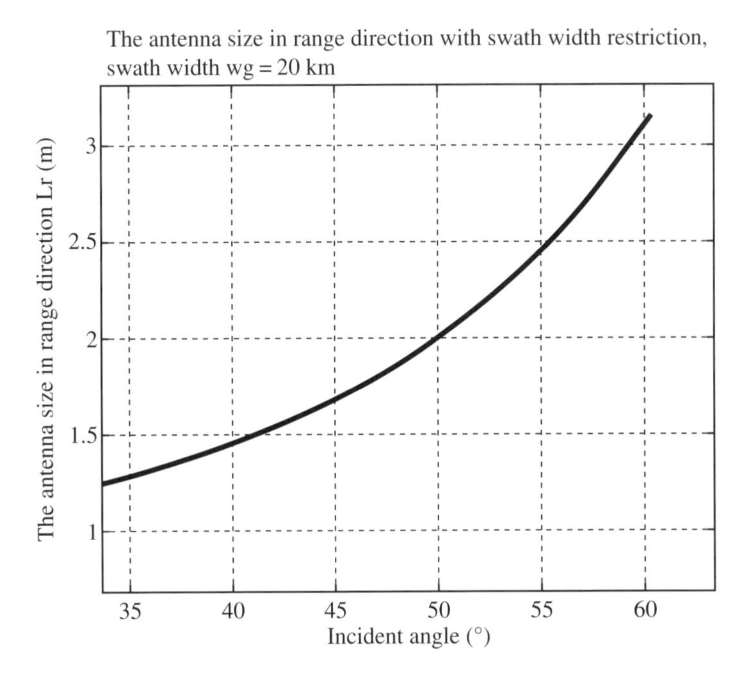


**Figure 3.4** Calculated antenna size in the range direction in strip-map mode.

In the case of resolution determination, if we want to improve the system sensitivity and reduce $NE\sigma_0$, we need to increase the average transmit power, the antenna aperture area, or the antenna efficiency. Since $NE\sigma_0$ is inversely proportional to antenna aperture area and the square of antenna efficiency, increasing the effective area of the antenna aperture or improving antenna efficiency can significantly improve the system sensitivity.

In summary, antenna aperture size is limited by such factors as ambiguity, resolution, swath width, system sensitivity, $NE\sigma_0$, and so on. To meet $NE\sigma_0$ and ambiguity requirements, the antenna aperture area should be large. However, the antenna is required to be small in the range and azimuth directions to meet the resolution and swath width requirements. They are contradictory. Therefore, it is difficult to design the antenna size satisfying the above four factors at the same time, which must be compromised.

### 3.2.3 Scanning Feature

There are many operation modes for spaceborne SAR, such as strip-map SAR, spotlight SAR, and ScanSAR. To meet the azimuth resolution and the swath width at different viewing angles and under different working modes, it is necessary to design the beam in the range and azimuth directions under different operation modes. In general, an antenna array has rectangular aperture or rectangular grid. When designing a beam in the range direction, it is required to consider not only beam width, beam shape, side lobe level, and gain but also ambiguity requirements. When designing an azimuth beam, in addition to considering the multiple modes required for the system, shape, width, side lobe level, and gain of the beam are also taken into account. Different operation modes require the antenna to have different beam widths. For example, spaceborne SAR requires 1~3 m strip-map resolution, where the azimuth beam width of the antenna with 1 m resolution is three times that of a 3 m resolution antenna. To keep the swath width the same at different viewing angles, the antenna beam in the range direction needs to be broadened.

When the antenna is transmitting, a saturated power amplifier is usually adopted. Therefore, only phase weighting is considered. When the antenna is receiving, normally both amplitude and phase weighting are adopted to obtain better side lobe features. Antenna beam synthesis is an optimization process of the antenna beam under specified conditions. The target of the range direction beam synthesis is to minimize electrical level fluctuation, to maximize gain, and to obtain an optimized side lobe (to minimize distance-to-ambiguity ratio) within the observation area. The target of the azimuth beam synthesis is to minimize beam width and the electrical level fluctuation of the beam width.

The optimization and synthesis method of antenna beams includes scattered focus, inverse feeding of adjacent cells at the edge, genetic algorithm, particle swarm optimization algorithm, etc. [1, 2] The former two methods are relatively simple, but their beam broadening effects are relatively bad, the ripple of the beam after broadening is large, and the side lobe level is deteriorated.

The genetic algorithm and the particle swarm optimization algorithm are relatively better methods that can optimize the amplitude and phase separately, or only the phase. The targets of the optimization are beam width, beam shape, side lobe level, gain, and so on. The genetic algorithm offers strong local search ability in optimizing the excitation coefficient of the antenna element. The adaptive operator is used to speed up the convergence rate. The mean value of the convergence result is also slightly better than that of the fixed mutation probability.

For example, assuming that the number of antenna elements is 120, and the spacing is 0.688 $\lambda_0$, the synthesis results of the antenna beam are shown in Figures 3.5 and 3.6.

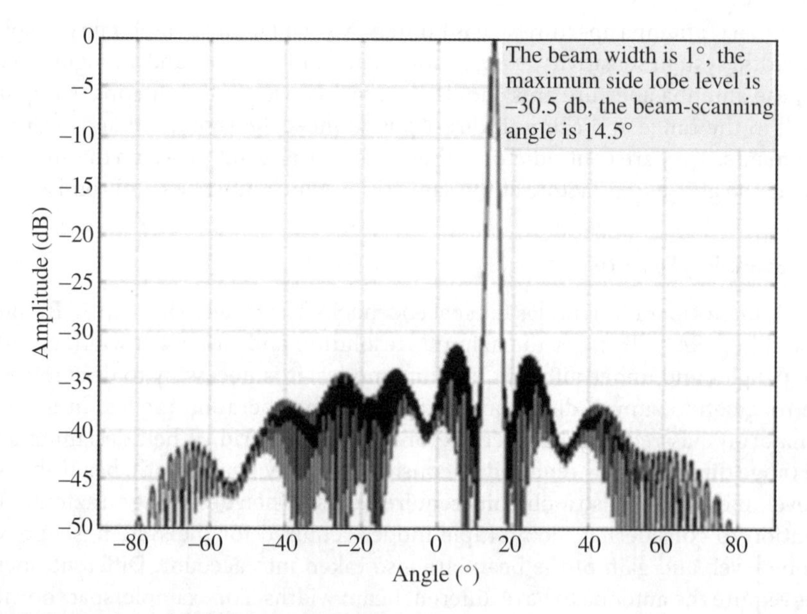


**Figure 3.5** Synthesis results of 1° antenna beam.

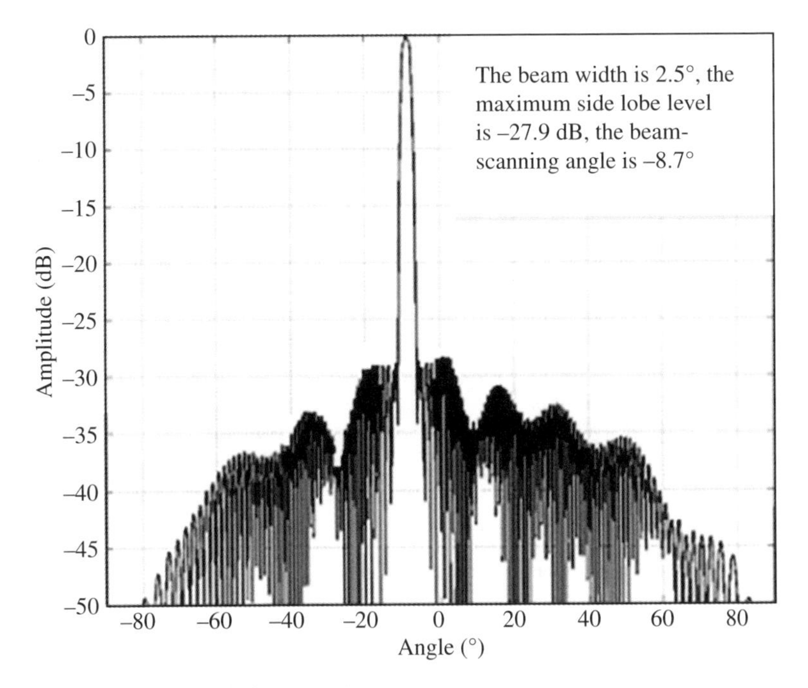


**Figure 3.6** Synthesis results of 2.5° antenna beam.

In Figure 3.5, the targets of beam synthesis are beam width 1°, side lobe level −30.5 dB, and beam-scanning angle 14.5°. In Figure 3.6, the targets of beam synthesis are beam width 2.5°, side lobe level −27.9 dB, and beam-scanning angle −8.5°.

A large azimuth (±60°) and a large elevation beam-scanning angle are usually required in ground-based and airborne phased array antennas. However, spaceborne SAR expects around ±20° scanning angle in the range direction under normal strip-map, spotlight, scanning, and ground moving target indication operation modes, to realize earth observation with a variable viewing angle. Only spotlight SAR, sliding SAR, and mosaic SAR require the beam to have certain azimuth scanning ability.

### 3.2.4 Internal Calibration

The instability of many parameters in a spaceborne SAR system may result in measurement error for the backscattering coefficient of the target. To realize precise measurement of a ground target, a spaceborne SAR system usually performs both internal and external system calibration. External calibration can mainly complete measurements of the transfer function in the SAR radar system and the radar antenna pattern. It can also measure the SAR transmitting power. The internal calibration mainly completes two functions: calibrating gain and power, monitoring the main work status of the radar. That means internal calibration corrects the radar system to ensure normal operation. It mainly concerns the absolute value of the channel change, fulfilling calibration of the gain in the receiver of the SAR radar and power of the transmitter. In the meantime, the signal characteristics of the SAR radar, such as range compression main lobe width, peak side lobe level, and integral side lobe, can be monitored.

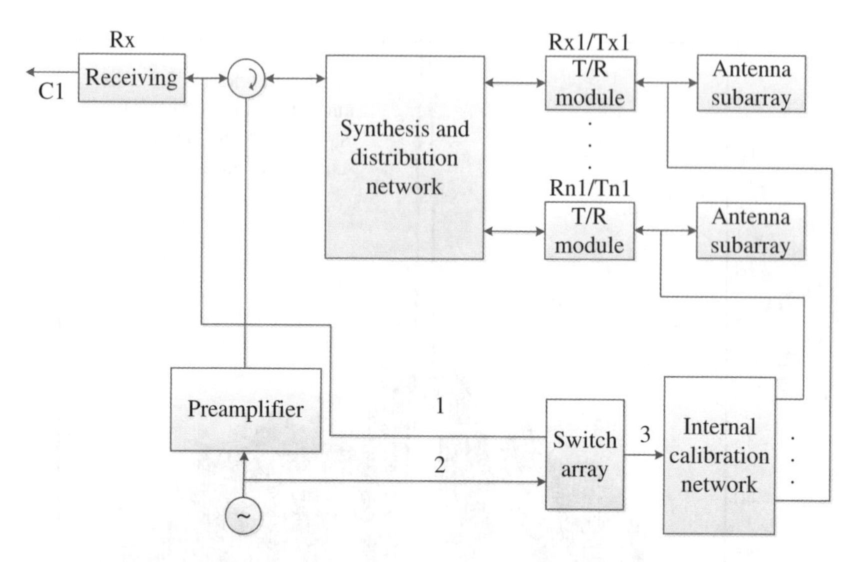


**Figure 3.7** Block diagram of internal calibration principle.

The internal calibration can calibrate the antenna channel as well. It can monitor the transmission characteristics of multiple channels in the active antenna system and perform the real-time compensation for maintaining the amplitude and phase relationship between each transmission channel. From the perspective of antenna correction, waveform correction based on the fast Fourier transform method is simple in hardware [3]. However, when considering the factors of internal calibration, equipment with a large amount of feedback correction network is needed. The amplitude and phase distribution of each feeding network can be obtained by using system calibration data or special working conditions (such as detecting components one by one), so as to realize detection of aging and failure in components and modules. The internal calibration principle is shown in Figure 3.7.

The transmit power and the receiver gain are measured by using time sharing. The procedure is divided into three steps. First, a coupling signal is produced at the transmitter/receiver (T/R) module output port and goes through the internal calibration network and switch array, whose ports 1 and 3 are connected at the moment. The signal reaches the data collector after passing the receiver, and the available power equation is given by

$$T_{x1}R_xk_1 = C_1 \tag{3.7}$$

where $k_1$ is the power transfer constant, $T_{x1}$ is the transmit power of the T/R module, $R_x$ is the receiver gain, and $C_1$ is the collected data.

Then a signal coupled from the output port of the signal source is input to the switch array, whose ports 2 and 3 are connected now, and the signal is coupled with the input signal at the output port of the T/R module after going through the internal calibration network. Then a monitor signal is obtained at the output port of the A/D, whose power equation is given by

$$\alpha R_{x1}R_xk_2 = C_2 \tag{3.8}$$

where $\alpha$ is the signal power, $k_2$ is the power transfer constant, $R_{x1}$ is the receiver gain in the T/R module, and $C_2$ is the collected data.

Finally, a signal coupled from output of the signal source is input to the switch array, whose ports 1 and 2 are connected now, the signal is input to the receiver through port 1, and monitor data is obtained by a data-forming device, with the power equation

$$\alpha R_x k_3 = C_3 \tag{3.9}$$

The transfer function of the receiving and transmitting system can be obtained by triple measurements, and it is given by

$$R_{x1} T_{x1} R_x = \frac{C_1 C_2 k_3}{k_1 k_2 C_3} \tag{3.10}$$

where $R_{x1}$ is the receiver gain in a single T/R module, $T_{x1}$ is the transmitter gain in a single T/R module, and $R_x$ is the main receiver gain. From Eq. (3.8), the output power change of each T/R module can be obtained. The receiver gain in the T/R module can be obtained from Eqs. (3.9) and (3.10).

For an X-band 32-element phased array antenna, if both phase shifter and attenuator in the T/R module are six-bit, an internal calibration with resolution better than 0.5 dB can be achieved by the above internal calibration method. And the calibration accuracy of the antenna amplitude and phase can be verified if the root-mean-square error in the amplitude is better than 0.12 dB, and the root-mean-square error in the phase is better than 1.8°.

## 3.3 Antenna Array

Antennas applied to SAR have many forms, such as the reflector antenna, the microstrip antenna array, the slotted waveguide antenna array, the printed dipole antenna array, and the open waveguide antenna array. The choice of antenna mainly depends on the system requirements, and it is optimized in terms of electrical properties, structure, size, and cost. In this section, several common antenna elements of planar array antennas are analyzed and discussed.

### 3.3.1 Microstrip Patch Antenna

A microstrip patch antenna is created by mounting a conductor sheet on dielectric substrate with a conductor ground plate. It can be fed by microstrip line, dielectric slab line, and coaxial probe. An electromagnetic field is excited between the conductor patch and the ground plate and is radiated out via the gap between patch and ground plate. Usually, the thickness of the dielectric substrate is very small compared with the wavelength. Therefore, compared with a conventional antenna, a microstrip antenna offers unique advantages: low profile, small size, and light weight. Because of the planar structure, it is easy to integrate with feeding networks and an active circuit. It is also easy to realize circular polarization, dual polarization, and dual-band working capacity. The main disadvantages of the single-layer microstrip patch antenna are narrow bandwidth and low efficiency, especially when used in subarray. The design process of a multilayer microstrip patch antenna with broadened bandwidth is complicated, and its thermal performance is moderate.

In a phased array antenna, the microstrip antenna element can be directly connected to the T/R module, so that the loss of the microstrip transmission line can be greatly reduced. For a microstrip antenna, if the input impedance meets the requirements of the frequency band, the other parameters are not a concern. Therefore, the frequency bandwidth of a patch antenna is normally defined as the frequency range where the voltage standing wave ratio is smaller than a given value. The relative frequency bandwidth of a conventional microstrip antenna is 1%~6%, and that of a multilayer or back cavity one is 20%~30% of the center frequency [4]. The most effective method to broaden the bandwidth of a microstrip antenna is to employ multiple structures. It is well known that in circuit theory, the bandwidth can be broadened when using a tightly coupled loop with a staggered tuning. According to a similar principle, the bandwidth of multilayer microstrip antenna can also be broadened. For this reason, a microstrip antenna in multilayer structure is realized by stacking layers, for example, the log periodic structure is an effective way to achieve broadband. The broadband performance is obtained at the expense of increasing the antenna thickness. Another more effective method is using back cavity. The supporting structure of a microstrip antenna is an air cavity mounted at the bottom of the patch, which reduces the effective $\varepsilon_r$ and expands bandwidth of the antenna, due to the cavity modes.

#### 3.3.1.1 Microstrip Antenna Analysis

Three types of microstrip antenna analysis methods are commonly used.

The first one obtains the current distribution of the microstrip patch antenna using a transmission line model, then obtains the antenna element aperture field in spectral domain, and finally transforms to the antenna pattern in the two main planes at point $P\ (R, \theta, \varphi)$ [4]:

$$F_\theta(\theta,0) = F_c(\theta,0)\frac{\cos\theta}{\left|\cos\theta - j\sqrt{\varepsilon_r - \sin^2\theta}\cot\left(K_0 h\sqrt{\varepsilon_r - \sin^2\delta}\right)\right|}\cdot$$

$$\left|\cos\theta + \frac{(\varepsilon_r - 1)\sin^2\theta}{\varepsilon_r\cos\theta + j\sqrt{\varepsilon_r - \sin^2\theta}\tan\left(K_0 h\sqrt{\varepsilon_r - \sin^2\delta}\right)}\right|$$

$$F_\phi\left(\theta,\frac{\pi}{2}\right) = F_c\left(\theta,\frac{\pi}{2}\right)$$

$$\frac{\cos\theta}{\left|\cos\theta - j\sqrt{\varepsilon_r - \sin^2\theta}\cot\left(K_0 h\sqrt{\varepsilon_r - \sin^2\delta}\right)\right|} \tag{3.11}$$

where $F_\theta(\theta,0)$ is the radiation pattern at $(\theta,0)$, $\varepsilon_r$ is the relative dielectric constant, $F_c(\theta,0)$ is the radiation pattern of the element at $(\theta,0)$, and $P_r$ is the radiation power, where

$$F_c(\theta,0) = \frac{\cos\theta\left(\frac{1}{2}K_0 b\sin\theta\right)}{\left(\frac{K_0 b}{\pi}\sin\theta\right)^2 - 1}$$

$$F_c\left(\theta, \frac{\pi}{2}\right) = j_0\left(\frac{1}{2}K_0 a \sin\theta\right) \tag{3.12}$$

It is shown in Eqs. (3.11) and (3.12) that the pattern of a microstrip antenna is a product of the Fourier transform factor of the patch current and the "microstrip factor" derived from the Green function. When $\theta = \pi/2$, the microstrip factor is always in the $\varphi = 0$ plane (E plane) and $\theta^{\phi} = \pi/2$ plane (H plane). Due to the existence of conductor ground plane, it is equivalent to a negative image of a zero current leading the electric field to be zero at $\theta = \pi/2$. As can be also seen from Eqs. (3.11) and (3.12), the current is not directly related to the substrate parameters, but the microstrip factor is derived from the Green function.

The second is the variable separation method, such as the cavity model method. It is based on the assumption of magnetic wall cavity. Either the expansion method or mode matching method is used to solve the internal field of the microstrip patch antenna. The external field is obtained by using equivalent magnetic current distribution of the antenna. When solving separation of variables, the cavity model is suitable for thin microstrip antenna with regular shape. For thick microstrip substrate, the assumption of magnetic wall cavity is not suitable.

The third method is based on the integral equation. The distribution of the equivalent magnetic current or the patch current can be obtained by using the cavity model or transmission line model. By integrating the Green function with the source distribution, the total field can be obtained. The advantage of this method is that the field source (or equivalent field source) is not derived from the solution of the integral equation; moreover, the microstrip and fringe effect can be considered.

For microstrip antennas with complex boundary conditions, the integral equation method is an ideal choice. The space domain or frequency domain Green functions satisfying the boundary conditions are derived, so as to establish the current integral equation of the microstrip structure, which is solved by the method of moments. The fundamental is an integral equation converted into matrix form, and a solution with high accuracy can be obtained numerically.

#### 3.3.1.2 Microstrip Antenna Design

The first step of designing the microstrip antenna array is to select a suitable radiation unit according to the characteristics, the specifications, and the difficulty in fabricating the array. If two kinds of linear polarizations are required, the antenna element must be two-dimensional (2D) symmetric. Obviously square and circular-shaped patches can meet this requirement. In addition, from the feeding mode point of view, dual-polarization microstrip antennas mainly include probe fed, coplanar microstrip fed, and slot coupled.

There is much literature on microstrip antennas. This chapter focuses on discussion and analysis of corner-fed, side-fed, and side-fed and slot-coupled-fed dual-polarized antenna elements.

***Angle-Fed Hybrid-Feed Dual-Polarized Microstrip Patch Antenna Element*** A corner-fed dual-polarized microstrip patch antenna array offers the advantage of 10 dB port isolation [5]. The double-point corner-fed square microstrip antenna shown in Figure 3.8 has the same advantages in a dual-polarization microstrip antenna array compared to a double-point side-feed microstrip patch antenna. Because of the need to arrange two

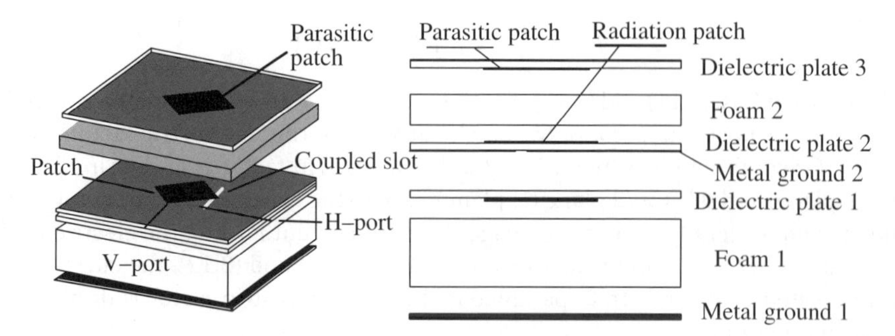


**Figure 3.8** Corner-fed double-layer microstrip patch antenna element.

sets of dual-polarization feeding networks in a small space in a linear array, a hybrid feeding structure is used in the antenna element [6].

The antenna elements are divided into six layers from bottom to top: metal ground 1, foam 1, dielectric plate 1, dielectric plate 2, foam 2, and dielectric plate 3. The radiating patch and vertical polarization feeder are located on the upper surface of dielectric plate 2. A metal film with a coupling slot is between dielectric plates 1 and 2. The horizontal polarization feeder is on the lower surface of dielectric plate 1. The parasitic patch that broadens the bandwidth is placed upside down on the lower surface of dielectric plate 3. Dielectric plate located above 3 can also function as a radome. To reduce the back radiation of the antenna, and to facilitate the antenna installation, a reflecting metal plate is arranged at a quarter wavelength below dielectric plate 1 and used as the antenna installation base as well, with low dielectric constant foam 1 placed between them.

The above-mentioned antenna element is a broadband dual-polarization microstrip patch element. The antenna size at X band after optimization is as follows: a = 9 mm, b = 9.7 mm, $W$ = 1 mm, $L$ = 8 mm, $S$ = 2.95 mm, and the thickness of foam 1 and foam 2 is 6.5 and 2.6 mm, respectively. The size of the reflecting metal plate is $30 \times 30\ \text{mm}^2$. All dielectric plates adopt the same material, and the relative dielectric constant is 2.94 with 0.508 mm thickness. Figure 3.9a gives the S-parameter simulation results of the antenna element. It can be seen that the bandwidth of less than −10 dB reflection coefficient at the port of horizontal polarization and vertical polarization is 14.5% and 15.9%, respectively, which covers 8.86∼10.25 and 8.74∼10.25 GHz, respectively. The isolation performance is not superior, mainly due to the horizontal polarization port passing through a slender slot coupling, and this coupling slot has changed the symmetry of the antenna, resulting in distortion of the symmetrical distribution in the electromagnetic field.

Figure 3.9b and c show the antenna radiation pattern, including main polarization and cross polarization. For the vertical polarization port, the cross-polarization component of the E plane is 20 dB lower than the major polarization component. The H plane is slightly worse, and the polarization is only −15 dB. For the horizontal polarization port, the cross-polarization components of the E and H surfaces are all 19 dB lower than the main polarization component. The calculated gain of antenna elements is about 9.9 dB. By comparing the radiation pattern in the main polarization and cross-polarization directions when the two ports are being fed, we can see that the coupling slot reduces the cross-polarization performance in the antenna.

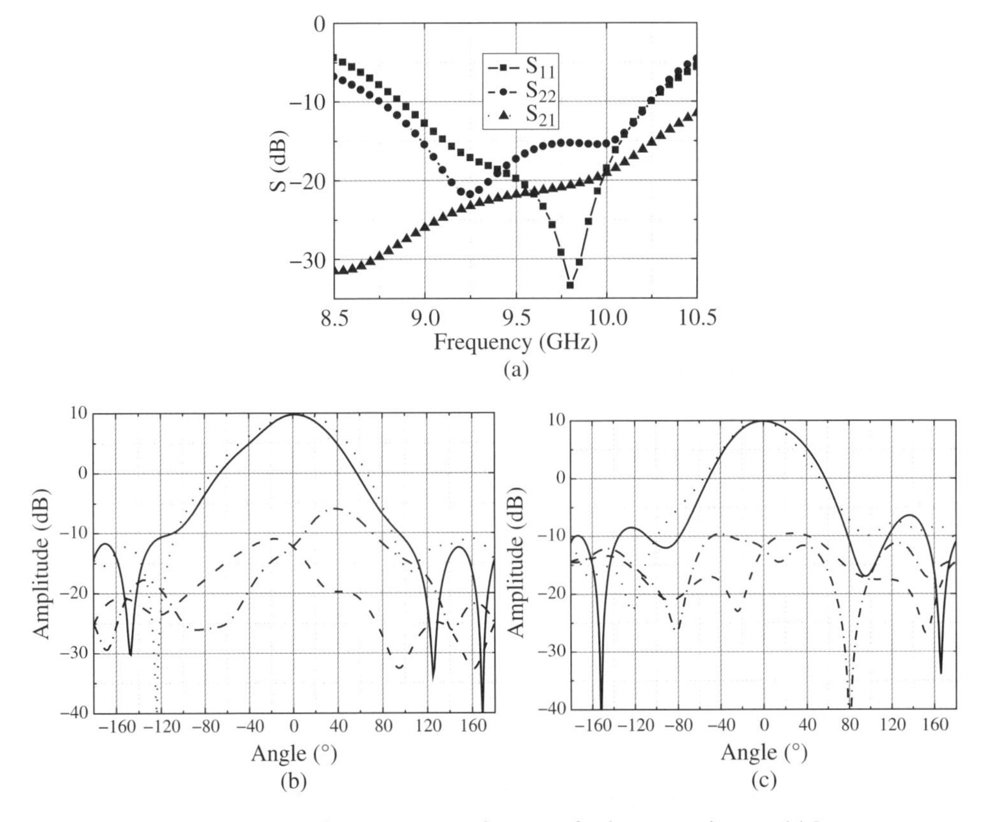


**Figure 3.9** S-parameters and radiation patterns of a corner-feed antenna element. (a) S-parameters; (b) radiation patterns in main polarization and cross-polarization directions at port V; (c) radiation patterns in main polarization and cross-polarization directions at port H.

***Side-Fed Hybrid-Feed Dual-Polarized Microstrip Patch Antenna Element*** When the corner-feed square microstrip antenna is used to construct an antenna array, the diagonal size is increased by $\sqrt{2}$ times relative to the edge length of the square patch, and the spacing between antenna elements is reduced, which is not conducive to the layout of a broadband parallel feeding network. In comparison, the side-feed square patch is advantageous. The horizontal antenna polarization is fed through slot coupling, and the vertical polarization port is fed through coplanar microstrip line. The antenna stack-up and dielectric materials are exactly the same as that of corner-feed square microstrip antennas, and both top and side views are shown in Figure 3.10. The basic antenna dimensions are as follows: parasitic square patch, $10 \times 10\,mm^2$; radiation patch, $9 \times 9\,mm^2$; $W_1 = 2\,mm$; $W_2 = 1\,mm$; $L = 7.4\,mm$; $S = 2\,mm$; the radiating slot is 3.4 mm away from the center; and the length of the coupled microstrip feed short line is 3.3 mm. The thickness of foam 2 is 2.7 mm. Figure 3.9a shows the coefficient of the S-parameters. $S_{11}$ denotes the coplanar microstrip feeding port, i.e. V-port. $S_{22}$ denotes the slot coupling feeding port, i.e. H-port. It can be seen that the bandwidth of less than -10 dB reflection loss at the two ports is 15.2% and 17.2%, respectively, which covers 8.75~10.19 and 8.79~10.44 GHz, respectively. The isolation between two ports can reach −20 dB. Compared with the corner-fed dual-polarized microstrip

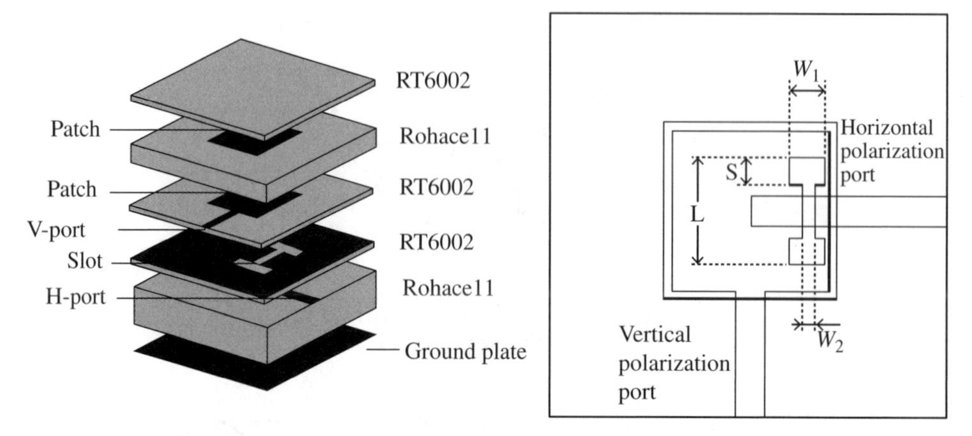


**Figure 3.10** Side-feed hybrid-feed dual-polarization microstrip patch antenna element.

antenna element discussed in the previous section, the polarization isolation is obviously improved in the main frequency band. The schematic of side-feed hybrid-feed dual-polarization microstrip patch antenna element is shown as Figure 3.10.

Radiation patterns are shown in Figure 3.11b and c. Figure 3.11b shows the radiation patterns in the main polarization and cross-polarization planes, when the V-port is excited. The cross polarization is less than −25 dB in the E plane and less than −17 dB in the H plane. For the H-port shown in Figure 3.11c, the cross-polarization components in the H plane and E plane are similar, 23 dB lower than that of the main polarization. These characteristics are the same as those of the corner-fed dual-polarized microstrip antenna, which is why the coupling slot breaks the 2D symmetry of the antenna.

***Side-Fed Slot-Coupled-Fed Microstrip Patch Dual-Polarized Antenna Element*** When the number of linear arrays is small, the feeding networks of the two polarizations can be arranged on the same layer, and slot coupling can be used for feeding both of the antenna elements. The advantage is that the radiation element and feeding network are separated from the ground plane by a coupling slot, which makes them relatively independent. It is also beneficial for optimization of the antenna, and the radiation of the feeding network is eliminated. Normally, the two coupling slots of the antenna are placed below the patch layer, and the two slots are in a T-shape distribution. The coupling slot and radiation patch have one-dimensional (1D) symmetry, as shown in Figure 3.12a. The stack-up structure of the antenna as a typical example used in X band is shown in Figure 3.12b. The main dimensions are as follows: the major slot width corresponding to port 1 is 0.3 mm, the two rectangles are $1 \times 0.75$ mm, and the H-shape slot is 5 mm long and 2 mm off the center. The corresponding dimensions at port 2 are 0.3 mm, $1\,\text{mm} \times 1\,\text{mm}$, 4.5 mm, and 2.5 mm. The size of the radiation patch is $7.7\,\text{mm} \times 7.7\,\text{mm}$, the parasitic radiation patch is $9.5\,\text{mm} \times 9.5\,\text{mm}$, and the distance between them is 2.8 mm. The simulated S-parameters of the antenna are shown in Figure 3.13a. The −10 dB reflection coefficient bandwidths at the two ports are 8.77~10.5 and 8.7~10.67 GHz, respectively. Port isolation in-band reaches −33 dB. The antenna radiation patterns in the main polarization and cross-polarization directions are given in Figure 3.13b and c. It can be seen that the cross polarization in the main lobe is lower than −30 dB.

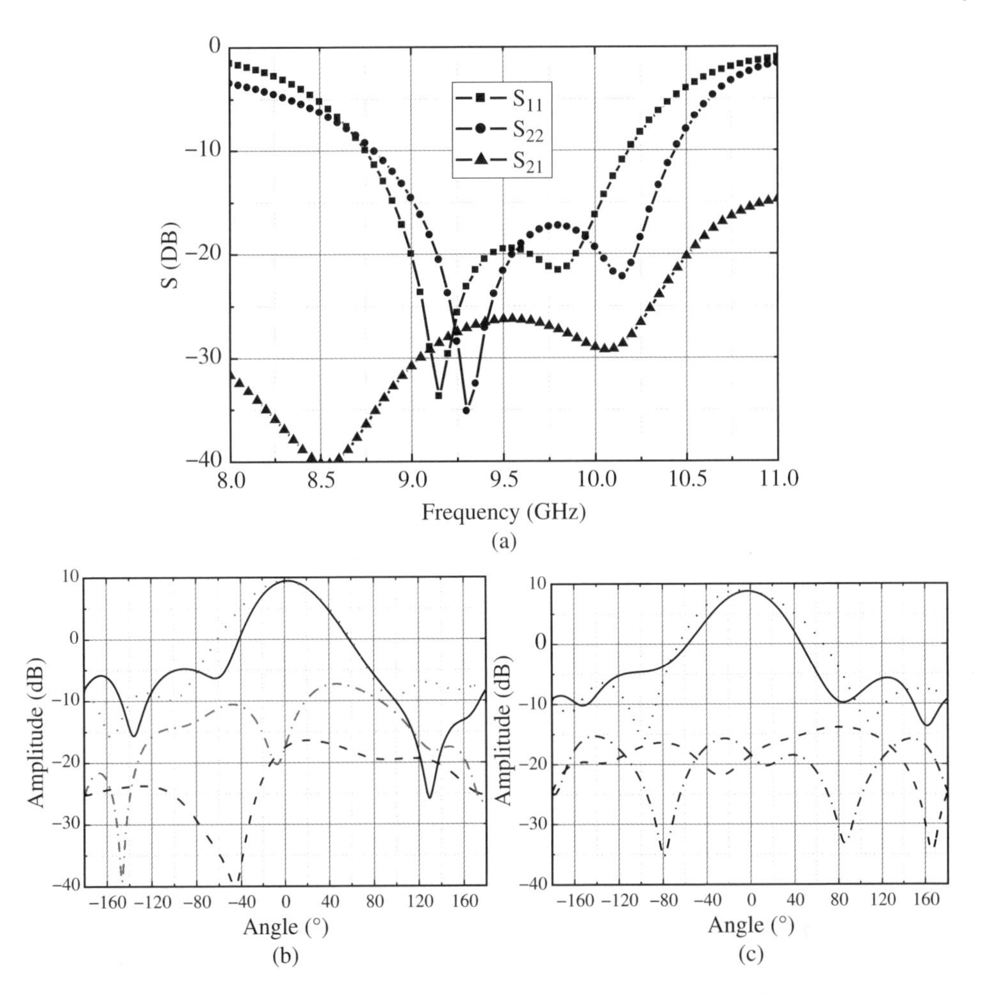


**Figure 3.11** Characteristics of a dual-polarization antenna element. (a) Calculated S-parameters; (b) radiation patterns in main polarization and cross polarization at port V; (c) radiation patterns in main polarization and cross polarization at port H.

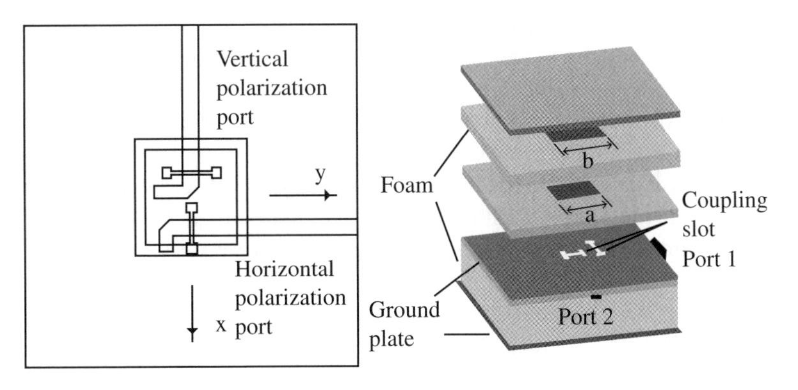


**Figure 3.12** Double-slot coupling dual-polarization microstrip antenna.

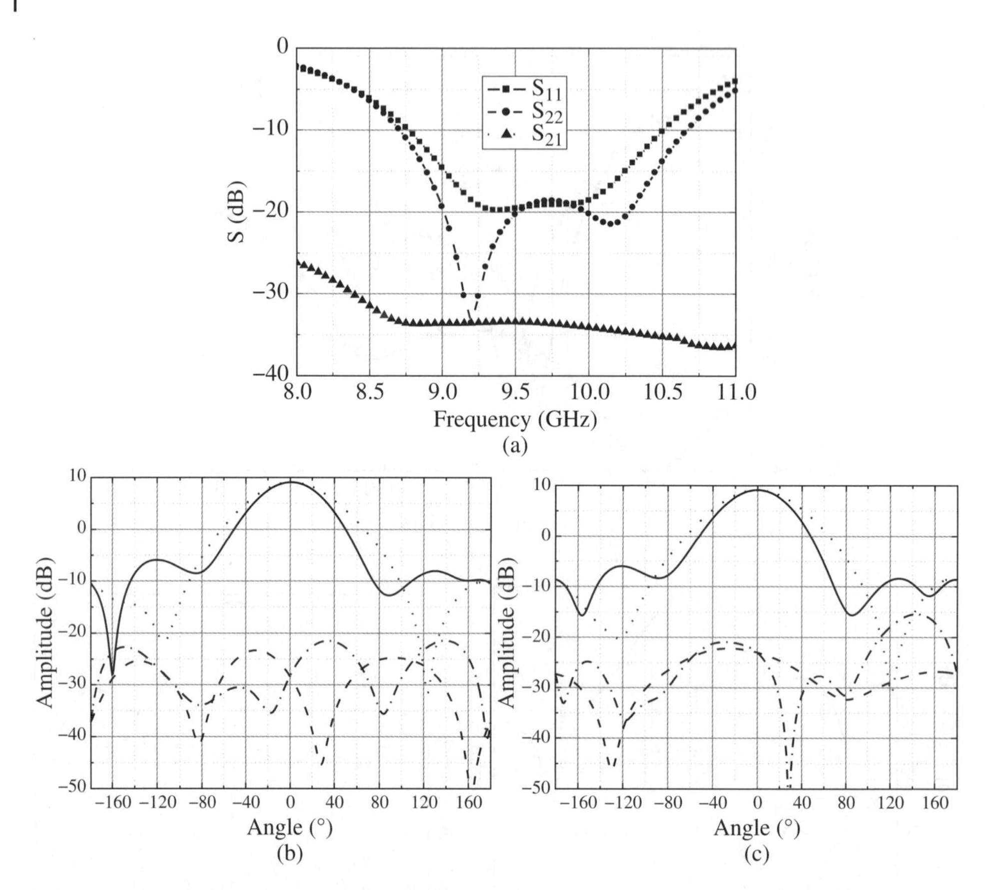


**Figure 3.13** Characteristics of a double-slot coupling antenna element. (a) Calculated S-parameters; (b) radiation patterns in main polarization and cross polarization at port V; (c) radiation patterns in main polarization and cross polarization at port H.

In view of the above microstrip antenna elements, the disadvantages of the corner-fed element are obvious, as it occupies a large space, and there is no space for the distribution of the feeding network. The side-feed element has advantages in port isolation and cross-polarization suppression. But it is difficult to deploy two sets of parallel feeding networks in the same plane because of small unit spacing. The side-feed slot-coupling element has relatively more dielectric layers. The dual-feeding performance is substantially affected by the manufacturing process. In addition, the dual-polarization network is located on both sides of the same dielectric plate, so the mutual coupling between two sets of networks will deteriorate the isolation between ports. Therefore, it is more appropriate to combine side-feed with slot-coupling-fed elements in subarray-level applications of broadband miniature dual-polarization phased array antennas.

### 3.3.2 Dipole Antenna

A typical representative of parallel-fed antenna arrays is the planar dipole antenna array, and the parallel-feed power division network usually uses microstrip lines, dielectric striplines, and air striplines. The advantages of the microstrip dipole antenna array are that the microstrip-feed balun and microstrip power divider can be integrated on the same surface of dielectric plate and the linear array is easily processed and convenient for volume production.

The integration of dipole antenna element and power division network is the key to improve the performance of the array antenna. One radio-frequency (RF) signal is fed to each antenna element according to the predetermined requirements, and the power distribution network is integrated after mutual coupling between antenna elements is analyzed. For a 1:$N$ power divider/combiner network, when $N$ is large, only circuit analysis instead of field analysis can be used. In this way, the influence between the single-stage power division networks and between the transmission lines cannot be considered, which leads to a certain difference between the theoretical simulation and actual test results. The integration design of dipole array and power division network usually uses an approach combining theory with experiment. The aperture field or near-field method is used to analyze the deviation of the amplitude and phase distribution of the antenna aperture from the ideal situation and to compensate for it.

The mutual coupling between planar dipole array antenna elements is relatively large. Mutual coupling will change the input impedance of the antenna element, making the antenna significantly mismatch, thus changing the distribution of effective aperture. Due to the fact that mutual coupling varies with frequency, it is very difficult to suppress it.

#### 3.3.2.1 Antenna Element Structure

Figure 3.14 shows the structure of a printed microstrip dipole antenna element. The input port of an antenna element is a microstrip line, which can be designed and processed together with a microstrip power divider. This type of dipole antenna element offers good broadband performance [7].

#### 3.3.2.2 Theoretic Analysis

To improve the efficiency of the antenna array and to reduce the side lobe level, the mutual coupling model of the array antenna needs to be established based on the theoretic analysis of the antenna element. According to the selected antenna element, a proper simplification is applied, and an accurate integral equation is established based

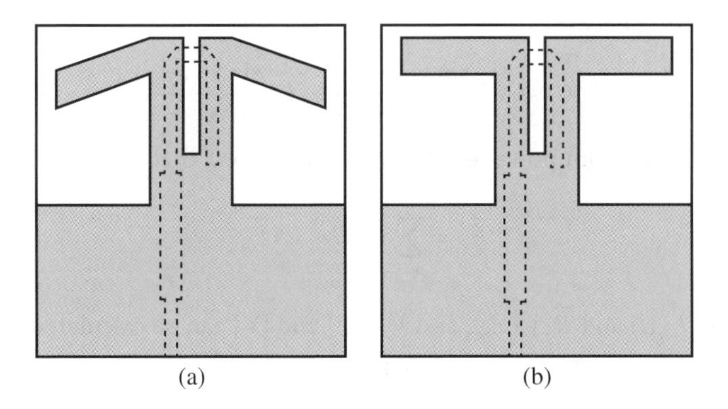


**Figure 3.14** The structure schematic of a printed microstrip dipole antenna element. (a) Umbrella-type dipole; (b) straight arm dipole.

on electromagnetic field theory and boundary conditions. The analysis of a printed dipole antenna array is briefly introduced below.

First, the microstrip dipole antenna element is equivalent to a thin cylindrical dipole antenna with a radius $r = (W + t)/4$, where $W$ is the width of the printed microstrip dipole, and $t$ is its thickness.

Suppose that the length of the dipole antenna element $L$ is much larger than its width $W$, and the influence on the horizontal current distribution of the dipole can be ignored. Considering the current distribution is along the dipole arm only, the integral equation of the whole antenna array can be established:

$$-\int_0^L \vec{I}(l) \cdot E_l^m dl = V_m \tag{3.13}$$

$$V_m = \int_v (J_i E^m - M_i H^m) dv \tag{3.14}$$

where $E_l^m$ is the $\vec{l}$ component of the test current field $E^m$, and $V_m$ is the response of the test field to an external current source.

By using the method of moments, current distribution is expanded as follows through a set of basic functions defined along the thin antenna:

$$\vec{I}(l) = \sum_{n=1}^{N} I_n \vec{F}_n(l) \tag{3.15}$$

where $I_n$ is the expansion coefficient. Let test source function $\vec{J}_m(l) = W_m(l)\vec{l}$.

Substituting Eq. (3.15) to Eqs. (3.13) and (3.14), the matrix format is given by

$$[\mathbf{Z}][\mathbf{I}] = [\mathbf{V}] \tag{3.16}$$

where

$$Z_{mn} = -\int \vec{F}_n(l) \vec{E}_l^m dl \tag{3.17}$$

$$V_m = \int_m W_m(l) \vec{l} \vec{E}^i(l) dl \tag{3.18}$$

If both expansion function and test function are chosen as piecewise sine functions, then the expression becomes

$$\vec{F}_n(l) = \vec{l}_1 \frac{\sin\gamma(l - l_1)}{\sin\gamma(l_2 - l_1)} + \vec{l}_2 \frac{\sin\gamma(l_3 - l)}{\sin\gamma(l_3 - l_2)} \tag{3.19}$$

Choosing an ideal point power excitation ($\delta$ excitation), that is

$$\vec{E}_i = \sum_{p=1}^{P} v_p \delta(l - l_p) \vec{l}_p \tag{3.20}$$

Substituting $\vec{F}_n(l)$ and $\vec{E}_i$ to $Z_{mn}$ and $V_m$, [Z] and [V] can be calculated by

$$[I] = [Z]^{-1}[V] = [Y][V] \tag{3.21}$$

The current distribution on the dipole can be calculated by Eq. (3.21), then the input impedance and the radiation pattern of the whole antenna array can be calculated. The above analysis shows that only the input impedance in the antenna element dipole (including reflector) and mutual impedance between each antenna element dipole (including reflector) are calculated. For the antenna element with balancer, it is only necessary to calculate with corresponding equivalent circuit.

#### 3.3.2.3 Typical Example

According to the above design ideas, an X-band antenna element with a relative bandwidth of 21% is designed. A 2D scanning active phased antenna array, composed of 72 columns × 16 rows of the antenna element, is shown in Figure 3.15. The antenna radiation patterns at typical frequency points are shown in Figures 3.16 and 3.17.

**Figure 3.15** X-band dipole antenna array.

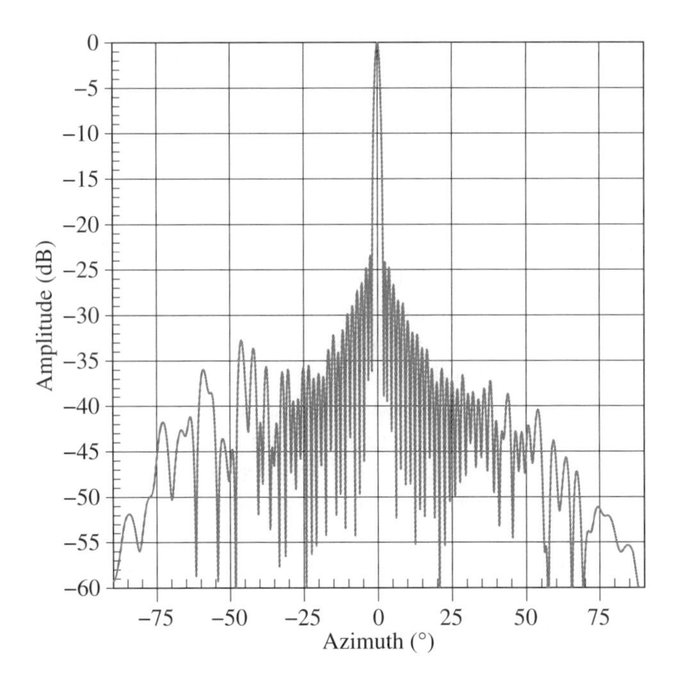


**Figure 3.16** Radiation pattern of antenna array in azimuth plane.

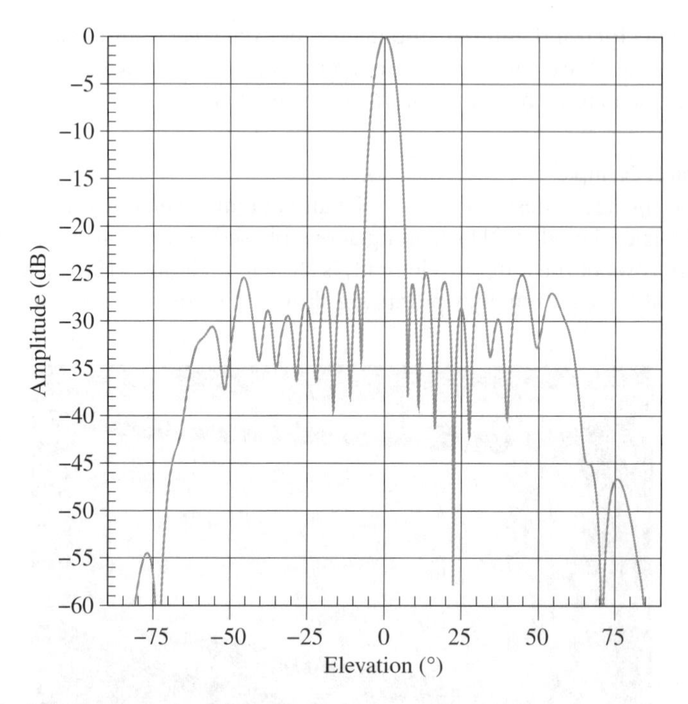


**Figure 3.17** Radiation pattern of antenna array in elevation plane.

### 3.3.3 Waveguide Slot Antenna

A waveguide slot antenna has the advantages of easily controllable aperture amplitude and phase distribution, high efficiency, simple and compact structure, and stable performance. In addition, the waveguide itself is the feeding system of the slot. The use of precision machining can ensure the fabrication accuracy of the antenna, making it easy to realize antennas with low or ultra-low side lobe. Therefore, it has been widely used in airborne and spaceborne SAR.

For standard rectangular waveguide that transmits $TE_{10}$ mode, current distribution on the waveguide wall is shown in Figure 3.18. The conventional slot forms of rectangular waveguide are shown in Figure 3.18 as 1 to 5. Slot 1 in the middle of the waveguide and slot 2 on the narrow wall will not produce radiation due to the lack of cutting surface current. The other three slots generate radiations as they cut the surface currents. Among them, broad wall longitudinal slot 4 and narrow wall tilted edge slot 5 are the most common slots. To reduce the size and weight of the antenna, a common method is to reduce the size of the narrow wall to half or a quarter of standard size. This waveguide is called *half-* or *quarter-height waveguide.* When the waveguide height is reduced, the resonance length of the longitudinal slot increases, which enlarges the mutual coupling effect on the longitudinal slot antenna. When the waveguide size is reduced to quarter height, in addition to the increase in slot length, the forward and backward scattering waves excited by the slot in the waveguide are no longer identical, so that the longitudinal slots cannot be equivalent to parallel units. For longitudinal slot on broad wall, its excitation intensity increases as the distance between the centerline of the slot and the

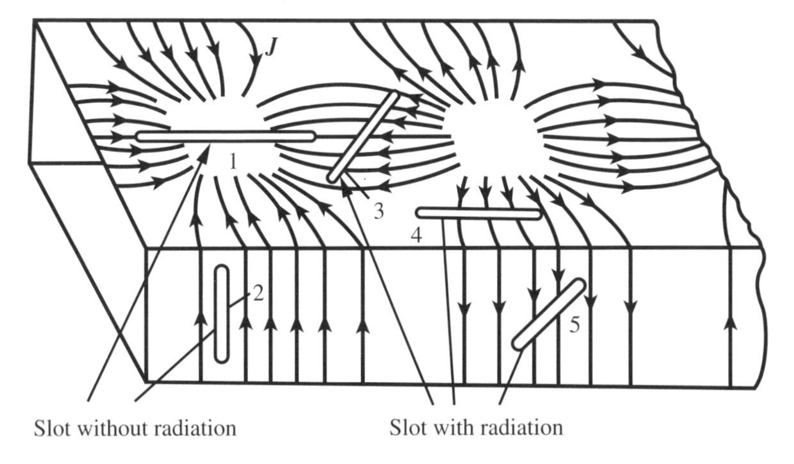


**Figure 3.18** Current distribution and slot forms of rectangular waveguide.

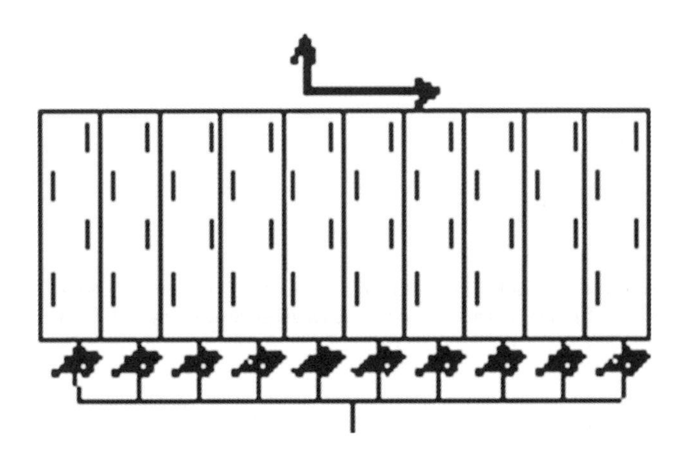

**Figure 3.19** Slot array distribution on waveguide broad wall.

centerline of the broad wall increases, hence the slots are distributed on both sides of the centerline of the broad wall, as shown in Figure 3.19.

When the scanning angle is relatively large or the dual-polarization array shares an aperture, the size of broad wall in the waveguide broad wall slot antenna array needs to be squeezed. A direct method is to use a ridge waveguide antenna. A single-ridge waveguide longitudinal slot antenna array is used for vertical polarization, whereas a waveguide narrow wall slot array is used for horizontal polarization. To reduce the cross-polarization component of a conventional oblique slot array, a metal rod [8] or diaphragm excitation noninclined slot or ridge waveguide inclined slot antenna may be used [9].

There are two forms of single-ridge waveguide longitudinal slot array, symmetrical and asymmetrical, as shown in Figure 3.20. Compared with conventional waveguides, the symmetrical single-ridge waveguide longitudinal slot array shown in Figure 3.20a has significantly smaller broad wall size. The size of the broad wall of a standard

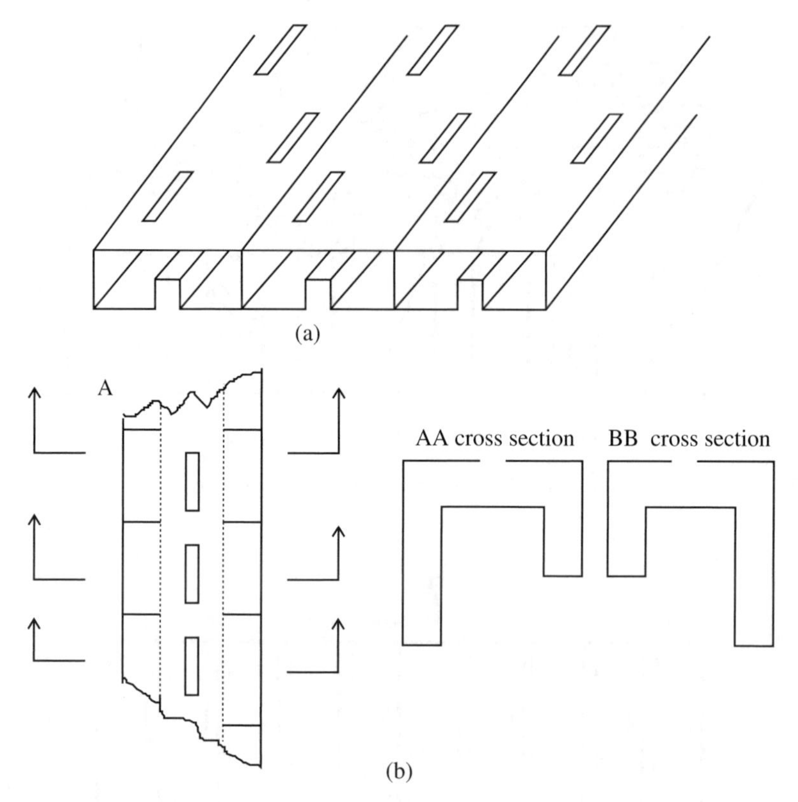


Figure 3.20 Single-ridge waveguide longitudinal slot array. (a) Symmetrical single-ridge waveguide longitudinal slot array; (b) asymmetrical single-ridge waveguide longitudinal slot array.

symmetrical single-ridge waveguide is around 0.58 $\lambda_0$. The asymmetric single-ridge waveguide longitudinal slot array is an ideal radiating element for an airborne 1D wide-angle scanning phased array antenna.

In a single-ridge waveguide longitudinal slot array, multiple longitudinal slots are made on the single-ridge waveguide, and all longitudinal slots are on the broad wall of the waveguide, as shown in Figure 3.20. An antenna with this kind of slot has many obvious advantages, such as parasitic-free side lobe, easy-to-realize low-level antenna side lobe, meeting the phased array antenna for wide angle (±60°) scanning, etc. Here an asymmetric single-ridge waveguide longitudinal slot array is analyzed as an example [10].

#### 3.3.3.1 Theoretical Analysis

The analysis of asymmetrical single-ridge waveguide longitudinal slot array with $m$ elements ($m$ slots) (shown in Figure 3.20), is mainly based on two design equations:

The first design equation is

$$\frac{Y_n^a}{G_0} = K_1 f_n \frac{U_n^s}{U_n} \tag{3.22}$$

where

$$K_1 = \frac{2K_t}{K_0}\sqrt{\frac{K_T^2}{\omega\mu_0\beta_{10}G_0}} \tag{3.23}$$

$$f_n = \frac{\left(\frac{\pi}{2}K_0 l_n\right)\cos(\beta_{10}l_n)}{\left(\frac{\pi}{2}K_0 l_n\right)^2 - \left(\frac{\beta_{10}}{K_0}\right)^2}\cdot\frac{1}{\omega}\cdot\int_{-\frac{\omega}{2}}^{\frac{\omega}{2}} g\left(\frac{d_n}{\lambda}\right)dx \tag{3.24}$$

$$K_T^2 = K_0^2 - \beta_{10}^2 \tag{3.25}$$

The second design equation is

$$\frac{Y_n^a}{G_0} = \frac{2f_n^2}{\frac{2f_n^2}{\frac{Y_n}{G_0}} + j\frac{\beta_{10}K_0^3}{4\pi K_T^4}\sum_{m=1}^{N}\frac{U_m^s}{U_n^s}g_{mn}} \tag{3.26}$$

where $Y_n^a$ is the slot admittance, $G_0$ is the waveguide admittance, $U_n^s$ is excitation voltage, and $U_n$ is the mode voltage of the waveguide. The mutual admittance is given by

$$g_{mn} = \int_{-K_0 l_m}^{K_0 l_m}\cos\left(\frac{z_m'}{4l_m/\lambda_0}\right)\cdot\left\{\frac{l}{4l_n/\lambda_0}\cdot\left[\frac{e^{-jK_0R_1}}{K_0R_1}+\frac{e^{-jK_0R_2}}{K_0R_2}\right]+\left[1-\frac{l}{\left(4l_n/\lambda_0\right)^2}\right]\right.$$
$$\left.\cdot\int_{-K_0 l_n}^{K_0 l_n}\cos\left(\frac{z_n'}{4l_n/\lambda_0}\right)\frac{e^{-jK_0R}}{K_0R}dz_n'\right\}dz_m' \tag{3.27}$$

#### 3.3.3.2 Computational Method

The above two equations have laid the foundation for the design of a symmetrical single-ridge waveguide longitudinal slot antenna. A design example is introduced in the following: for an X-band asymmetrical single-ridge waveguide longitudinal slot array with $n$ elements, the aperture of the linear array is the Taylor distribution, and for the design parameters, the side lobe level is −28 dB and the number of equal side lobes is 5. The specific design and calculation methods are as follows:

(1) According to the aperture distribution, the slot voltage distribution $U_n^s$ is directly calculated by the corresponding formula. The initial parameters of a group of $n$ slots are prespecified as $\left(\frac{d_n}{\lambda}, l_n\right)$. Without considering mutual coupling, a set of appropriate slot parameters, obtained directly from the first design equation $U_n^s$ [Eq. (3.22)] and self-admittance of the isolated slot, are used as initial parameters $\left(\frac{d_n^0}{\lambda}, l_n^0\right)$ for the iteration.
(2) It can be seen from the second design equation [Eq. (3.26)] that to make the voltage of all slots in the same phase, the active admittance of each slot needs to be zero. The denominator of Eq. (3.26) is a real number, and the imaginary part of the denominator is zero. The following equation can be obtained:

$$\frac{\frac{Y_p^a}{G_0}}{\frac{Y_n^a}{G_0}} = \frac{f_p}{f_n}\cdot\frac{U_p^s}{U_n^s}\cdot\frac{U_n}{U_p} \tag{3.28}$$

Equation (3.28) is the ratio of the $p$th slot after applying the first design equation and $n$th slot after applying the first design equation. According to Eq. (3.28), $2n-1$ equations can be obtained, together with the matching constraint equation:

$$\sum_{n=1}^{N} Y_n^{\alpha} = N_0 \tag{3.29}$$

Thus, $2n$ nonlinear equations are established. The basic principle of the selection of $N_0$ is that $N_0 = 1$ for end feeding, and $N_0 = 2$ for center feeding to achieve a good match. Of course, $N_0$ can also be determined in the actual process according to specific circumstances.

(3) A quasi-Newton method can be used to solve the nonlinear equations. Compared with other numerical methods, this method is fast with good stability.

(4) After iteration, the obtained slot parameters $\left(\frac{d_n^1}{\lambda}, l_n^1\right)$ are compared with previous calculated slot parameters $\left(\frac{d_n^0}{\lambda}, l_n^0\right)$; if the deviation is greater than the design requirement, a second iteration is performed until design requirements are met.

#### 3.3.3.3 Typical Example

According to the above steps, a planar waveguide slot array antenna applied to a small platform is designed. Figure 3.21 shows a half-height rectangular waveguide slot array antenna. Figure 3.22 shows the measured radiation pattern of the planar antenna array. The theoretical design is −30 dB Taylor weighting. As shown in Figure 3.22, the measured side lobe voltage level is better than −28 dB. To increase the scanning angle of the antenna beam, a ridge waveguide slot array is normally used. Figure 3.23 shows a large-scale planar waveguide slot phased array antenna based on module assembly. The aluminum alloy slot waveguide antenna can be achieved successfully by one-time design

**Figure 3.21** A small-scale waveguide slot antenna.

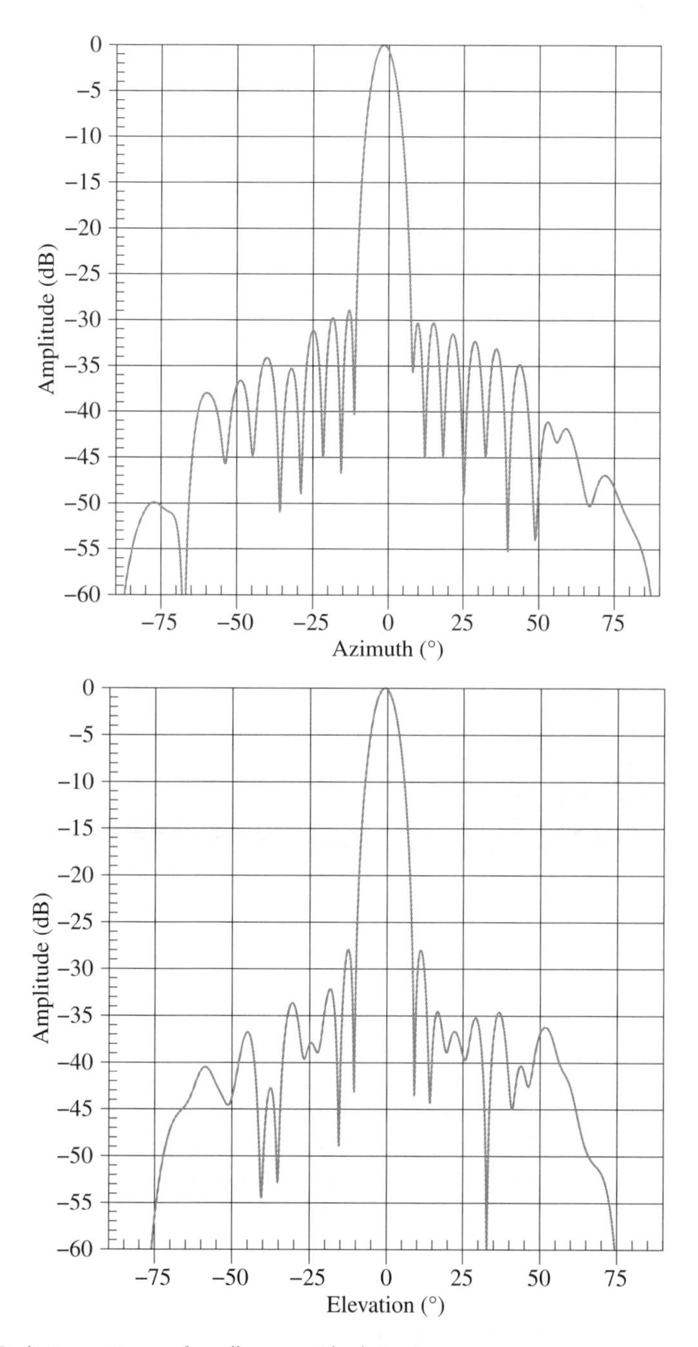


**Figure 3.22** Radiation patterns of small waveguide slot antenna.

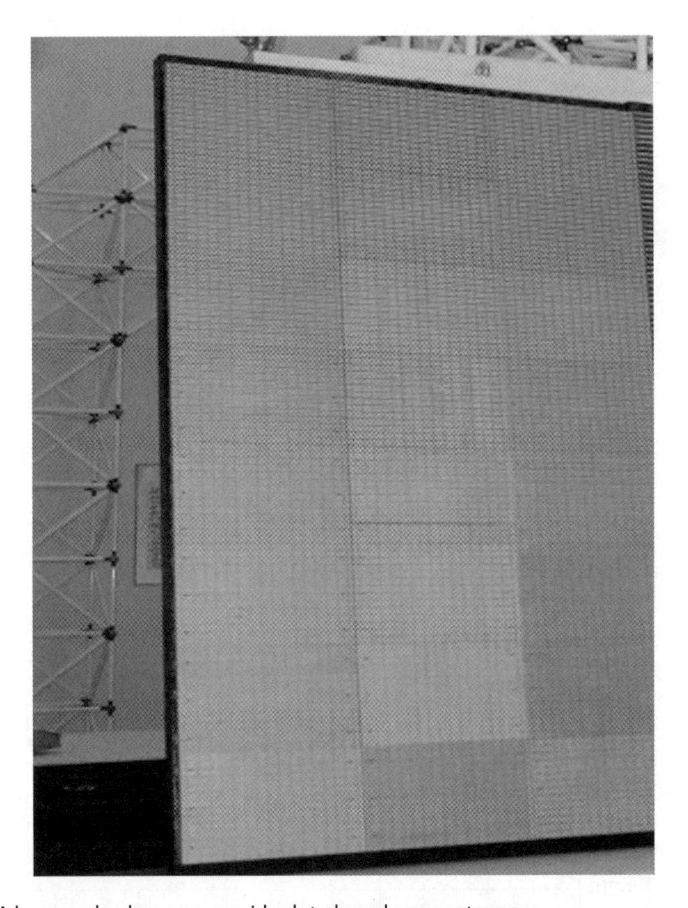

**Figure 3.23** A large-scale planar waveguide slot phased array antenna.

and processing, because the material is stable and reliable, and the processing technology is mature.

For high-frequency phased array antenna above C band, the slot waveguide antenna is of great advantage. With improvement of the SAR system performance, not only high resolution but also wide swaths are required. Research on 2D wide-angle SAR antennas is ongoing. In addition to microstrip patch antennas and dipole antennas, open waveguide is also preferred, especially in frequencies above X band. Open waveguide has the advantages of broadband [11], including high strength, easy to integrate electromechanical functions, etc. The 40%~50% relative bandwidth can meet the requirement of high-resolution SAR.

## 3.4 Airborne Antenna Structure

Antenna structure design is important to the electrical performance of airborne radar. It is also an important part of the environmental adaptability design of airborne radar. Environmental adaptability means the equipment is capable of achieving all intended function and performance without being destroyed under different environments within its lifetime. Environmental adaptability is an important quality characteristic of airborne SAR. For drones, helicopters, and manned aircraft, because of the different characteristics of the aircraft, the flight altitude and speed are different, and the equipment environment adaptability requirements for SAR systems are different. Therefore, the environmental adaptability of the antenna structure needs to be specifically designed.

Under normal circumstances, airborne SAR must meet the requirements of high temperature, low temperature, damp heat, mold, salt spray, acceleration, vibration, and impact in addition to the environmental conditions specified by the aircraft. There must also be sufficient high intensity radiated field (HIRF) protection to prevent damage and malfunction of the radar equipment due to direct or indirect HIRF effects on the aircraft avionics system and to prevent the radar's radiation field from affecting the aircraft's avionics system, which may lead to an accident.

The air pressure of airborne vehicles is different at different altitudes. Generally, low-pressure operation needs to be considered in SAR system design. The problems caused by low pressure are mainly the difficulty of heat dissipation of electronic equipment, high voltage, and high-power ignition, especially the power capacity problem of concentrated-feed SAR antennas, which needs to be considered at high altitude. For phased array radar, due to the use of a T/R module, generally there is no high-voltage equipment and high-power microwave transmission line. The major problem caused by low pressure is its influence on the cooling of the electronic equipment. If liquid cooling is used for the power device in a radar system, it will not be affected by low pressure. For transmission line and passive RF devices that use natural heat dissipation, because the natural heat dissipation capability is decreased at high altitude and in a low-pressure environment, the heat consumption is small. The temperature at high altitude is low, if the natural heat dissipation module is well designed; it can meet the requirements of a low-pressure environment.

### 3.4.1 Airborne Antenna Environment Condition

In addition to the electromagnetic environment, an airborne antenna environment includes the mechanical and climate environments. The mechanical environment includes vibration, shock, and acceleration. The airborne vibration frequency range is generally 5∼2000 Hz, the maximum amplitude range is 0.06∼3.0 mm, and the acceleration range is 1∼10 g. The value is determined by the type of aircraft and the location and method of antenna installation. The impact reaches 15∼30 g acceleration in six directions along three vertical axes, and the pulse duration is about 10 ms. In aircraft maneuvers, sometimes due to sudden changes in airflow, such electronic equipment as antennas can withstand acceleration, ranging from 1∼18 g. Structure design should prevent the following faults of the antenna: components cracking, coating peeling off, mechanical damage and deformation, fixed joints cracking, components falling off, and separation of equipment and mounting rack as well as structural deformation, relative displacement, cracking, and connecting point displacement of the component.

Climatic conditions include temperature, altitude, humidity, dust, mold, salt mist, etc. The regional temperature difference in China is between −57 °C and 48 °C. The temperature difference between summer and winter can reach 60 °C. This difference can be 80 °C when an aircraft is up in the air. The fastest temperature ramp can reach

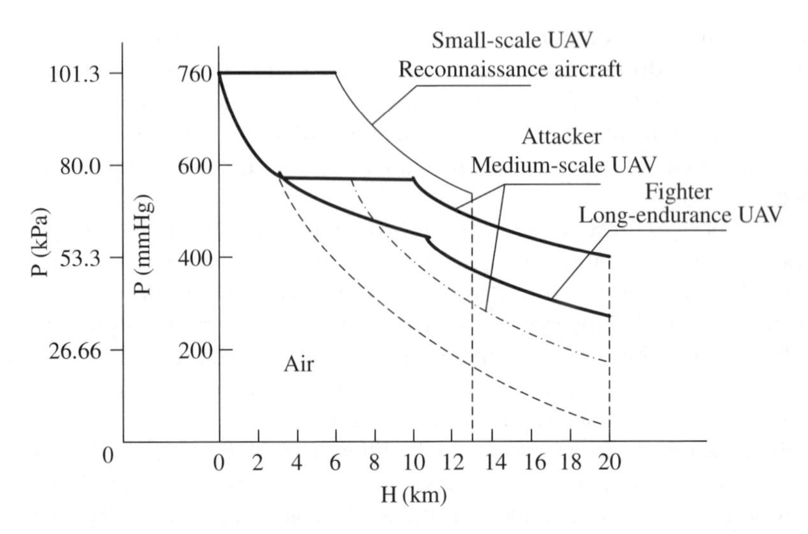


**Figure 3.24** Relationship between the flying height and the pressure.

$10\,°\mathrm{C\,min}^{-1}$. When the aircraft rises from sea level to 10,668 m (35,000 ft), the pressure reduces from 101.32 to 24.84 kPa, and the electronic equipment located in the nonpressurized area will be affected by the low pressure, as shown in Figure 3.24. As a result of the large maneuvering range of aircraft platforms, the airborne electronic equipment also needs to survive such harsh environments as humidity, mold, salt fog, and so on.

### 3.4.2 Airborne Antenna Structure Design

The mechanical vibration and shock during aircraft takeoff can cause permanent structural deformation of the phased array antenna frame, which reduces the connector's elasticity and leads to cumulative fatigue failure, such as fracture of connecting bolts, and temporary or permanent performance reduction of the airborne antenna. The general phased array antenna system equipment is rigidly installed on the carrier. To improve the shock resistance capacity of the structure, the structure stiffness needs to be improved. High stiffness also has high strength; therefore, the impact resistance is increased accordingly. If structural damping is used, the mechanical energy of the shock will be changed into thermal loss of the structure or material, which may attenuate all resonant peaks in a very wide spectrum.

The anti-shock/impact ability is generally improved by increasing the structure rigidness. Under the principle of minimizing the system weight, the following methods can be used: properly arranging the structure stiffener, increasing the overall structure rigidity, avoiding using cantilever structure, shortening the spacing between support points, and increasing the number or strength of the fastening points.

If the antenna structure is large and lightweight, array surface deformation will easily appear in an airborne vibration condition. The design should not only guarantee the precision of antenna assembly but also ensure shock resistance of the antenna array. High-modulus carbon fiber (CF) composites can be used in the overall structure of the antenna, which may improve the overall rigidity and reduce the weight. If the antenna structure is small and the precision demands for the antenna array are high,

high-strength aluminum alloy is generally chosen in the structural frame, which increases the stiffness by increasing the truss height and the moment of inertia. At the same time, structural stability design should be enhanced to improve the overall rigidity and shock resistance.

Based on requirements for the mechanical environment of the antenna structure design and its evaluation criteria, it is required to perform antenna mechanical analysis, including strength, stiffness, acceleration, vibration and aircraft cruising condition, and the safety margin of the antenna structure is calculated based on the analysis results.

A SAR system is normally large in size and contains many setups. It is strictly limited in radar weight, size, space, and power consumption. The difficulty of designing a radar structure is primarily because of the following: the general antenna has a larger aperture that highlights the contradiction between weight reduction and stiffness to obtain a large antenna power product; for phased array antennas, the stand-alone network (such as antenna array surface, a magnitude of T/R modules, beam-forming network, time-delay amplifier module, correction network, switch matrix, data distributor, liquid cooling distributor), high-/low-frequency cable network, and liquid cooling network are mounted on the antenna frame, which makes high-density mounting difficult.

To solve the difficulty of lightweight antenna design, the following technical methods are generally adopted to reduce the system weight: during the design of an antenna structure, the stiffness and strength analysis on the structure are important. The shape of components/devices in multiple layouts, installation position, fixing forms, interface types, and maintainability should be analyzed and compared. Integrated design is used to reduce the number of parts, workload and assembly difficulty, and system weight.

Lightweight materials are used as much as possible, such as aluminum alloy, magnesium alloy, titanium alloy, or CF composite material. A module box with such metal material as lightweight and high-strength aluminum alloy (aluminum–magnesium alloy and carbon–silicon–aluminum alloy) is strongly recommended. Antirust aluminum or stainless-steel tube should be used as the liquid cooling transmission pipeline.

A large proportion of the electronic equipment weight is the weight of the components. The weight ratio of various components (including printed circuit boards) accounts for 60%. Therefore, it is very important to strictly control the weight of components.

## 3.5 Spaceborne Antenna Structure

The SAR system is one of the most important satellite payloads that plays a leading role in satellite design. Each subsystem of the satellite platform is designed to give priority to the requirements of the payload on the satellite platform. The payload is also designed to take the capability of the satellite platform into account. In satellite design, efforts should be made to coordinate the payload and the platform for overall optimization.

The spaceborne SAR system is composed of the antenna and central electronic equipment (receiving channel, frequency source, waveform generation, data acquisition, data compression, etc.). The antenna is mounted on the satellite body, which is considered extravehicular equipment. The central electronic equipment is installed inside the satellite cabin. Compared to the central electronic equipment, the operating environment of the antenna is much harsher. An example of the active phased array antenna is given below for discussion.

### 3.5.1 Spaceborne Antenna Environment Requirements

Similar to other subsystems of the satellite, the active phased array antenna must be able to adapt to both the mechanical environment during launch and the space environment after entering orbit.

#### 3.5.1.1 Mechanical Environment

Active phased array antennas are required to adapt to the mechanical environment during launch, which includes as vibration, shock, acceleration, noise, etc. The antenna structure should therefore be designed with sufficient strength and stiffness. A modal analysis is required to avoid resonance, which may damage the active phased array antenna. To operate in the mechanical environment during launch, the phased array antenna unfolding mechanism needs to be locked until the satellite is in orbit.

#### 3.5.1.2 Weightlessness

The satellite in orbit is operating under weightlessness, and the active phased array antenna is in a gravity environment when being debugged on the ground. Although a floating platform is used to simulate weightlessness of an active phased array antenna, the performance of the plane accuracy of the antenna differs with or without gravity. In addition, the repetition precision of unfolding an antenna also brings about differences in flatness. Appropriate measures are required during the design or the ground debugging phase so as to ensure that the performance of an active phased array antenna in orbit can meet requirements under weightlessness.

#### 3.5.1.3 Vacuum State

For phased array antennas, due to the extensive use of T/R modules and delay lines, some microwave absorbing materials are used to overcome the cavity effects and electromagnetic interference of the T/R components and delay lines. The T/R modules and delay lines use some composite glue in the manufacturing process. These materials will appear out of gas and evaporate under vacuum, which will pollute the chips used in the T/R components and delay lines and affect their lifetime. Vacuum discharge may damage certain circuit components, and the secondary electron multiplication effect may damage microwave components. In the design and manufacturing of the phased array antenna, these problems need to be addressed.

#### 3.5.1.4 Temperature Variation

A phased array antenna is composed of antenna array, T/R module, delay line, beam controller, secondary power source, and many other devices, among which the T/R module has very strict temperature requirements. External and internal heat flux of the satellite will lead to a variation in the temperature field of the phased array antenna, so the thermal control method needs to be taken into account to ensure the appropriate temperature variation.

The thermal design is carried out at the same time as the electrical performance and structural design, so that the thermal resistance of the active heating element to the components or heat dissipation surface is small. Among them, some parts need to be electrically insulated but thermally conductive. Therefore, special measures are required, such as the use of beryllium oxide material, which can ensure good electrical insulation and thermal conductivity, whereas electrical conductor is prohibited in some parts. Then special measures should be taken to dissipate the heat. For example, for thermal conduction in the inner conductor of a transmitting power division network (air dielectric substrate line network) of center-feed antenna, measures of blackening internal and external conductor surfaces (at the cost of increasing the loss) or boosting a $\lambda/4$ short bar are taken.

#### 3.5.1.5 Space Radiation Environment

High-energy electrons, protons, and heavy ions may damage the surface of phased array antenna and electronic components. Single-event upset and lock can cause the complementary metal–oxide–semiconductor circuit to be faulty. In the design and manufacturing of a phased array antenna, these factors should be carefully considered and appropriate measures adopted to meet the requirements of the space radiation environment.

The payload generally consumes a lot of the energy of the satellite, so very high requirements are proposed on quality, size, power consumption, and reliability. An active phased array antenna includes a large number of T/R modules. The efficiency of the final stage power amplifier in a T/R module is critical for RF power conversion efficiency in the phased array antenna. Widely used GaAs monolithic amplifiers have an efficiency of around 40%. The conversion efficiency of the new generation of semiconductor GaN monolithic amplifiers will be further improved, reaching 50%~60%. For a phased array antenna, except electromagnetic energy radiated by the antenna, all other energy is converted into thermal energy, which increases the burden to control the thermal energy. Therefore, improving the overall efficiency of the antenna in spaceborne SAR is continuously pursued.

### 3.5.2 Antenna Structure and Mechanism Design

Structure and mechanism are important to the phased array antenna. The following aspects are of concern in the design: (1) environmental conditions, including the launch environment (mainly mechanical environments, such as overload acceleration, vibration, shock, and noise), orbital environment (vacuum, alternating temperature, particle and ultraviolet radiation, atomic oxygen in low earth orbit, orbit maneuver load); (2) the satellite configuration design, including shape, maximum envelope dimension (allow space inside the fairing), antenna layout, mass center, moment of inertia, and interface relationship between various machines (satellite platform), electric design (electromagnetic compatibility, insulation, electrostatic prevention, and grounding), thermal (thermal isolation and thermal conductivity); (3) performance requirements, including stiffness, intrinsic frequency (avoiding coupling, decreasing deformation), strength (static and dynamic stress response does not exceed the allowable value), and deformation (antenna thermal deformation limit, platform precision requirements); (4) quality; (5) movement of the mechanism; and (6) reliability (especially the mechanism reliability).

Under the premise of meeting demands for the above-mentioned environmental conditions and satellite configuration and performance, the strength safety margin in the

**Table 3.1** Minimum safety margins in the structural design.

| Material | Parameter | Minimum safety margin |
|---|---|---|
| Metal | Yield strength | 0.0 |
| | Ultimate strength | 0.12 |
| | Stability | 0.25 |
| Composite material | First layer failure | 0.25 |
| | Bearing strength | 0.25 |
| | Stability | 0.30 |

structure design is

$$\text{strength safety margin} = \frac{\text{allowable load or stress}}{\text{designed load or stress}} - 1 \tag{3.30}$$

The designed load = the used load × design safety factor, design safety factor is normally set to be 1.5.

The minimum safety margins in the structural design are shown in Table 3.1.

The safety margin of the mechanism design is

$$\text{static moment margin} = \frac{\text{driving moment} - \text{driving moment needed to produce acceleration}}{\text{resisting moment}} - 1 \tag{3.31}$$

$$\text{dynamic moment margin} = \frac{\text{driving moment} - \text{resisting moment}}{\text{driving moment needed to produce acceleration}} - 1 \tag{3.32}$$

Both static and dynamic moment margins in all development stages are shown in Table 3.2.

#### 3.5.2.1 Analysis of Structure and Mechanism Design

The purpose of design analysis is to predict the structure and mechanism performance (stiffness, strength, deformation, and safety margin) and to optimize design. The analytical model should be capable of representing an actual situation of the structure or mechanism. The key parts of the model should be verified by experiments. During analysis,

**Table 3.2** Minimum safety margins in the mechanism design.

| Type of margin | Development stage | Minimum safety margin |
|---|---|---|
| Static moment margin | Conceptual design review | 1.75 |
| | Initial design review | 1.30 |
| | Critical design review | 1.25 |
| | Identification test | 1.0 |
| Dynamic moment margin | | 0.25 |

the input parameters of the mechanism and physical performance are derived from statistical test results.

The goal of analysis and optimization of the antenna structure and mechanism is to look for for tradeoffs between performance safety margin and quality. Therefore, it is very important to correctly define the boundary condition of the structure or mechanism. In addition, optimized design is based on structural analysis, including dynamic analysis and static analysis. The most important methods of dynamic analysis include modal analysis, sinusoidal vibration, random vibration, and response analysis of noise and impulse. Static analysis mainly covers analysis and calculation of stress, strain, deformation, and stability.

Mechanism kinematics and dynamics analysis should be performed on the design of phased array antennas to determine the duration of mechanism unfolding, unfolded motion trajectory, angular velocity when locked, impact of unfolding, etc.

#### 3.5.2.2 Structure and Mechanism Materials

There are three major types of materials used in antenna structures and mechanisms: metal, composite, and auxiliary. The metal materials commonly used in antenna structures and mechanisms are aluminum alloy, magnesium alloy, titanium alloy, steel, beryllium, beryllium alloy, etc. Composite material can be classified into structural composite material and functional composite material according to its application; structural composite material currently dominates. Structural composite materials are mainly used in bearing and secondary bearing structures, requiring light weight, high strength, and high rigidity, and they can undergo thermal shock. In some cases, low expansion coefficient and good thermal insulation are required. The structural composite material is primarily composed of reinforcement and matrix. The reinforcement is used to bear various loads. The matrix plays the role of bonding and enhancing the reinforcement system and forming and transferring stress and toughness.

Composite material is mainly made of organic polymer, a small amount of metal, ceramic, and carbon (graphite). Reinforcement is a key part of high-performance structural composites and is instrumental in increasing the strength and improving the performance. At present, the novel reinforcements widely used include inorganic fibers – such as CF, alumina fiber, silicon carbide fiber, and special glass fiber – and organic fibers – such as aramid fiber (Kevlar), PBO fiber (poly-p-phenylene benzobisoxazole), and ultra-high molecular weight polyethylene fiber (UHMWPE).

The composite material features high specific strength and high specific stiffness. Fiber reinforcement materials used for antenna structure and mechanism are mainly glass fiber, CF (graphite fiber), Kevlar fiber, boron fiber, and silicon carbide fiber. High modulus CF/epoxy resin composite is currently an important material with low relative density, high elastic modulus, high strength, and high corrosion resistance. The longitudinal thermal expansion coefficient is approaching zero or negative, which guarantees the dimensional stability and flexibility of the material in a wide temperature range. Parameters of typical high-modulus CF/epoxy resin unidirectional composites are shown in Table 3.3.

Other auxiliary materials include adhesives, film, and foam rubber. Adhesive is used for connecting various structural parts. Film is used for gluing the sandwich panel, the sandwich core, and the embedded parts. Foam rubber (also called *splicing glue*) is used

**Table 3.3** Engineering constants for typical high-modulus CF/epoxy resin unidirectional composites (reference value).

| Parameter | M40j/epoxy | M60j/epoxy |
|---|---|---|
| Longitudinal tensile modulus (GPa) | 223 | 331 |
| Transverse tensile modulus (GPa) | 6.98 | 8.01 |
| Shear modulus (GPa) | 3.87 | 4.16 |
| Poisson ratio | 0.255 | 0.32 |

to splice the pre-embedded parts and the postembedded parts with the honeycomb core as well as to splice the honeycomb core itself.

#### 3.5.2.3 Structure and Mechanism Testing

The test includes structural parts and the mechanism unfolding component. The test of structure parts mainly consists of a static test and a dynamic test. The dynamic test includes vibration (modal, sinusoidal vibration, random vibration, and shock simulation), noise, and shock.

The unfolding-locking mechanism test includes a running-in test, driving torque test under normal temperature and pressure, friction torque test under normal temperature and high–low vacuum temperature, torsional stiffness test, dynamic equivalent stiffness test, and space environment simulation test of the motor. The press and release mechanism test includes function identification, bearing capacity, environment and function of flame cutter, shear capability of high friction coating cover under pressing force, etc.

The complete antenna test includes the unfolding test, noise test, disturbance state modal test, sinusoidal vibration test, random vibration test, and unfolded state modal test.

## References

1 Chen, G., Wang, X., Zhuang, Z., and Wang, D. (1996). *Genetic Algorithm and Its Application*. People's Posts and Telecommunications Press.
2 Qi, M., Wang, W., and Jin, M. (2008). Optimization of antenna array pattern based on particle swarm optimization algorithm. *Radar Science and Technology* 6 (3): 231–234.
3 Lu, J., Wu, M., Chen, S., and Fang, Z. (2000). FFT based calibration method of phased array radar. *Journal of Radio Science* 15 (2): 221–224.
4 Zhong, S. (1991). *Microstrip Antenna Theory and Application*. Xi'an Electronic and Science University Press.
5 M. Bonadiman, R. Schildberg, and J.C. da S. Lacava (2004). Design of a dual-polarized L-band microstrip antenna with high level of isolation for SAR applications. *Proceedings of the IEEE Antennas and Propagation Society Symposium*, Monterey, California (June 20–25, 2004). Piscataway, NJ: IEEE.
6 Wang, W., Liang, X.-L., Zhang, Y.-M., and Zhong, S.-S. (2007). Experimental characterization of a broadband dual-polarized microstrip antenna for X-band SAR applications. *Microwave and Optical Technology Letters* 49 (3): 649–652.

7 Lu, X., Zhang, Y., and Ni, W. (2004). Ultra-wideband UHF antenna for airborne SAR. *Radar Science and Technology* 2 (1): 57–60.
8 Hashemi-Yeganeh, S. and Elliott, R.S. (1990). Analysis of untitled edge slot excited by tilted wires. *IEEE Transactions on Antennas and Propagation* 38 (11): 1737–1745.
9 W. Wang, J. Jin, J.-G. Lu, S.-S. Zhong (2005). Waveguide antenna array with broadband, dual-polarization and low cross-polarization for X-band SAR applications. *Proceedings of the 2005 IEEE International Radar Conference,* Arlington, VA (May 9–12, 2005). Piscataway, NJ: IEEE.
10 Green, J. and Shnitkin, H. Asymmetric ridge waveguide radiating element for a scanned planar array. *IEEE Transactions on Antennas and Propagation* 38 (8): 161–165.
11 Hao, L., Jiaguo, L., Nianyun, H., and Desen, F. (2003). Research of broadband coaxial-to-rectangular waveguide end launcher antenna element. *Acta Electronica Sinica* 31 (9): 1365–1367.

# 4

# Transmit/Receive Module

## 4.1 Overview

The transmitter, which transforms AC or DC electric power into microwave energy, generates a radio-frequency (RF) signal, as required by the radar system, which is then transferred by the feeding system to an antenna radiating into space. Different radar systems have different requirements with regard to transmitters, which differ mainly in instantaneous bandwidth, signal waveform, and signal stability. For example, pulse compression radars generally use long pulse widths for increasing the range of action and overcoming the limitation of high peak power. To obtain high-range resolution and improve the recognition of target characteristics, wide instantaneous bandwidth signals are usually used. Moving target display radar requires the transmitter's RF signal phase to have high stability between pulses. Generally speaking, the basic requirements of a synthetic aperture radar (SAR) system for the transmitter are:

(1) *Wide instantaneous bandwidth*. It is well known that a SAR system is based on the synthesis aperture of a coherent pulse sequence radiated by an antenna. Every pulse in the pulse sequence is a wide pulse with the same linear frequency modulation. The range resolution is obtained by compressing it into a narrow pulse through a matched filter in the receiver. The range resolution is defined by the instantaneous bandwidth. Meanwhile, it is required that the microwave signal generated by the transmitter should have good amplitude and phase stability.

(2) *Wide working pulse width* $\tau$. The method of the SAR system to improve the range resolution is to compress the received wide pulse radiated by the antenna into a narrow pulse $\tau_0$, where $\tau_0 = 1/B$ or $\tau_0 = 1/\Delta f$. Usually $\tau_{max} \approx 20{\sim}30$ μs. For a transmitter with instantaneous bandwidth of 1 GHz and $\tau_0 = 1$ ns, the pulse compression ratio can reach $(2{\sim}3) \times 10^4$ times.

(3) *High pulse repetition frequency (PRF)*. The PRF in a SAR system is related to velocity and height of the moving platform and aperture of the antenna $L_a$. A high PRF is usually adopted to avoid orientation ambiguity caused by spectrum folding. The PRF needs to be changed if the observation angle in the SAR system is altered.

(4) *High efficiency*. SAR is usually applied in the scenario of moving payloads, such as aircraft, missiles, satellites, etc. These payloads are precisely placed with scarce energy, and in many circumstances only batteries are available for the power supply. Therefore, it is very important to improve the transmitter efficiency by any means.

*Design Technology of Synthetic Aperture Radar*, First Edition. Jiaguo Lu.

For a SAR system, conventional transmitters can be categorized as electric vacuum tube transmitters, solid-state transmitters, and distributed transmit/receive modules (T/R modules). The former two types, which have been discussed in numerous papers and monographs, are not discussed in this book. This chapter focuses on the design of T/R modules.

## 4.2 Basic Demands

In comparison with conventional radars, SAR has specific requirements for the T/R module in terms of accuracy and consistency of amplitude and phase in a wide bandwidth, adaptation, and reliability required by the space platform.

### 4.2.1 Amplitude and Phase Accuracy

The amplitude and phase accuracy are used to characterize a single T/R module. They are mainly determined by the digital phase shifter and digital attenuator in the T/R module.

The phase accuracy $E_{\Phi}$, which is related to the channel phase $\Phi$, is given by

$$E_{\Phi} = \sqrt{\frac{\sum_{i=1}^{n} (\Phi_i^r - \Phi_i^h)^2}{n}} \tag{4.1}$$

where $\Phi_i^r$ is the actual channel phase, $\Phi_i^h$ is the expected channel phase, $n$ is the number of control states, and $i = 1, 2, \ldots, n$.

The amplitude accuracy $E_A$, which is related to the channel gain $A$, is given by

$$E_A = \sqrt{\frac{\sum_{i=1}^{n} (A_i^r - A_i^h)^2}{n}} \tag{4.2}$$

where $A_i^r$ is the actual channel gain, $A_i^h$ is the expected channel gain, $n$ is the number of control states, and $i = 1, 2, \ldots, n$.

In engineering design, the amplitude accuracy is often directly expressed as the attenuation accuracy required for digital attenuators.

### 4.2.2 Amplitude and Phase Consistency

The amplitude and phase consistency are used to characterize the amplitude and phase relationships between multiple T/R modules. They are not only determined by the corresponding components inside the T/R module but also intimately related to various discontinuities inside channels. The phase consistency $C_{\Phi}$ is given by

$$C_{\Phi} = \sqrt{\frac{\sum_{i=1}^{m} (\Phi_i - \Phi_a)^2}{m}} \tag{4.3}$$

where $\Phi_i$ is the channel phase, $\Phi_a$ is the average of $\Phi_i$, i.e. $\Phi_a = (\Phi_1 + \Phi_2 + \cdots + \Phi_m)/m$, $m$ is the number of channels, and $i = 1, 2, \ldots, m$.

Similarly, the amplitude consistency is related to channel gain, which is given by

$$C_{\mathrm{A}} = \sqrt{\frac{\sum_{i=1}^{m} (A_i - A_a)^2}{m}} \tag{4.4}$$

where $A_i$ is channel amplitude, $A_a$ is the average of $A_{i,}$ i.e. $A_a = (A_1 + A_2 + \cdots + A_m)/m$, $m$ is the number of channels, and $i = 1, 2, \ldots, m$.

Because the T/R module generally works within a certain bandwidth, it is further required to define the channel amplitude and channel phase fluctuating within the bandwidth when calculating the amplitude and phase consistency. If $m$ pieces of the T/R module's channel phase and gain fluctuating within band are $\Delta\Phi_i$, $\Delta A_i$ ($i = 1, 2,\ldots,$ m), the phase consistency $\Delta C_{\Phi}$ and the amplitude consistency $\Delta C_A$ are defined as

$$\Delta C_{\Phi} = \sqrt{\frac{\sum_{i=1}^{m} (\Delta\Phi_i - \Delta\Phi_a)^2}{m}} \tag{4.5}$$

$$\Delta C_{\mathrm{A}} = \sqrt{\frac{\sum_{i=1}^{m} (\Delta A_i - \Delta A_a)^2}{m}} \tag{4.6}$$

where $\Delta\Phi_i$ is the channel phase fluctuation within the bandwidth, $\Delta A_i$ is channel amplitude fluctuation within the bandwidth, $m$ is the number of channels, and $i = 1, 2, \ldots, m$, $\Delta\Phi_a$ and $\Delta A_a$ are the average of $\Delta\Phi_i$ and $\Delta A_i$, respectively, given by

$$\Delta\Phi_a = (\Delta\Phi_1 + \Delta\Phi_2 + \cdots + \Delta\Phi_m) \text{ and}$$

$$\Delta A_a = (\Delta A_1 + \Delta A_2 + \cdots + \Delta A_m)/m$$

The above-defined $\Delta C_{\Phi}$ and $\Delta C_A$ are sometimes referred to as the *nonlinear phase and amplitude specifications* in the T/R module.

### 4.2.3 Assembly Adaptability of Antenna Arrays

The assembly adaptability of the antenna array is mainly considered from two aspects: selection of single channel or multichannel configuration and selection of feeding network and assembling method.

#### 4.2.3.1 Single-Channel and Multichannel Configurations

At low-frequency applications, due to the long wavelength, large element spacing of the antenna array, and limited space available for the T/R module offered by the system, a single T/R module configuration is generally adopted. In higher-frequency radar, especially in the two-dimensional (2D) phased array radar system, the antenna spacing is small. Due to the large size of the connector, the mounting hole, and so on, the size of the conventional single-channel T/R component cannot be further reduced, which brings layout difficulty to the antenna array structure design. Multichannel form is often used in T/R component design to further improve the integration of T/R components. The feed network of the antenna array system should be simplified to reduce the size

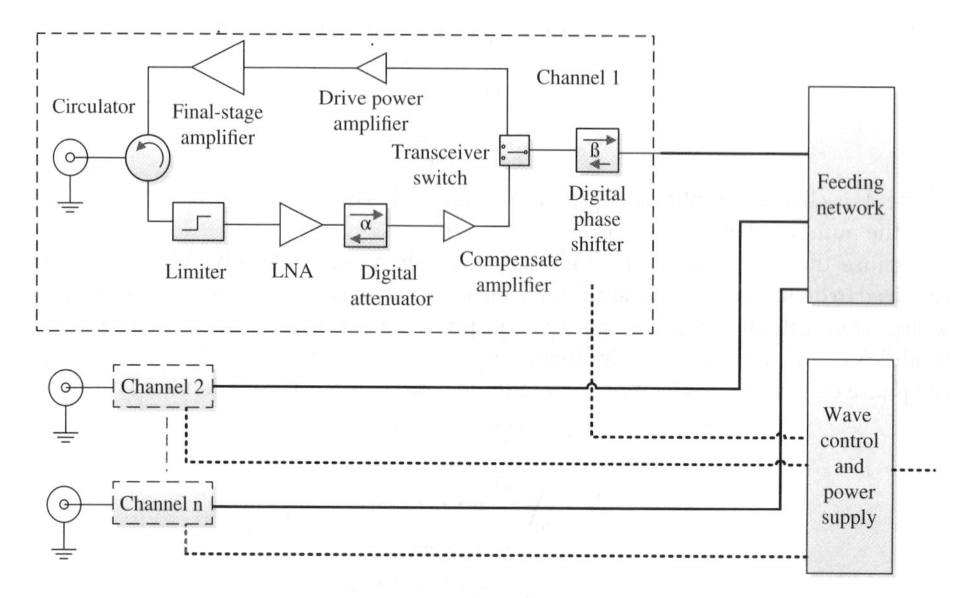


Figure 4.1 Schematic of a multichannel T/R module.

and weight of the system. At the same time, the multichannel T/R component form is adopted, which also reduces the assembly complexity of the active antenna array.

Figure 4.1 shows the schematic of a multichannel T/R module. Due to the fact that multiple T/R channels and their feeding networks are integrated inside the T/R module, and all T/R channels share one control and power supply interface in this configuration, the integration of the T/R module can be effectively improved. The channels number is generally set to be the *n*th power of 2 to integrate the feed network inside the module.

The more channels integrated in the T/R module, the larger the circuits inside the module, the higher assembly process demands, and the lower production yield is. Therefore, the channel number inside the T/R module is determined wholly based on the array structure design, the assembly process capability, and the acceptable cost of volume manufacture.

#### 4.2.3.2 Feeding and Assembling Method

The feeding network and assembling method of the T/R module are significantly influenced by the following aspects of an active antenna array: structure assembling method, thermal design, and feeding layout of the antenna elements. Therefore, the concurrent design can be made by combining the T/R module with the antenna system.

For one-dimensional phased array systems, cables are generally used to connect the feeding network in the T/R module. The advantage is that the assembling position in T/R modules does not require one-to-one correspondence to the antenna elements, and the module assembling is seldom affected by the antenna element spacing and the antenna aperture, though the size and weight of the antenna system may be compromised.

For a 2D phased array system, in the design without the aperture transformation, the T/R modules require one-to-one correspondence to the antenna elements. Due to a large number of cables, inflexibility of connecting method, and difficulty to repair, blind socket connectors are commonly used to connect T/R modules and antenna elements. For connections between T/R modules and feeding networks, either cables or blind

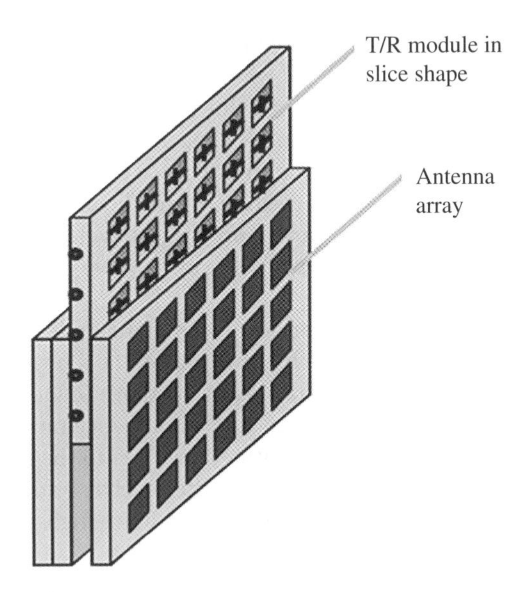


**Figure 4.2** Schematic of a tile antenna system.

socket connecters can be selected depending on the array assembling situation. In addition, in highly integrated SAR systems, the whole antenna system is stacked in tile shape according to different functions (as shown in Figure 4.2). The T/R module should be designed in a slice shape. This is also an important development trend for inactive antenna systems [1].

### 4.2.4 Reliability

The reliability of a T/R module should be improved in terms of design analysis, procedure control, experimental screening, etc. The following aspects need to be considered generally:

(1) *A certain amount of redundancy needs to be included in the module design when specifying electrical properties at top-level planning*. For example, when considering output power of the transmitting power transistor, the power capacity of the load resistance of the attenuator, and the power capacity of the limiter switch, the practical working power level should be lower than the nominal power level of the devices, retaining a certain margin.
(2) *Simplified design is the first choice in reliability design*. Main approaches include optimizing layout in the functional modules, designing and selecting the simplest circuit structures conforming to electrical specifications, and eliminating dispensable components by analyzing the circuit chain features. A typical example is that two T/R modules share one drive circuit.
(3) *Tolerance design in the modular circuit*. The concept of tolerance design should be highlighted in designing each unit circuit, electric circuit, and the structure in the T/R module, to avoid malfunction or failure in the module due to manufacture error variation and temperature gradient variation.
(4) *Protective design of transient states*. In the design of the T/R module, harmful transient states appear in two cases. One is the moment when the power is on or off.

Another is the moment when the excitation of the power amplifier is on or off. The transistor and the circuit may fail when the collector (or drain) voltage fluctuates without being stable enough. For the former case, the protection methods are applied by introducing an overshot proof capacitor and an electromagnetic compatibility (EMC) network of canceling interference deburring, and selecting components with large redundancy. For the latter case, the protection methods include formulating and following a strict power-on program when the module is powered on and operated, and switching on and off the microwave excitation signal when the power supply is stabilized.

(5) *Failure safety design.* During the service and maintenance of the T/R module, it is required that when one certain circuit or device fails or malfunctions, a chain reaction in the other circuits that may result in failure influence spreading should be avoided. Special attention should be paid to influence on the receiver from the leakage power of a high-power transmitter as well as to the influence on the final stage of the higher-power transmitter from the receiver and the antenna. In general, the module reliability can be improved by adding a high-power isolator behind the final-stage transistor in the transmitting channel.

(6) *Thermal design.* Temperature is a main factor that influences the function of the electronic components and the T/R module. Usually, the temperature is exponentially related to the life of the electronic components. Therefore, the thermal design of the component is a very important part. The core of thermal design is to maintain the components to work under reliable conditions. For instance, the case temperature of a transmitting high-power transistor should be kept below 80 °C. For this purpose, the circuit simulation needs to be optimized.

(7) *Procedure control is performed after the telecommunications and structure design are completed.* This specifies requirements of quality control and operations standards for the following procedures: selecting, outsourcing, inspecting, burn in, screening, and assembling of the components; fabrication of microstrip lines, printed circuit board (PCB), and shield packages; and, finally, the assembling and testing of the T/R module. The development and production of large-scale components should be carried out in stages according to the special prototype experiment, small batch production and commissioning, mass production, automatic testing, and debugging production.

(8) *Environmental testing is an important method to improve module reliability.* Electrical performance aging, screening, and environment adaptability testing are effective methods to decrease the failure rate and to eliminate early failure of components. Therefore, for mass production, performance of 100% aging and screening tests on T/R modules is needed. If necessary, a reliability growth test is performed to expose reliability issues in T/R modules and to inspect the design quality so as to improve SAR system reliability.

## 4.3 T/R Module Design

T/R modules need to meet not only the performance specifications but also the reliability specifications. Therefore, the design of a T/R module is a compromise procedure, instead of overemphasizing one index. Under the prerequisite of satisfying system requirements, the circuit, structure, EMC, reliability, manufacturing, etc. are taken into account comprehensively to find a tradeoff among these critical parameters, such as compromise between bandwidth and efficiency of the power device, and compromise between electrical properties, size, and weight.

The stability of the module determines whether the design is successful or not. The transceiver is a closed loop. Noise always exists in the circuit even though duplex is adopted. If the isolation between the transmitter and the receiver is not high enough, once the module is powered on, a positive feedback of the system noise can cause a self-oscillation, and the module is unstable. The yield of the module is another key factor that determines whether a design is successful or not. The yield influences not only the cost of the module and the system but also the uniformity and reliability of the module. The yield of the module is closely related to the structure, the processing technology, and so on. In the early stages of module design, material thermal properties, welding technology, and assembling method need to be considered.

## 4.3.1 Electrical Design

### 4.3.1.1 Receiver

A typical T/R module receiver is shown in Figure 4.3. The main function of the receiver is to amplify the signal received by the antenna. In the meantime, the noise factor should be kept low enough to meet sensitivity specifications of the receiver. In the receiver, a digital attenuator is usually added to adjust the receiver gain digitally. In a SAR system with duplex transceiver, a digital phase shifter is usually located in the common path, where the signal phase for both receiver and transmitter can be digitally adjusted.

In practice, the receiver is destabilized by two factors. One is the leakage and reflection of the power in the transmitter channel. The worst case happens when the antenna is open, and all output power of the transmitter goes into the receiver. Another factor is high-power echo under normal operation. When the radar system reflects large areas at close range (such as clouds, ground echoes, radome surface ice, etc.), it will return echoes with very high signal levels. If no protection is adopted, the receiving channel will not work properly or may even be burnt out. Therefore, the receiving channel input terminal generally needs to be protected by the upper limiter. The limiter has a low insertion loss when passing a small signal normal echo. A large amount of attenuation is produced when the signal input is increased to the slice level, and the slice level output is maintained when the signal is greater than the slice level input. The limiter level of the

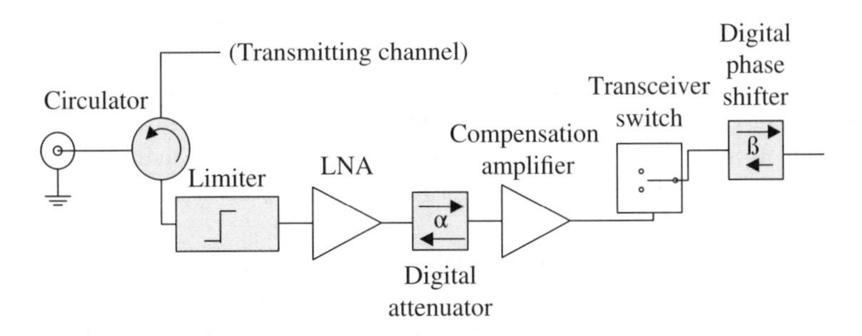


**Figure 4.3** Typical T/R module receiver.

limiter should be chosen to be lower than the absolute maximum input signal power allowed of the first-stage low noise amplifier (LNA) to ensure that the leakage signal after clipping does not burn the first-stage LNA.

Noise factor is a key factor in the receiver design. The noise factor in a cascaded system is given by

$$F = F_1 + \frac{F_2 - 1}{G_1} + \frac{F_3 - 1}{G_1 G_2} + \ldots + \frac{F_i - 1}{G_1 G_{2\ldots} G_{i-1}} + \ldots + \frac{F_n - 1}{G_1 G_{2\ldots} G_{n-1}} \tag{4.7}$$

where $F_i$ is the noise in the $i$th stage in the cascaded system, $G_i$ is the magnification of the $i$th stage amplifying circuit, $n$ is the number of stages, and $i = 1, 2, \ldots, n - 1$.

It can be seen from Eq. (4.7) that the noise factor in the receiver is dominantly defined by the loss of front passive components as well as the noise factor and gain in the first-stage LNA. Therefore, in the design of the receiver, the first-stage amplifier needs to be selected with low noise factor and high in gain, and the loss before the first-stage LNA needs to be decreased as much as possible.

The dynamic scope in the receiver determines the intensity variation range in allowable echo of the system. Two parameters are employed to characterize the dynamic scope: minimum detectable signal amplitude $Si_{min}$ and input compression point power $P_{-1(in)}$ at 1 dB.

For the receiver, the minimum detectable signal level lies within the signal bandwidth. At room temperature, the equivalent white noise power level in the receiver at the input port is given by

$$Si_{min}(dBm) = -114 + 10lg(BW) + NF + 10lg\left(\frac{S_o}{N_o}\right)_{min} \tag{4.8}$$

where $BW$ is the signal bandwidth in MHz, $NF$ is the noise factor in the receiver, and $\left(\frac{S_o}{N_o}\right)_{min}$ is the identification factor, which usually is 1.

When the amplitude of the input signal is going up to the 1 dB compression point, the amplitude of the input signal is determined by the gain of cascaded LNAs and the size of output $P_{-1}$. The parameters in the LNA require proper selection in practical design. In addition, under the same component combination, the 1 dB compression point is strongly influenced by the order of the connecting components inside the receiver. After the digital attenuator in the typical T/R component receiving channel shown in Figure 4.3 is moved behind the compensation amplifier, a receiving channel topology as shown in Figure 4.4 is formed. Both designs contain identical devices. However, when the digital attenuator is put behind the compensation amplifier, the amplitude of input signal in the compensation amplifier will not vary with insertion loss of the digital attenuator, causing gain compression to easily occur in the second-stage amplifier of the receiver; consequently, the 1 dB compression point power in the receiver will be reduced. Taking the X-band T/R component as an example, under the same device selection, the input 1 dB compression point power of the receiving channel can be increased by about 3.5 dB after the digital attenuator is placed between the two-stage amplifiers.

As the SAR system resolution goes up higher, demands are put on the bandwidth of the T/R module, which challenges gain flatness in the receiver. The in-band gain flatness in the receiver is influenced by the factors of amplitude-frequency characteristic, standing wave and frequency characteristic of insertion loss in the transmission line, etc.

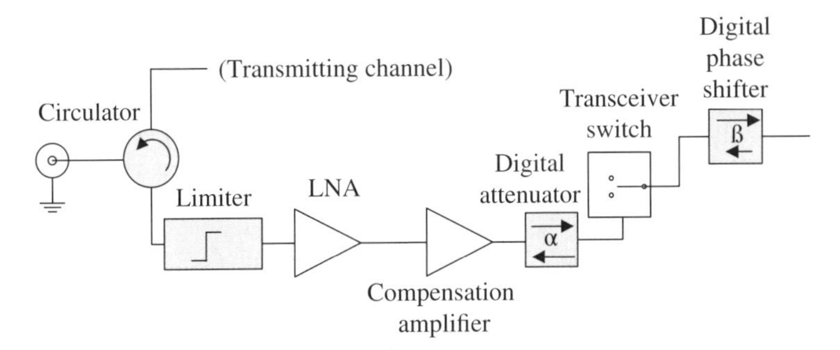


**Figure 4.4** Topology of a T/R module receiver with reduced pin-1.

The loss of conventional transmission line mainly consists of three parts [2]: dielectric loss, conduction loss, and radiation loss. The dielectric loss in simplified formula is given by

$$\alpha_d = 27.27\frac{\varepsilon_e - 1}{\varepsilon_r - 1}\left(\frac{\varepsilon_r}{\varepsilon_e}\right)\frac{tan\ \delta_e}{\lambda_g} \tag{4.9}$$

where $\varepsilon_r$ is relative dielectric constant, $\varepsilon_e$ is equivalent dielectric constant, $tan\ \delta_e$ is loss tangent angle, and $\lambda_g$ is guide wavelength.

As shown in Eq. (4.9), dielectric loss decreases with the increase of wavelength and increases as frequency. It means the loss of a passive circuit is characterized with negative slope and goes up.

Considering the fact that the insertion loss of a passive device has a negative slope characteristic, positive slope gain compensation, or balance circuit compensation, is usually is designed in an LNA to make the receiver gain tend to be flat. In practical applications, it is not always required that the flatter the gain the better the result, because transmission loss of the atmosphere also shows a negative slope characteristic, and the transmitter power and the receiver gain are designed to have a positive slope in band, which sometimes can have a certain compensating effect.

In addition, thermal stability in the receiver is also important. During practical operation of an active antenna array, the temperatures of T/R modules in the array differ widely with ambient temperature, operation mode, and thermal design. This difference is mainly manifested in two ways: one is the overall increase or decrease in the array temperature, the other is the different temperatures of T/R modules in different positions. Experimental results indicate that the gain of a GaAs amplifier decreases as temperature goes up or increases as temperature goes down, the gain temperature characteristic of a typical GaAs amplifier being characterized with −0.001 dB/°C per 1 dB variation [3]. For example, for an LNA with typical gain of 25 dB, its gain decreases 0.25 dB (0.001 dB/°C × 10 °C × 25) for every 10 °C ramping up. For the T/R module receiver, which normally consists of two- or three-stage amplifiers, the overall gain is usually higher than 50 dB. Without compensation design, variation of the receiver gain will exceed 0.5 dB for every 10 °C variation. To improve thermal stability in a T/R module, a popular method is to design a biased network with temperature compensation for the amplifier or to adopt a temperature compensation attenuator in the receiver. The temperature compensation attenuator is easy to use – an absorption microwave attenuator is featured, with the attenuation varying with temperature. Products with a series of different temperature coefficients (characterization of the variation of the attenuation with the temperature) can be selected. According to temperature characteristic of a GaAs device, the selected temperature compensation attenuator should increase attenuation at low temperatures and decrease attenuation at high temperatures. As a result, the temperature variation of the gain in the receiver is maintained at a relatively low level.

#### 4.3.1.2 Transmitter

A typical T/R module transmitter is shown in Figure 4.5. The main function is to amplify the excited signal from feed networks. The phase of the transmitted signal during operation can be adjusted by controlling a digital phase shifter in the common channel, so as to realize beam scanning of the antenna.

For the transmit channel, the input and output matching network design of the power amplifier is very important. The matching network of the amplifier can be optimized by load-pull characterization and simulation, to obtain both high output power and efficiency. Second, the transmission line causes direct loss of the transmit power, due to insertion loss between the power amplifier and the antenna radiation element. In addition, the influence of the antenna on the transceiver load cannot be neglected. The standing wave at the antenna port varies with the scanning angle during wide-angle scanning of the active antenna array. For the final stage, the load is no longer 50 Ω. The output port is mismatched, which changes the output power into a function of the scanning angle and even makes the final-stage amplifier unstable. For example, at a scanning angle of 75°, the input impedance of the antenna varying in the range of $50 \pm 20j$, the standing wave ratio can be larger than 3 : 1, which will cause a serious load-pull. Therefore, it should be considered during the T/R module design to suppress the load-pull to a certain extent.

Conventional transmitter configurations are shown in Figure 4.5. Figure 4.6a shows a switching method of a single-pole double-throw switch for transmitter and receiver. Due to the fact that the switch cannot isolate the signal reflected from the antenna port, the final-stage amplifier is directly influenced by the port mismatch during antenna scanning, showing an obvious load-pull phenomenon in the transmitter. This configuration is generally applied in systems with high integration and small antenna scanning angle.

Figure 4.6b shows a configuration where the single-pole double-throw switch is replaced by a circulator. No additional control signal is required to switch the transceiver, which simplifies circuit design in the module. In the meantime, low insertion loss in the circulator can offer better system specifications. However, similar to the switching method, a pure circulator also cannot alleviate the output mismatch problem.

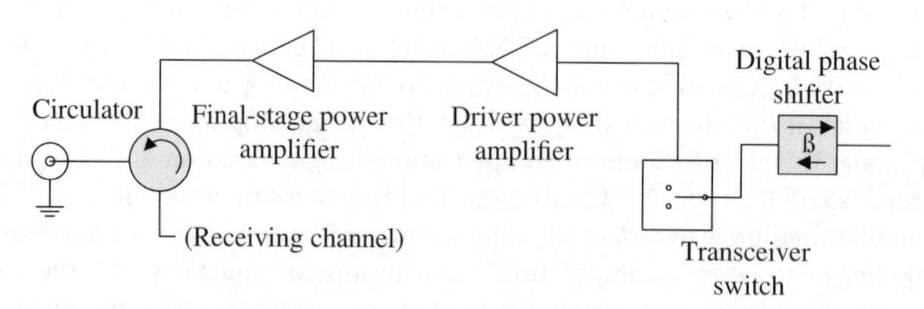


Figure 4.5 Typical T/R transmitter.

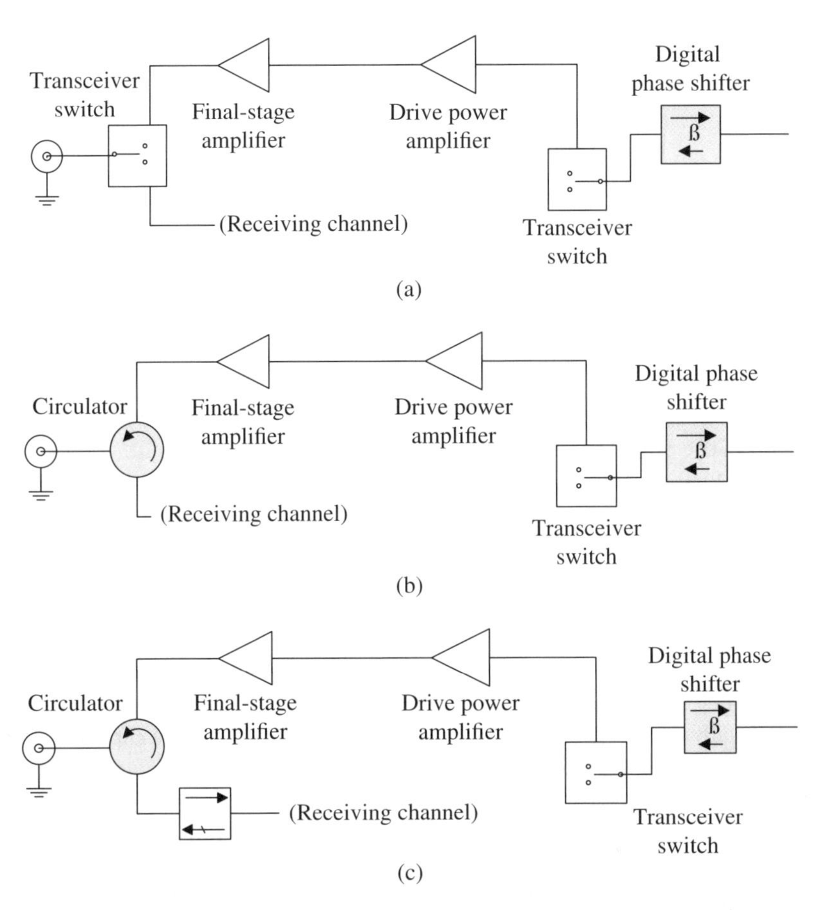


**Figure 4.6** Conventional transmitter configurations. (a) Switch type; (b) pure circulator; (c) circulator combined with isolator.

Figure 4.6c shows a way of combination with the lowest risk and the highest safety. When the transmitting channel is working, the reflection caused by the antenna mismatch will be absorbed by the isolator of the receiving channel without returning to the output of the final amplifier of the transmitting channel. Therefore, the influence of the mismatched amplifiers at the final stage of the antenna beam scanning is avoided. The disadvantage in this way is that the isolator increases insertion loss in the receiver, which deteriorates the noise factor to some extent.

In the design of a high-frequency T/R module, GaAs- and GaN-based class A and class AB power amplifiers are mainly adopted. The peak current of the transmitter is high, together with the resistors associated with low-frequency connector cables and internal circuit lines, causing a high circuit loss, and consequently reducing the module efficiency.

In a SAR system, the transmitter in the T/R module usually works in pulse mode. The quality of pulse waveform directly influences the radar imaging quality, in which the pulse waveform drop is a key factor that dominates the quality of the radar image. Generally speaking, for Class C amplifiers, the pulse drop is mainly determined by internal

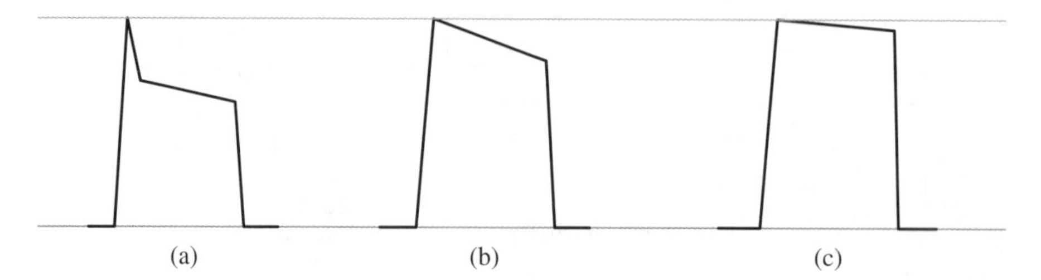


**Figure 4.7** Influence of energy storage capacitor on the pulse waveform drop. (a) Without storage capacitor; (b) with insufficient storage capacity; (c) with sufficient storage capacity.

thermal effect of the transistor itself and specific circumstances of the external circuits. For Class A or Class AB amplifiers, the pulse waveform drop is mainly created by pulse modulation in the power supply. In practical design, a storage capacitor is placed inside the module. During receiving, the storage capacitor is charged by the power supply. During transmitter operation, both the storage capacitor and the power supply provide energy simultaneously. The capacity amount in the storage capacitor is significantly instrumental to the pulse waveform, as shown in Figure 4.7. In Figure 4.7a, without a storage capacitor during pulse operation, the equivalent inductance, generated by the circuit between power supply and power amplifier, causes the output power to decrease, and an overshoot spike appears in the power waveform. In Figure 4.7b, a storage capacitor with insufficient capacity is added. Now, voltage at the power amplifier during transmitting gradually decreases with energy consumption in the storage capacitor, showing a big drop in the pulse waveform. In Figure 4.7c, with adequate storage capacity, voltage drop in the capacitor is trivial during transmitting, and consequently, the pulse waveform drop is very small and the effective output power within the pulse is larger.

The capacity in the storage capacitor can be estimated based on specifications of pulse waveform drop in the transmitted pulse signal and operation pulse width. The relationship between the stored energy and voltage is given by

$$W = \frac{1}{2}CU^2 \tag{4.10}$$

Assuming that all power in the power amplifier comes from the capacitor discharge during transmitting, then

$$P\tau = \frac{1}{2}(CU_1^2 - CU_2^2) \tag{4.11}$$

where $P$ is the power consumption in the amplifier from the power supply, $\tau$ is pulse width of the transmitter, $C$ is the capacity in the storage capacitor, and $U_1$ and $U_2$ are the voltage of the power supply at the start and stop point of the pulse, respectively, which determines the pulse waveform drop.

The required capacity in the storage capacitor can be calculated based on the above equation. Due to the fact that the power supply provides power as well, the capacity calculated with the above equation has a certain redundancy, which can be relaxed in a practical design.

#### 4.3.1.3 Beam Steering and Electric Interface

The interface in the T/R module is defined according to the control means and function in the SAR system. It can be divided into a transistor–transistor logic (TTL) interface and a differential interface as per interface voltage level, and serial control and parallel control as per control mode. The TTL interface only requires one control line for every signal. The number of physical pins in the modules can be made small, but it is not suitable for long-distance transmission without good signal anti-interference capability. It is normally used in small-scale antenna arrays. A pair of wires is used in the differential interface to replace a single wire. Both signals with equal amplitude and opposite phase are transferred on each wire, which increases complexity in the interface. However, due to high interference resistance, the differential signal is highly immune to external electromagnetic interference (EMI). The differential signal generates much less EMI than the TTL signal, except for its insensitivity to external interference. The differential signal is often adopted as the interface in the T/R module, in systems with long-distance transmission and a complex electromagnetic environment.

There are many device control bits inside the T/R component, which requires a larger number of control lines. In most cases, parallel interface is not suitable to connect the control interface externally. A dedicated beam-steering chip is normally used to transfer parallel interface to serial interface to reduce the number of interface pins. This type of beam-steering chip usually consists of two-stage latch, control and protection circuits of the power sequence in the transceiver, built-in test equipment detection under working status, protection circuits in the power supply, etc. The logic functional diagram in a typical beam-steering chip is shown in Figure 4.8.

#### 4.3.1.4 Power Supply and Time Sequence

For a T/R module that adopts GaAs and GaN devices as power amplifiers in the transmitter, attention should be paid to the overall consideration of the power sequence in the whole module. Both gate and drain in these types of devices need to be biased during operation. The static working point is controlled by biasing gate with reverse voltage. If the gate bias is off while the drain bias is still on, the clamping of drain current will be lost, and the device will be burned by the undergoing high drain current. Therefore, when powering on these types of power devices in the T/R module, one should follow the following time sequence: first, bias the gate, then the drain. And vice versa when powering off, that is, first, disconnect the drain bias, then the gate bias.

In system applications, there exists a certain risk when protection relies on an external power on/off sequence in the T/R module. If the external protection fails by accident, it may burn a large area of the T/R module on the array. Therefore, it is necessary to have a negative voltage protection circuit inside the T/R module. Figure 4.9 shows a relatively simple negative voltage protection circuit in which the transmitted modulated signal TRT is blocked, and the drain bias is switched off by the power modulation circuit inside the module to protect the power device.

The T/R module in a SAR system usually transmits and receives in time division. The transceiver in the T/R module shares the transceiver switch, the circulator, etc. Due to the limited isolation from the transceiver switch, when the transmitter and the receiver are powered on simultaneously, the amplification gain in the transmitter and the receiver is much higher than isolation of the switch, producing a positive feedback loop and generating a transceiver self-excitation. Therefore, the time sequence of the power supply in the transceiver should be strictly controlled to make sure that the receiver and transmitter power supply voltages are not applied simultaneously. The time sequence of a typical T/R module is shown in Figure 4.10.

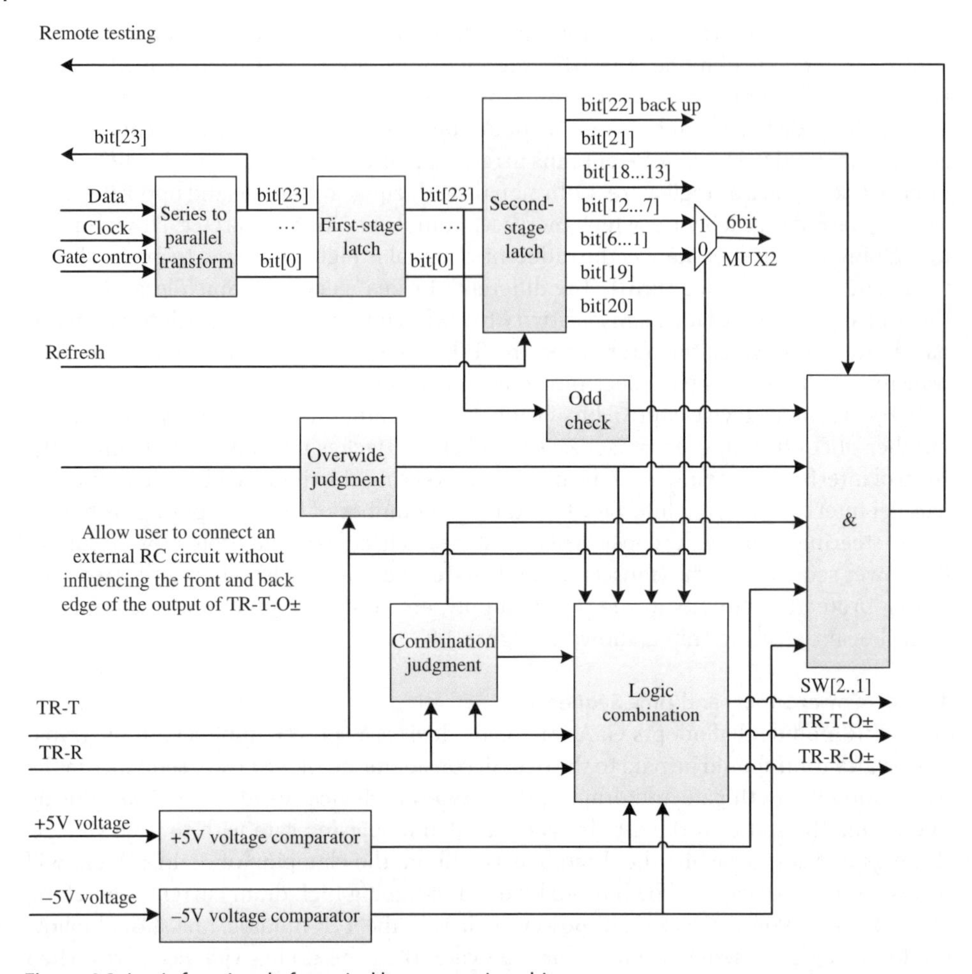


**Figure 4.8** Logic functional of a typical beam-steering chip.

## 4.3.2 Structure Design

### 4.3.2.1 Physical Interface

Limited by the size of the antenna aperture, an integrated design of the structure layout in a T/R module is generally used to meet the requirements of array assembling when the weight and size are considered. Physical interface is an important part of T/R module design, which affects both performance and quality of the radar. Physical interface influences not only the signal transfer characterization but also the assembling method. Microwave signal, control signal, and assembling interface need to be considered during the design, together with an overall consideration of the circuit layout.

### 4.3.2.2 Thermal Dissipation

Thermal dissipation is another important part of T/R module design. Good thermal dissipation design can improve stability and reliability in the T/R module. First, the cooling method of the device and the module is defined in the thermal dissipation design.

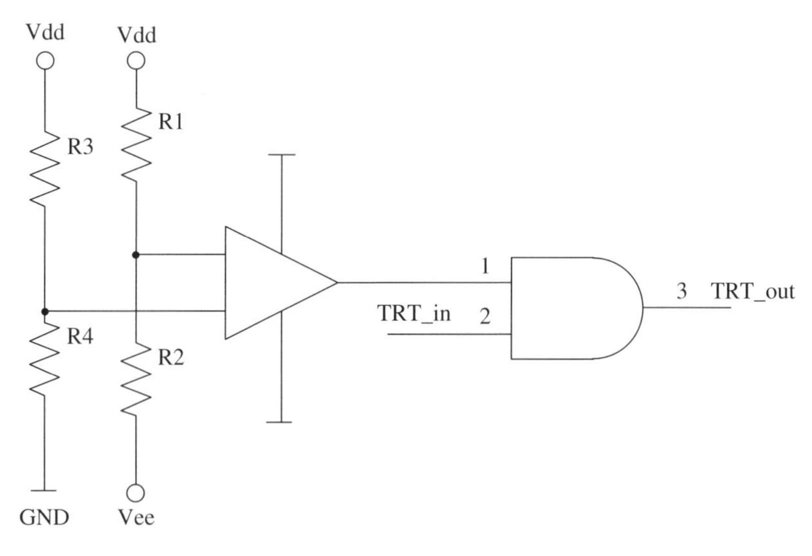


**Figure 4.9** Negative voltage protection circuit.

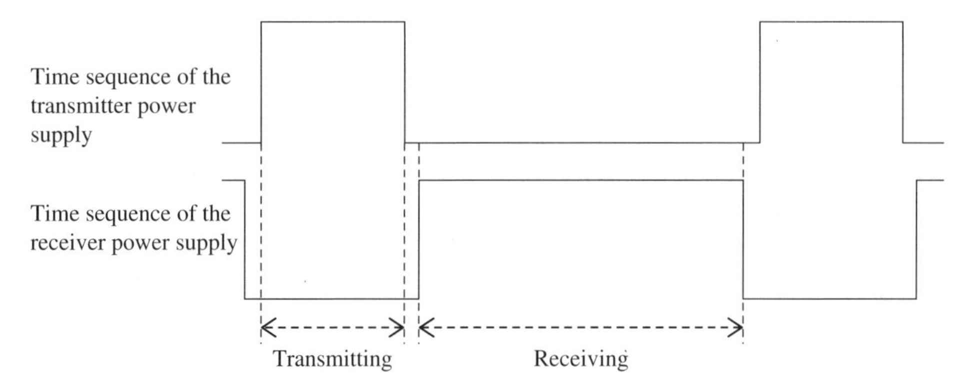


**Figure 4.10** Time sequence of a typical T/R module.

The choice of cooling method directly influences the selection of the device and the assembling method in the module. It even influences reliability, weight, and cost of the T/R module. There are currently widely used cooling methods, such as natural cooling, forced air cooling, forced liquid cooling, etc. To enhance the thermal dissipation effect, the thermal conductivity of the package and the internal devices in the T/R modules should be taken into account. The thermal resistance in every segment of the heat transfer path should be minimized to create a low thermal resistance path. For example, one can choose high thermal conductivity materials to provide a proper heat sink for high-power devices.

The thermal source in the T/R module is mainly concentrated in the transmitter, and generally the final-stage power amplifier in the transmitter is the major thermal source. The heat dissipation design in the T/R module is based on analysis and simulation of the practical working conditions, including mainly the following three aspects.

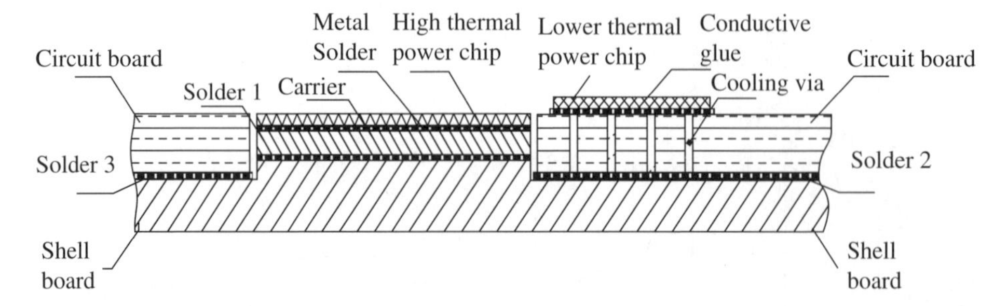


**Figure 4.11** Diagram of T/R module assembly.

***Assembling Method and Material Characteristic of the T/R Module*** The assembling technology is chosen according to the thermal power in the device. Generally speaking, for high thermal power devices in the transmitter, the devices are usually mounted on a high thermal conductivity carrier with gold–tin solder. Then the carrier is soldered on the T/R module package by using solders with a low melting point. For devices with low thermal power in the receiver, they are generally bonded and sintered directly on the surface of the circuit board, and the heat is dissipated through the cooling vias on the board, as shown in Figure 4.11.

For every material in the heat flow path, thermal conductivity of the material needs to be known to properly estimate the thermal resistance. Table 4.1 lists the thermal conductivity of several materials commonly used in T/R modules.

***Analysis of Thermal Resistance in the T/R Module*** According to the assembly method of the device, the size and thickness of every material in the thermal path, from thermal device to external bottom in the module package, can be calculated. Taking into account that the inhomogeneous diffusion distribution of the heat flux, in which most of the heat flux is concentrated in the 45° conical region under the contact surface (as shown in Figure 4.12), diffusion characteristics in the contact area need to be considered when calculating the size of materials.

When the physical dimension of a component and the thermal conductivity of a material are known, thermal resistance introduced by this component can be calculated by

$$\theta = \frac{H}{KS} \tag{4.12}$$

where $\theta$ is the thermal resistance (°C W$^{-1}$), $H$ the is material thickness along the direction of heat flux (m), $K$ is the thermal conductivity (W m$^{-1}$ K), and $S$ is the cross-sectional area of the heat flux (m$^2$).

Based on Eq. (4.12), the thermal resistance introduced by every material between the bottom of the thermal device and the bottom of the T/R module package can be calculated. For example, for the power device in Figure 4.11, the thermal resistance generated by the solder, the heat conductive carrier, and the bottom plate of the housing can be separately calculated.

The thermal resistance in the device is normally provided by the manufacturer. Under certain working conditions, the real operation junction temperature and the temperature at the backside of the device can be measured by infrared thermal camera.

**Table 4.1** Thermal conductivity of conventional materials in a T/R module.

| Material description | Thermal conductivity (W m$^{-1}$ K) |
|---|---|
| Lead–tin solder (63Sn37Pn) | 50 |
| Gold–tin solder (80Au20Sn) | 251 |
| Gold–germanium solder (88Au12Ge) | 232 |
| Indium–tin solder (50In50Sn) | 200 |
| Gold (Au) | 300 |
| Silver (Ag) | 420 |
| Copper (Cu) | 394 |
| Tungsten (W) | 150 |
| Molybdenum (Mo) | 140 |
| Aluminum (Al) | 216 |
| Silicon (Si) | 150 |
| Alumina ($Al_2O_3$) | 27 |
| Molybdenum–copper alloy | 165 |
| Oxygen-free copper | 390 |
| Kovar alloy | 14 |
| Plastic molding compound | 0.75 |
| Aluminum nitride (AlN) | 170–200 |
| Beryllium oxide (BeO) | 230 |
| Aluminum silicon carbide (AlSiC) | 160 |
| GaAs | 59 |
| Epoxy glue | 0.79 |
| Thermal grease silicone | 170 |
| Microstrip board | 0.63 |
| FR4 laminate | 0.32 |

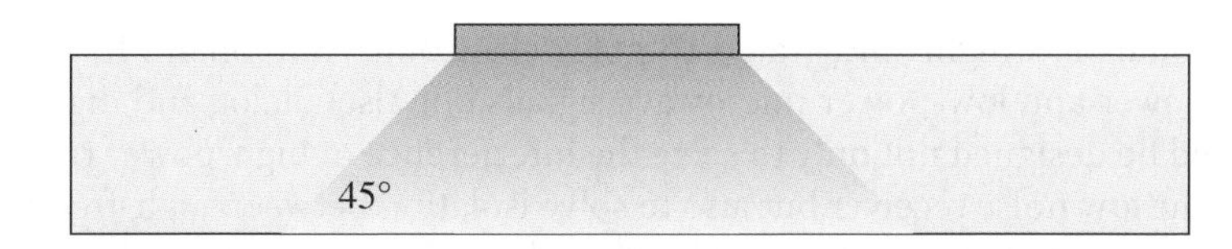


**Figure 4.12** Diffusion distribution of the heat flux.

The thermal resistance in the device can then be extrapolated according to the operating conditions.

***Analysis of the Junction Temperature of the Chip*** As per the thermal resistance $\theta$ of every material obtained by the above-described steps, combined with thermal power $Q$ in the device during operation, the temperature ramp $\Delta T$ between the thermal junction of the device and the bottom of the T/R module package can be given by

$$\Delta T = Q\theta_{\Sigma} \tag{4.13}$$

where $\theta_{\Sigma}$ is sum of the thermal resistance in every material. The maximum package temperature $T_c$ can be obtained from the ambient temperature and the analysis of thermal design. Then the junction temperature $T_j$ of the power device in the T/R module is given by

$$T_j = T_c + \Delta T \tag{4.14}$$

Based on the junction temperature calculated from Eq. (4.14), one can judge whether the thermal design of the T/R module satisfies reliability requirements in the SAR system or not.

#### 4.3.2.3 Protection of the T/R Module

The T/R module is very sensitive to harsh environments. If the module is not protected properly, it may cause the aging of the device, substrate, and solder used in assembly, resulting in gradual degradation of the circuit performance, such as frequency dropping, gain dropping, noise increasing, bandwidth narrowing, output power drop, etc. Therefore, the protection of the T/R module has become an important research area. It is not just a process technology implementation but also involves many aspects of raw materials, devices, circuits, structures, processes, and comprehensive technical research.

The following aspects are generally considered in T/R module protection: all kinds of surfaces in the structure should be designed simply and smoothly, and the transition should be smooth enough to prevent local heating and stress concentration. Vents and drains should be added where seepage and retained moisture exist. Grooves, blind holes, and cracks should be avoided to prevent retention and aggregation of a corrosive medium. The existence of blind holes makes it difficult to clean corrosive liquid inside the hole, which not only pollutes the plating solution but also has harmful effects on the coating quality. Try to use the same metal or metals with a small potential difference to make direct contact to prevent galvanic corrosion. A metal gasket or a coating layer can be used as a transition to contact both metals. Cathode materials should be chosen for critical components.

### 4.3.3 EMC

The T/R module is a highly integrated complex electronic component. It contains not only high-power and low-power microwave signals but also analog and digital signals. EMC should be designed not only to solve the interference of high-power transmission signals on the low noise receiver but also to solve isolation between high-frequency signal and digital signal. In the meantime, the mutual interference between the T/R module and other electronic devices in the SAR system must be taken into account. The EMC design in the T/R module needs to ensure that the T/R module functions properly in a practical electromagnetic environment, without performance degradation and failure. It should also ensure that any EMI that influences proper function of any other modules and components inside the electromagnetic environment is suppressed. Currently, no EMC design criteria dedicated to the T/R module has been established. Instead, EMC specifications of GJB (Chinese standard) are generally used as the basis of module evaluation.

It is well known that there are three necessary conditions for EMI: interference source, transmission path, and sensitive receiving unit. EMC is designed to destroy one or more of these three conditions. Generally used methods are filtering technology, layout and routing technology, shielding technology, grounding technology, sealing technology, etc. For EMC design of the T/R module, one should pay attention to the following five aspects.

#### 4.3.3.1 Self-Oscillation and Cavity Effect

The existence of self-oscillation prevents signal transmission, wastes a lot of energy, and affects normal operation of the components and even the SAR system. The cavity effect is an important consideration in the EMC design of the module. The existence of cavity resonant frequency will lead to instability of the module and even failure in the entire module. In most cases, the cavity effect can cause self-oscillation of certain circuits inside the module. In the module design, one should try to increase the cavity resonant frequency, keeping the resonant frequency outside the working bandwidth and far away from the working frequency of the module. At high frequencies, the cavity resonant frequency inevitably falls inside the working bandwidth. In this case, one should optimize the cavity structure of the module to reduce the $Q$ value of the in-band cavity resonant frequency.

For rectangular cavity structure shown in Figure 4.13, the resonant wave number is given by

$$k_{mnl} = \sqrt{\left(\frac{m\pi}{a}\right)^2 + \left(\frac{n\pi}{b}\right)^2 + \left(\frac{l\pi}{d}\right)^2} \tag{4.15}$$

where $a$, $b$, and $d$ are the length, width, and height of the rectangle, respectively, and $m$, $n$, and $l$ are the half-wave number of the field in direction $x$, $y$, and $z$, respectively. The resonant frequency of $TE_{mnl}$ mode and $TM_{mnl}$ mode are given by

$$f_{mnl} = \frac{c}{2\sqrt{\mu_r \varepsilon_r}}\sqrt{\left(\frac{m}{a}\right)^2 + \left(\frac{n}{b}\right)^2 + \left(\frac{l}{d}\right)^2} \tag{4.16}$$

Assuming $b < a < d$, then the fundamental mode of TE mode is $TE_{101}$, and the fundamental mode of TM mode is $TM_{110}$. The lowest resonant frequency of the cavity is resonant frequency corresponding to $TE_{101}$. All resonant frequencies of the cavity can be calculated according to Eq. (4.16).

Quality factor $Q_0$ characterizes the relationship between the stored energy and the dissipated energy inside the cavity and influences the EMC of the module. $Q_0$ is given by

$$Q_0 = 2\pi\frac{W}{W_T} = 2\pi\frac{W}{\mathrm{P_L T}} = \frac{2\pi}{\mathrm{T}}\frac{W}{P_L} = 2\pi \mathrm{f_0}\frac{W}{P_L} = \omega_0\frac{W}{P_L} \tag{4.17}$$

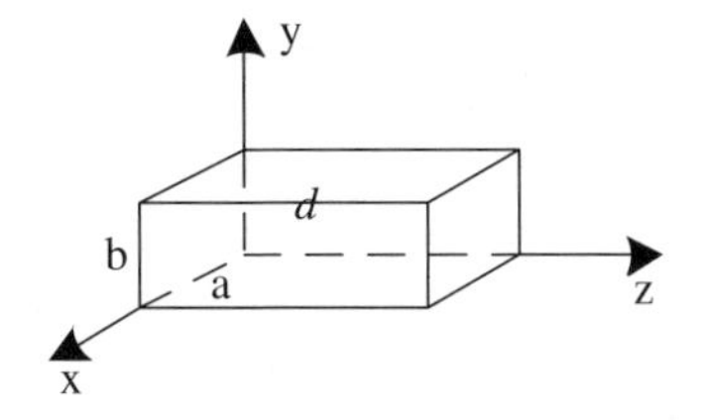


**Figure 4.13** Rectangle cavity structure.

where $\omega_0$ is the resonant angle frequency, $W$ is the energy stored in the cavity, $W_T$ is the dissipated resonant energy in one period, and $P_L$ is the average dissipated power in one period. $W$ and $P_L$ are given by

$$W = W_e + W_m = \frac{1}{2}\int_V \mu|H|^2 dv \tag{4.18}$$

$$P_L = \frac{1}{2}R_s \oint_S |H_t|^2 ds \tag{4.19}$$

where $R_S$ is the surface resistivity, $H_t$ is the tangent magnetic field, and $\delta$ is skin depth. From Eqs. (4.17) to (4.19), the quality factor $Q_o$ is given by

$$Q_0 = \frac{2}{\delta}\frac{\int_V \mu|H|^2 dv}{\oint_S |H_t|^2 ds} \tag{4.20}$$

If the tangent magnetic field at the inner wall of a resonant cavity is larger than the magnetic field inside the cavity, $Q_0$ is approximated to

$$Q_0 \approx \frac{1}{\delta}\frac{V}{S} \tag{4.21}$$

where $V$ is cavity volume and $S$ is the internal surface area of the cavity.

It can be seen from Eq. (4.21) that $Q_0$ in resonant cavity is proportional to the volume of the cavity and is inversely proportional to the internal surface area. Thus, in T/R module design, when resonance cannot be prevented, one should try to reduce $Q_0$ of resonant frequency by decreasing volume of the cavity and increasing the internal surface area, through adding metal lugs and bonding absorbing materials inside the cover.

#### 4.3.3.2 Power Integrity

The power instability of the T/R module is mainly caused by two aspects: one is that transient AC current is too high when the device is in high-speed switch status; the other is the existence of the inductance in current loop. From the performance point of view, it can be divided into three types: simultaneous switching noise (SSN), sometimes referred to as *Δi noise,* and ground bounce phenomenon can also be classified as such; nonideal source impedance; and resonance and edge effect.

In feeding network design, an inductor coil can be used at the input port of the power supply to suppress high-frequency common mode interference. A high-frequency rectifier EMI filter circuit can be used at the output port to not only prevent the influence of external EMI on the T/R module but also to suppress the conduction leakage of the T/R module through the power bus. It can also prevent the noise of the power supply from being added into the feed network when it is providing energy.

The interference caused by charge and discharge of the power supply during pulse operation should be taken into account in designing the filter of the power supply in a T/R module. A parallel decoupling capacitor should be added at places where the power enters into modules and places where the power is supplied to a key active device, to remove high-frequency noise from the power supply effectively. The filter for the power supply is generally designed to follow the following principles: filtering at both low frequency and high frequency should be considered simultaneously at places where the power is on; and at least one high-frequency filtering capacitor should be added for every (group of) power/ground of the devices. An inductor or a magnetic bead should be added at the system power supply side of the corresponding high-frequency filtering capacitor, when the working frequency of the device or the component is relatively high. The high-frequency filtering capacitor should be placed as close as possible to the power/ground pins.

#### 4.3.3.3 Grounding and Slot Coupling

From the EMC point of view, grounding is designed to restrain the coupling interference caused by current flowing through one common ground. For this purpose, a good grounding requires that unnecessary ground loop should be avoided by all means. According to the principle of three-ground separation, DC ground and shield ground (safety ground) should be separated to prevent the interference caused by signals passing through a common ground system. Ensure that the ground is well connected when assembling the device, and all ground wires should be as short as possible. For signal loop, power system loop, bottom plate, and package, a separate ground should be used, and all these loops should be connected to a common reference point. Every unit circuit inside the module should contact the package in a large area. Try to reduce the number of via holes and slots in structure design to prevent signal crosstalk due to coupling; when via holes and slots cannot be avoided, their size should be minimized by using a circular via hole and a narrow slot as much as possible.

#### 4.3.3.4 Electrostatic Prevention

Electrostatic discharge (ESD) is caused by current flow between two objects with different potentials. ESD can damage the semiconductor device, with the probability of immediate failure of the device being about 10% (short circuit, open circuit, no function, parameter change), and probability of potential damage being more than 90%. Although electrical parameters of the device still meet the specified requirements after ESD damage, the capability to resist electrical overstress and the device's reliability are reduced.

ESD protection of a T/R module is mainly designed in three aspects: reinforcing ESD immunity of the device itself, grounding, and circuit-level protection. More details are summarized below.

(1) *Reinforcing ESD immunity.* A protection circuit is designed to transfer high current, thus improving the ESD threshold of the device itself, or adding a Zener diode or a bypassing capacitor at the input port of the power supply.
(2) *Grounding design.* Grounding is the most fundamental and effective method to avoid ESD, by providin a discharge loop. The purpose of grounding is to provide a uniform structure plane or a low impedance path for ESD current, so as to avoid the buildup of the potential difference between two interconnected metals.
(3) *Circuit-level protection.* To reduce interference between low-frequency devices and high-frequency devices inside the module, low-frequency and high-frequency circuits should be isolated in the structure. A large contact surface is used for grounding when assembling circuits, making sure the contact resistance between the circuits and the package is kept low and shields the ESD to circuits. PCB is very sensitive to the magnetic field produced by electrostatic discharging current; therefore, the area of all loops on the circuit board should be kept as small as possible, and the loop area can be reduced by using a ground grid. The layout of the wires on the circuit board is a key factor to resist transient shock, and parasitic

inductance on the trace creates voltage spikes for transient shocks that may exceed the limits of the chip pins. Therefore, the parasitic inductance of the wires should be minimized, such as shortening the line, increasing the line width, etc.

#### 4.3.3.5 Electrical Wiring

The layout of wires directly affects the integrity of board-level signal. A good layout is an important approach to ensure integrity of the signal. For the layout of microwave multilayer circuit board in a T/R module, the points in the following paragraphs are considered.

The backside of substrate is usually designed as the ground plane for large-area soldering. This structure is particularly suitable for RF circuits, where it can provide the smallest loop inductance and a low ground surface impedance at high frequency. Another advantage of a large-area grounding surface is that it can minimize the radiation loop. This ensures that the circuit board has a minimal differential mode radiation and sensitivity to external interference.

Decoupling capacitors are used to filter out interference caused by common impedance on the power line, preventing parasitic feedback from causing interstage coupling or even oscillation [4]. A good high-frequency ceramic decoupling capacitor can be effective up to 1 GHz. Adding a 0.01 μF decoupling capacitor between integrated circuit components and the power source in microwave active devices can, on one hand, provide and absorb the charge and discharge energy produced during device switching transients. On the other hand, the high-frequency noise of the device can be filtered. Meanwhile, in the vicinity of an active device, short wires between the decoupling capacitors and the pins of the power source are required. The decoupling capacitor is connected directly to a nearby ground through a via hole.

Anti-EMI should be considered when arranging the positions of components on the substrate. One principle is that the short wires between the components should be kept to as few as possible. Pulse analog signal and RF signal should be separated reasonably in the layout to minimize mutual coupling.

The thickness of wires in the power supply network inside the substrate should be increased. The wires will be classified according to frequency and current switching characteristic. And the distance between noise components and non-noise components should be appropriately increased. The control signal wires should be kept far away from the substrate edge. An abrupt change in line width should be avoided. Use a 45° fold line instead of 90° fold line to reduce high-frequency radiation and coupling. Parallel arrangement of noise-sensitive lines and high-current, high-speed switching lines should be avoided.

### 4.3.4 Environment Adaptability

The environmental requirements of spaceborne and airborne T/R modules are largely different, due to the significant difference between a space environment and an air environment. Generally speaking, the environmental requirements for a spaceborne T/R module basically cover those for an airborne T/R module. Therefore, a spaceborne T/R module is taken as an example for environment adaptability analysis, mainly including mechanical environment during launch, electromagnetic environment, thermal environment, space radiation environment during normal operation, etc.

#### 4.3.4.1 Mechanical Environment

The tolerance to the mechanical environment of a spaceborne T/R module is designed based on a preliminary structural analysis model constructed from the satellite configuration and structure. Both the static force analysis and mode analysis of the structure are performed to select and define the appropriate structures, materials, connections of the module, and the experimental verification projects. The purposes of the design are as follows: first, to ensure that the structure has enough stiffness, strength, and safety, taking into account the thermal design, antiradiation, and EMC design; and second, to ensure that there is sufficient structural damping in the module.

To verify the reliability of the mechanical environment design, mechanical simulation of the T/R module structure is required, including modal analysis, acceleration analysis, random vibration response analysis, and sinusoidal vibration analysis. Thermal stress analysis is performed on the temperature range of the working and storage states of the T/R module. Currently, the finite element method is commonly used in mechanical simulation for T/R modules. It is used to observe response of modules under overload and vibration conditions. If necessary, experiments and tests can be used to verify whether the T/R module meets safety and reliability requirements.

#### 4.3.4.2 Thermal Environment

The thermal design of the T/R module is directly related to the reliability and the lifetime. With the continuous improvement of the integration level, the power density and thermal density of the T/R module also increase. Because the device failure is often closely related to its operating temperature (in general, the failure rate is doubled for every 10 °C increase of the operating temperature in the device), improper thermal design will lead to a series of problems, such as local overheating, excessive thermal stress, and uneven temperature distribution. Particular attention should be paid to the thermal management of the power chip. Considering that a number of T/R modules are often used in an antenna array at the same time, the thermal design of T/R modules must be combined with the environmental control of the array, to ensure that a T/R module is not only in a proper thermal environment but also in thermal balance in the array.

The heat is dissipated in three ways: conduction, convection, and radiation. The conduction is more effective than the radiation in the temperature scope allowed by the components. Usually the following three aspects are used to dissipate heat: first, a proper layout of the overall structure is required. Without affecting operation of the other components, a microwave power device with the highest dissipated power and its associate power supply should be put at the bottom of the T/R module package to shorten the thermal dissipation path. Second, materials with high thermal conductivity should be chosen. Third, the thermal impedance between contacting surfaces should be reduced. This requires a high smoothness and flatness in both planes in contact. If necessary, thermal pad or thermal silicone grease can be inserted between the contact surfaces to improve surface contact.

#### 4.3.4.3 Total Dose

The satellite is in the space environment with particle radiation, solar radiation, and electromagnetic radiation during operation. The charged particles in the space environment may lead to malfunction or failure in the electronic devices, which seriously affects the reliability and lifetime of the spacecraft [5], thus requiring the designer to comprehensively consider the SAR system adaptability to this radiation during the early stages of the design. Among them, the most important is to consider total radiation dose of the devices in orbit.

In selecting components and raw materials, for components without credible and authoritative data (for example, the data provided by vendors, the authoritative database, similar flight experience, etc.), their total radiation dose tolerance should be confirmed by experimental tests. When the total radiation dose tolerance of the selected components and the raw materials cannot satisfy the requirements at places where they are in use (including the margin required in radiation design), adequate protection should be designed until the requirements are met.

According to general specification requirements, electronic components and raw materials used in a stand-alone satellite should be capable of withstanding a total radiation dose of $2 \times 10^4$ rad (SI), and the key components need to withstand a total dose of $3 \times 10^4$ rad (SI).

The following methods can be used to withstand total radiation dose: radiation-resistant reinforcement components are used, and radiation shielding measures are adopted in the design so that components with radiation resistance lower than the total radiation dose can meet the requirements. Radiation-resistant reinforcement technology includes component-level radiation reinforcement, equipment-level shielding design, optimized design, etc. – for example, a series current-limiting resistor and a parallel filter capacitor are connected to every circuit board to prevent the destruction of the device caused by radiation. Mounting tantalum sheet on the surface of key devices. Assembling all components inside a shielding case. Choosing wires with high radiation tolerance, such as ARF252-type fluorine plastic wires.

Regarding components inside the T/R module, a GaAs RF chip has a radiation tolerance of $1 \times 10^5$ rad (SI), which is radiation-tolerant by nature, and no special radiation reinforcement is needed. For analog and digital controlling chips based on SI, with radiation reinforcement, their radiation tolerance can reach $1 \times 10^5$ rad (SI). Devices designed on silicon-based new integrated circuit material structure (such as silicon-on-insulator) offer better radiation resistance, and their total radiation tolerance can reach above 1 M Rad (SI) [6]. Such components as connectors, resistors, and capacitors are not sensitive to radiation.

#### 4.3.4.4 Micro-Discharge

The micro-discharge effect is a vacuum discharge resonance phenomenon that usually occurs between two metal surfaces or on the surface of a single dielectric. Micro-discharge is normally excited by the RF electric field, which accelerates electrons and is produced by secondary electrons coming from acceleration electrons bombarding the surface. The conditions for micro-discharge to occur vary according to micro-discharge types. As for micro-discharge between metal surfaces, its occurring conditions include that the mean free path of the electron must be larger than the distance between two metal surfaces, and the mean transition time of an electron between these two surfaces should be odd times of the half cycle in the RF electric field. For micro-discharge occurring on the surface of the dielectric, the DC current field produced by surface charges must be capable of accelerating electrons to go back to the dielectric surface, so that the secondary electron can be generated.

The micro-discharge sensitive areas on the surface of conventional metals are shown in Figure 4.14. The space between inner and outer conduction metals should be

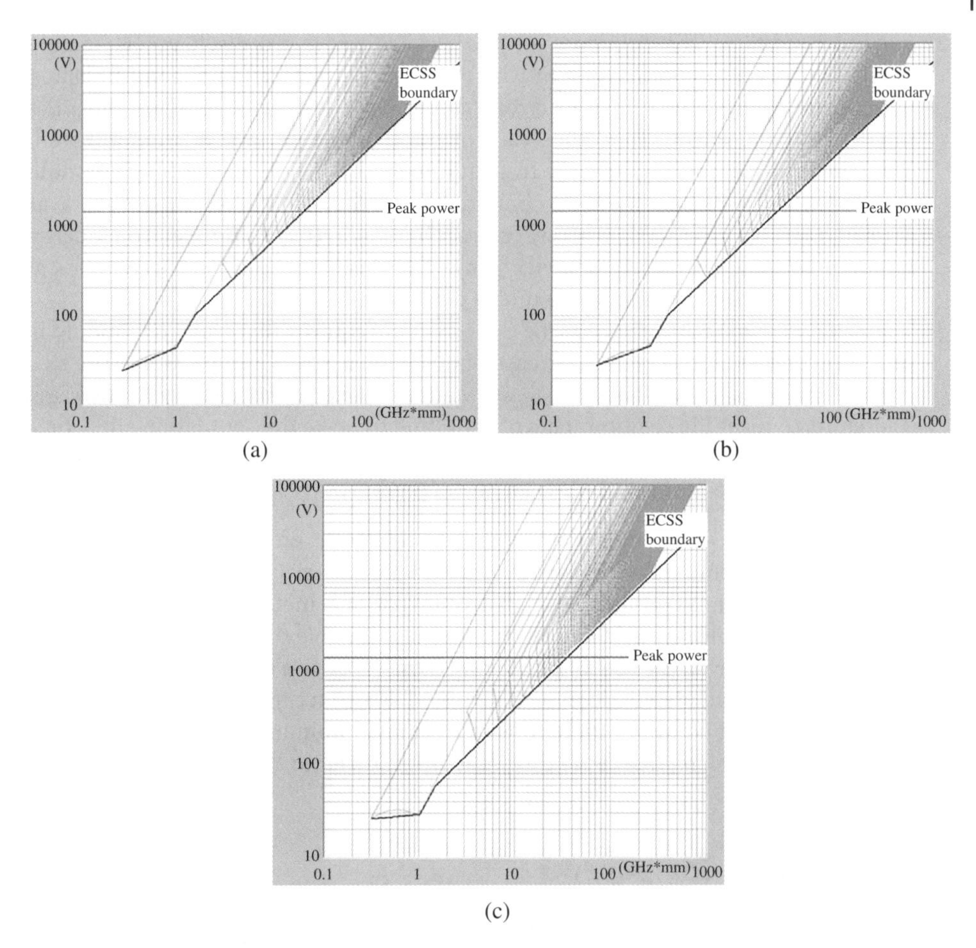


**Figure 4.14** Micro-discharge sensitive areas on the surface of conventional metals. (a) Gold surface; (b) silver surface; (c) aluminum surface. ECSS: European Cooperation for Space Standardization.

increased in the design, so that the maximum operating voltage is outside the sensitive areas.

There are many factors forming the micro-discharge effect and many influences that cause it, mainly in the following areas:

(1) *Dielectric*. The filling dielectric between two electrodes can reduce the mean free path of the electron, equivalently enlarging the spacing size. For most dielectrics, electrons with a relatively low energy can make the secondary electron emission coefficient of the surface $\delta > 1$, so that the initial discharge electron source can be obtained. Therefore, micro-discharge can easily occur between dielectric and dielectric, or between metal and dielectric. Moreover, more energy is generated by the dielectric rather than by metal when secondary electrons are emitted. Due to the fact that most dielectrics have a low thermal conductivity, the heat is difficult to dissipate from the discharging port of dielectrics, causing local overheating and damage to the device. In the meantime, the dielectric easily releases gas when it is heated. A large amount of gas accumulated in the device can cause the local gas pressure increase in the device and breakdown.

(2) *Surface condition.* The influence of the surface on micro-discharge is mainly divided into two categories, one of which is that the surface itself is not polished with burrs. A large number of electrons gathering at the burrs easily produce the micro-discharge effect. The other category is that the surface is contaminated. The contamination in high-power transmitters is the major cause of T/R module micro-discharge. When there is contamination, the micro-discharge occurs at a relatively low threshold, because contamination reduces the electron energy required to generate secondary electrons. The discharge can also lead to local ionization discharge in addition to micro-discharge.
(3) *Ventilation condition.* A spaceborne SAR system requires that the T/R module is properly ventilated to avoid an increase of mechanical stress and pressure. The size of the vent holes should be as small as possible to avoid unnecessary RF leakage. Therefore, to meet the requirements of EMC and to make the ventilation effect better, the size of the vent holes should be as large as possible. In practice, the method of opening a number of vent holes can be adopted.

Different materials have different secondary electron emission coefficients. If the second electron emission coefficient of a certain material is relatively small, the micro-discharge threshold is relatively high. In conventional metals, the secondary electron emission coefficients of gold and silver are small, suitable to be used as the surface of a component. However, the cost of employing these expensive metals is too high. Normally, the surface is gold plated, silver plated, or special surface treatment on aluminum. For example, the most common method is to treat an aluminum surface with alodine. After treatment, the micro-discharge threshold is improved to a relatively high value.

## 4.4 T/R Module Components

A T/R module is shown in Figure 4.15. The main devices of a T/R module include power amplifier, LNA, phase shifter, attenuator, switch, etc. The development of T/R module devices includes discrete components, hybrid integrated microwave circuits, and microwave monolithic integrated circuits (MMICs). Generally, the power amplifier, drive amplifier, and limiter are first defined in design according to the output power of the module. Then the LNA is determined according to the module noise figure requirements, and the LNA compensation is determined according to the module receiving gain. After that, phase shifters, attenuators, and beam-steering chips are determined according to amplitude and phase control resolution. Finally, switch chips are determined according to the size of the transceiver isolation.

### 4.4.1 Amplifier

#### 4.4.1.1 LNA

Noise is a major factor limiting the sensitivity of the T/R module receive channel. Therefore, low noise design is very important for the receiver. The noise figure is generally defined as

$$N_f = \frac{S_{in}/N_{in}}{S_{out}/N_{out}} \tag{4.22}$$

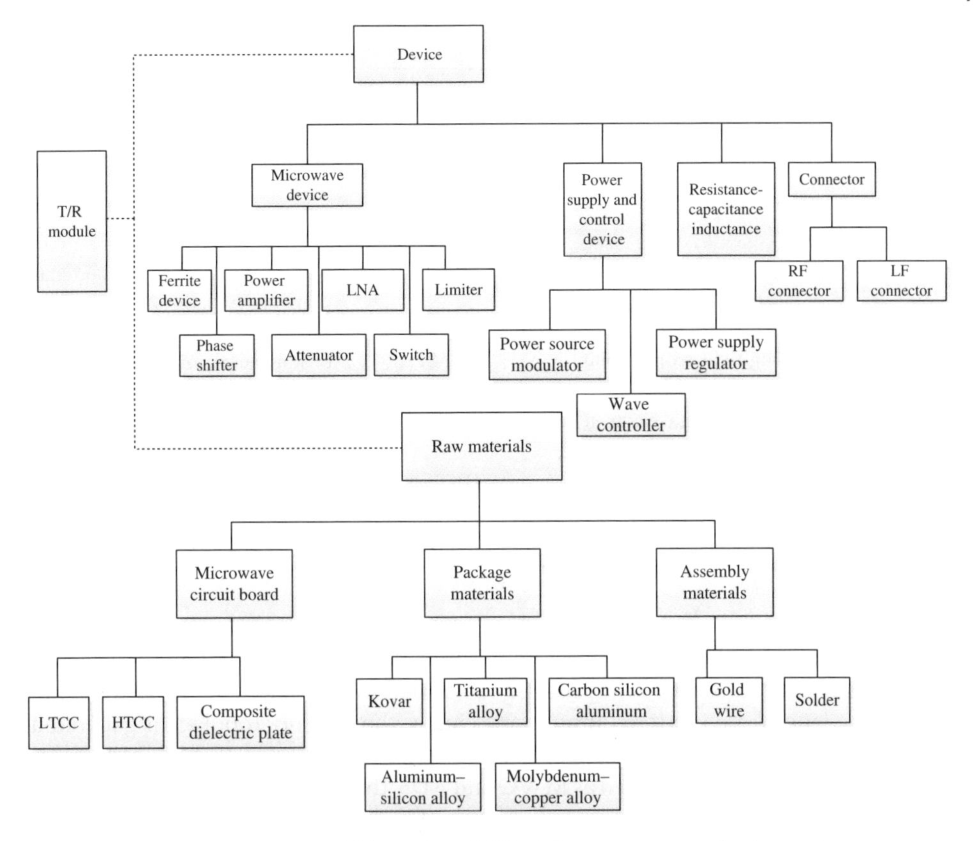


**Figure 4.15** Device composition of T/R module. HTCC: high-temperature cofired ceramic; LF: low-frequency.

where $N_f$ is the noise figure of the microwave device, $S_{in}$ and $N_{in}$ are the signal power and noise power at the input port in the microwave device, respectively, and $S_{out}$ and $N_{out}$ are the signal power and noise power at the output port in the microwave device, respectively.

It can be seen from Eq. (4.22) that the physical implication of the noise figure is that the signal-to-noise ratio (SNR) is changed due to noise generated by the device after the signal passes through the device, and the decrease multiplier of the SNR is the noise figure. For a cascade system, such as the T/R module receiver, the total noise figure can be calculated according to Eq. (4.7).

It can be seen from Eq. (4.7) that numerous items approach 0 if the first-stage amplifier $G_1$ has very high gain. In this case, the system cascade noise mainly depends on the noise figure in first-stage amplifier $F_1$. Therefore, the most effective method to realize a high-sensitivity receiver in receiver design is to select an LNA with low noise and high gain as the first-stage amplifier. The types and characteristics of the main microwave transistors used in LNAs are listed in Table 4.2.

The Smith chart is a convenient and efficient way to design an LNA. First, based on the S-parameters of the selected transistor, a stability circle, noise circle, and gain circle can be drawn in the Smith chart. After that, based on specific parameters of the design,

**Table 4.2** Types and characteristics of LNAs.

| Circuit type | Microwave semiconductors | Symbols | Applications and characteristics |
|---|---|---|---|
| LNAs | GaAs FET | GaAsFET | The most commonly used transistor, low noise, applicable in centimeter wavelength up to 18 GHz |
| | Bipolar junction transistor | BJT | The input impedance is moderate, easy to match, low price, suitable for frequencies below 3 GHz |
| | High-electron mobility transistor | HEMT | Best noise performance, can be used for mm-wavelength LNA |
| | Pseudo-high-electron mobility transistor | PHEMT | New type ultra-low noise transistor |
| | Heterojunction bipolar transistor | HBT | Used in mm-wavelength three-port device |
| | Double-gate FET | DGFET | Used in automatic gain control |

input impedance can be determined, and this impedance is then matched to the center in the Smith chart. The output impedance can be determined and matched by mapping input impedance from input to output port. Finally, the network topology is generated, and the design is optimized in microwave computer-aided design (CAD) software. In conventional CAD software, such as MWO and ADS, there are mature and standardized examples for reference, which are very convenient to use.

#### 4.4.1.2 Power Amplifier

The power amplifier converts DC power from the power supply into signal power under the action of the input signal, which amplifies signal power to a sufficient level and transmits the amplified signal to the antenna radiating to the space. Table 4.3 lists types and characteristics of the main microwave transistors used in power amplifiers.

**Table 4.3** Types and characteristics of microwave power amplifiers.

| Circuit type | Microwave semiconductor | Symbol | Applications and characteristics |
|---|---|---|---|
| Power amplifiers | Pseudo-high-electron mobility transistor | PHEMT | Mature in processing, high power and high gain, long-term stability |
| | Heterojunction bipolar transistor | HBT | High-current gain, dominating in high-frequency power amplifiers |
| | Bipolar junction transistor | BJT | Suitable for low-frequency, low-cost, and high-power applications |
| | Gunn diode | Gunn | Negative resistance injection lock-in amplifier, used in mm-wave or higher frequencies |

In addition to power, efficiency is another key parameter of a power amplifier, which is given by:

Drain efficiency:

$$\eta = \frac{P_{out}}{V_d I_d} D \tag{4.23}$$

Power-added efficiency:

$$\eta = \frac{P_{out} - P_{in}}{V_d I_d} D \tag{4.24}$$

where $V_d$ is drain voltage, $I_d$ is drain current, $P_{in}$ is input power, $P_{out}$ is output power, and $D$ is duty ratio. It should be noted that, due to very small gate current (a few mA), the power consumption of the gate is generally neglected in practice.

For high-power amplifier components, the load impedance of the final device needs to be carefully considered to maximize power output and efficiency. However, it must be noted that the load-pull experiment shows that the maximum output power matching and the maximum efficiency matching are different. The matching point should be designed as per the specific requirements.

With continuous progress in technology and materials, power amplifiers are usually divided into the following three generations [7].

***First Generation: Silicon Bipolar Power Transistors*** Below 4 GHz, the output power in silicon power devices is higher than that of other solid-state devices, which are dominant. Since the 1980s and 1990s, the structure and technology of this type of device have become mature. No more investment in fundamental research is going on; the focus is on the development of specific products. At present, there are many different types of products, which can be used both in pulse and continuous wave. The development of silicon power devices is as follows:

(1) As power combining technology has developed from engineering-oriented research on silicon power transistors below S band, reductions in cost, and improvements in power and efficiency, silicon power transistors have been used in solid-state transmitters and T/R modules. Compared with microwave vacuum tubes, they are more competitive in the entire life cycle.
(2) The operating frequency, power, and efficiency in silicon transistors are approaching their theoretical values. Therefore, no major breakthrough is anticipated in microwave characteristics of future silicon power transistors. Their development direction is to improve the matching technology of the transistor or introduce new internal matching circuits to increase output power in individual transistors.
(3) The output power of silicon power devices can be effectively improved by replacing a single silicon power device with a microwave power module consisting of multiple power chips. This type of microwave power module has already been widely used in practical products.
(4) Silicon field-effect transistors (FETs), especially longitudinally diffused metal–oxide–semiconductor (LDMOS) and vertically diffused metal–oxide–semiconductor (VDMOS), have higher operating temperatures and efficiency than silicon bipolar transistors. Therefore, LDMOS and VDMOS have become the mainstream of silicon power devices in many applications.

***Second Generation: GaAs FETs*** The electron mobility of GaAs is seven times higher than that of silicon. GaAs also offers high drift velocity. Therefore, GaAs has better frequency characteristics than silicon as well as better linearity, lower noise, stronger radiation resistance ability, wider operating temperature range, easier integration, and is more adaptable in harsh conditions. In the frequency range of 5~30 GHz, GaAs devices are the first choice for solid-state microwave power devices. The development of GaAs power devices is as follows: GaAs power devices above 4 GHz are currently the most important. GaAs and silicon power devices below 4 GHz coexist in a competitive state. It is expected that the power of a single GaAs transistor can provide 50~100 W at L and C band, 30~60 W at X band, and 20~40 W at Ku band in the near future.

In wideband and high-integration applications, the development of GaAs MMICs has been focused to realize low profile, light weight, high reliability, and low price. Currently, there are various types of MMIC products that cover the frequency from 1 to 100 GHz, providing many options for T/R modules.

***Third Generation: Wide-Bandgap Semiconductor Devices*** Wide-bandgap semiconductor devices are mainly silicon carbide FETs and GaN power transistors, and those solid microwave devices with higher frequency, power, and efficiency tend to be more developed, compared with the second-generation semiconductor devices. Solid-state amplifiers with better characteristics in size, reliability, and life cycle than silicon and GaAs can be developed, to meet higher requirements for output power, output power density, efficiency, linearity, operation voltage, and temperature.

The third-generation semiconductor devices are featured with high breakdown voltage, high output power, and high cutoff frequency and can realize a high-frequency, wide bandwidth. Their high thermal conductivity and strong radiation resistance ability enable them to obtain high output power, high junction temperature, and high thermal stability. The relationship between the second- and third-generation semiconductors is shown in Figure 4.16.

### 4.4.2 Microwave Control Device

#### 4.4.2.1 Attenuator

An attenuator is an adjustable or fixed resistive/capacitive or resistive network, which produces only amplitude attenuation for the input microwave signal without obvious phase shift or frequency distortion. In a SAR system, it is often necessary to weight the amplitude of the received signal. Therefore, digital attenuators are usually used in T/R modules to adjust the receiver gain. Taking into account the requirements of being conveniently controlled by program, small size, fast speed, low driving level, and easily integrated with planar circuits, digitally controlled attenuators are normally used in T/R modules. A digitally controlled attenuator is usually composed of a thin-film fixed attenuator and a FET-type single-pole double-throw switch, with a simple working principle. The attenuation value is a combination of a series of discrete fixed attenuation values. The combination is a cumulative process realized by the switch on–off. When all switches are turned to the fixed attenuators, maximum attenuation value is achieved and vice versa.

Figure 4.17 shows the diagram of a digitally controlled attenuator with 4 bit and 2 dB step. It can be seen that the attenuator is in the minimum state when all switches are

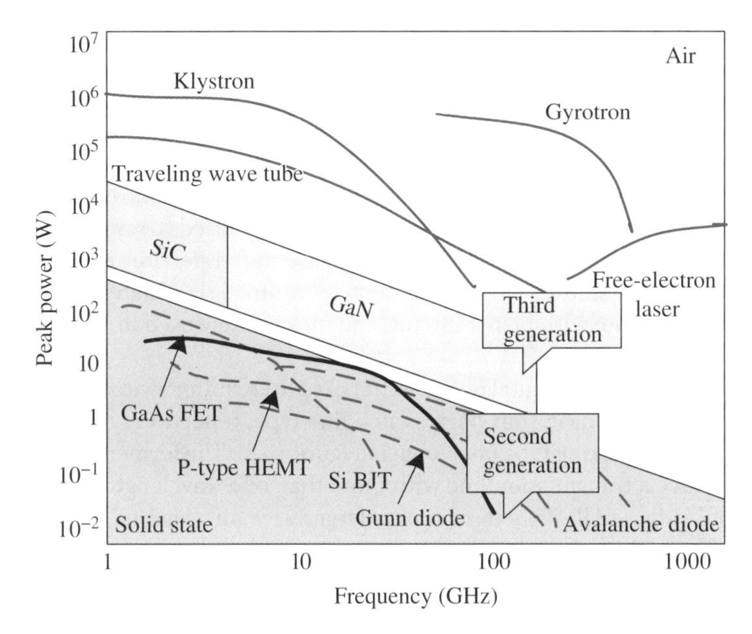


**Figure 4.16** Frequency-power distribution in different types of power transistors. BJT: bipolar junction transistor; HEMT: high-electron mobility transistor.

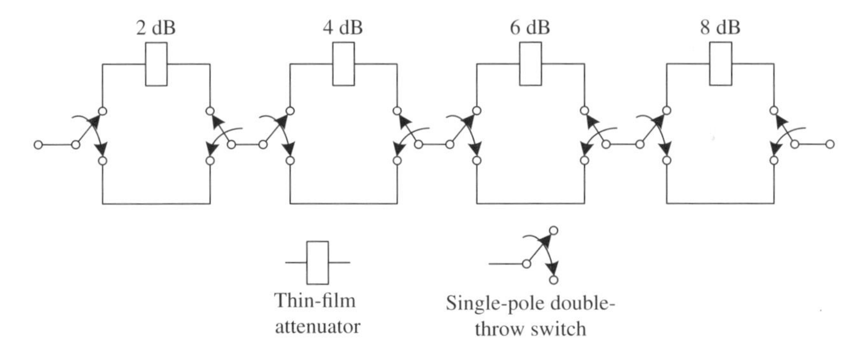


**Figure 4.17** Block diagram of a digitally controlled attenuator.

turned to the no attenuation state. In this case, no extra attenuation is added except the losses caused by the switches and transmission lines inside the attenuator. When the leftmost switch is turned to 2 dB attenuator film, the attenuator is in the 2 dB state. In this case, the total loss is the insertion loss plus the 2 dB attenuation. When all switches are turned to the attenuation film, the attenuator is in the maximum attenuation state. In this case, the total loss is insertion loss plus an additional 30(2 + 4 + 8 + 16) dB attenuation. In the same way, by combining different combinations of the switches, we can get $2^4(= 16)$ kinds of different attenuation states, including both minimum and maximum states.

#### 4.4.2.2 Phase Shifter

A phase shifter is a two-port network that realizes the desired phase relationship between output voltage (or current) and input voltage (or current) by adjusting one certain control signal. Phase shifters are mainly divided into two categories: analog and digital. An analog phase shifter is a device that produces continuous phase shift by using the control signal. A digital phase shifter is a device whose step value is predefined and fixed according to design requirements. For example, the step value of a 5 bit 360° digital phase shifter is 11.25°, while that of a 6 bit 360° digital phase shifter is 5.625°. Digital phase shifters are convenient for program control due to a digital signal to control. Therefore, they are widely used in phased array antenna systems. There are two main types of microwave digital phase shifter: ferrite and semiconductor. Semiconductor phase shifters have the advantages of small size, high speed, low driving power, easy to integrate with planar circuit, and more widely used in T/R modules than ferrite phase shifters.

Usually, FETs are used in digital phase shifters as phase shifter switches. There are four common types in the phase shift part: switch line type, reflection type, load line type, and high–low-pass filtering type (as shown in Figure 4.18). The former three types either need to introduce a transmission line with more than one wavelength or need to insert a number of bridges. All of them occupy a larger area and lead to narrow-operation bandwidth. The high–low-pass network phase shifter uses lumped elements, with the low-pass network increasing electrical length and the high-pass network decreasing electrical length, alternately changing transmission length, and producing different phase shifts, which is particularly suitable for monolithic integration.

The diagram and working principle of a phase shifter are similar to those of a digital attenuator, except that the attenuation film in the attenuator is replaced by the phase shift circuits shown in Figure 4.18a–d. Then different phase shift combinations can be obtained by integrating the switches differently.

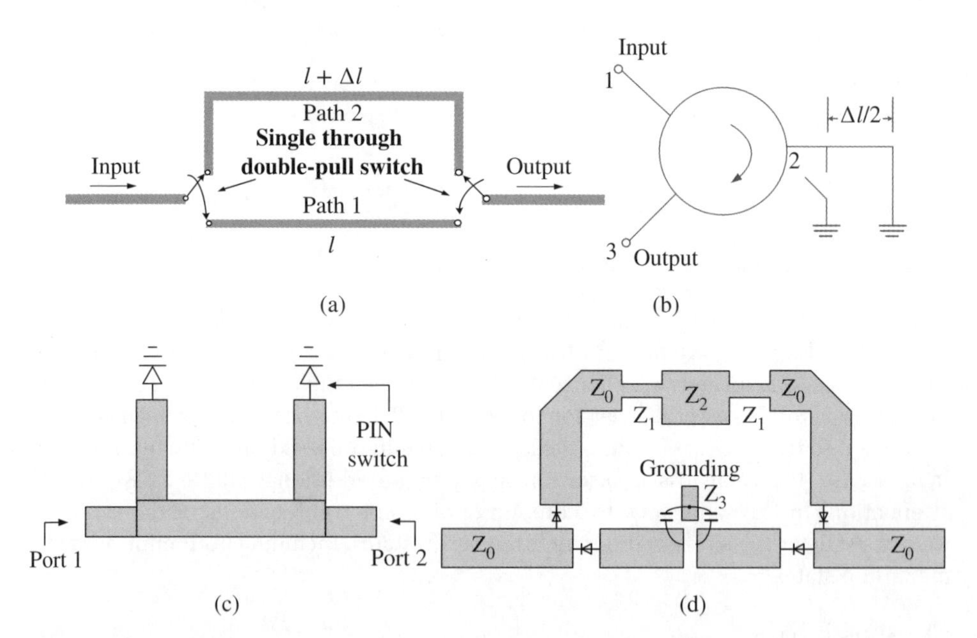


**Figure 4.18** Four types of phase shift circuits. (a) Switch line type; (b) reflection type; (c) load line type; (d) high–low-pass filtering type.

The accuracy in the phase shifter reflects performance of the T/R component to some degree. It is required that a conventional phase shifter in a T/R module not only provide accurate 360° phase control in several bits within the operation bandwidth but also offer characteristics of low insertion loss, low-amplitude modulation between different phase states, low input and output voltage standing wave ratio (VSWR), a high third-order intercept point, short switching response time, the capability to suppress false noise, etc.

A PIN diode chip is used in a traditional digital phase shifter as the switch realized by hybrid integrated circuit technology. The conventional circuit types are switching type, load type, reflection type, mixed type, etc. However, with the rapid development of MMIC, a variety of digital phase shifters based on GaAs technology have quickly occupied the market, becoming the mainstream of mature products. In terms of low cost, similar products based on the silicon–germanium process have begun to appear.

#### 4.4.2.3 Transceiver Switch

A transceiver switch is a switching circuit that changes between the transmitter and the receiver in the T/R module. The switching circuit usually connects one or more PIN diodes or FETs in parallel or in series to transmission lines. The PIN diodes or FETs are either in on-state (transfer energy) or off-state (reflection energy) controlled by DC bias, which transmitted or reflected the microwave power on transmission lines accordingly. Consequently, the function of transmitting/blocking or steering controlling microwave signal is realized. The switching circuit is generally divided into two types: series and parallel. The advantage of the series type is wide bandwidth, but PIN diodes or FETs will endure all power of the load; the parallel type endures lower power, with acceptable insertion loss and isolation.

The basic element of a PIN switch is a PIN diode, which consists of a highly doped P layer, N layer, and intrinsic semiconductor I layer sandwiched between these two layers. Under forward bias, the impedance of a PIN diode is small, approximating short circuit, whereas under reverse bias, its impedance is very high, approximating open circuit. This PIN diode characteristic makes them widely used in various microwave switch control circuits.

FET is primarily composed of source (S), drain (D), and gate (G), with gate controlling signal change between source and drain. The FET types include metal–oxide–semiconductor FET and gallium arsenide FET (GaAsFET). The difference is that the former is isolated by a PN junction, the latter is isolated by a Schottky junction, and GaAsFET is generally used as a microwave switch. When GaAsFET is used as a switch, the RF signal path is through the source and the drain, and the gate is used as a control terminal for the signal channel to be turned on or off. When the gate voltage is 0 V, the impedance between drain and source is low, and the GaAsFET is on. When the gate voltage is lower than the pinch-off voltage of the GaAsFET, the impedance between drain and source is high, and the GaAsFET is in cutoff state. In contrast to the PIN switch, the GaAsFET switch almost needs no current to control. Therefore, it is superior in power consumption and switching speed. The channel length of the GaAsFET basically defines the on-state of the switch and specifies insertion loss of the switch. The parasitic capacitance between drain and source determines the off-state of the switch, and therefore defines isolation of the switch.

A GaAsFET switch is generally used as transceiver switch in T/R modules.

#### 4.4.2.4 Limiter

A limiter is mainly used to protect the highly sensitive front-end receiver, to prevent power leakage of the transmitter from burning the LNA of the receiver in the first stage. It is normally required that the threshold level of the limiter is smaller than the maximum allowable power of the LNA in the first stage of the receiver. When input power is smaller than the threshold voltage of the limiter, the loss of the limiter is very small, approximately to a through condition. When input power is higher than threshold voltage, the limiter produces a very large attenuation. After that, the input power continues to increase, and output power of the limiter remains approximately constant. Of course, the limiter itself has maximal power durability. When input power is higher than maximal power durability of the limiter, the limiter will be burned to fail.

If the limiter relies on an external control signal to realize the limiting function, the limiter is called an *active limiter*. If it relies on RF self-detection to realize the limit function, it is called a *passive limiter*. Sometimes active and passive limiters can be mixed, which is called a *hybrid limiter*. To simplify the design, a passive limiter is usually used in T/R modules.

Figure 4.19 shows the diagram and equivalent circuit of a parallel multistage microstrip-based PIN limiter.

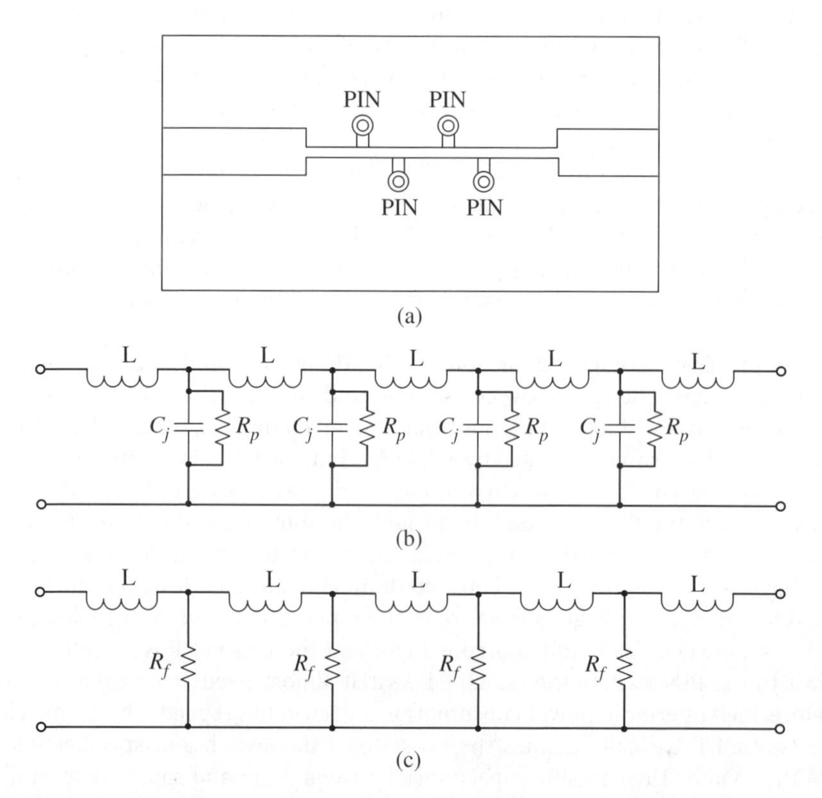


**Figure 4.19** A multistage PIN limiter. (a) Microstrip circuit board; (b) low-voltage equivalent circuit; (c) high-voltage equivalent circuit.

When microwave signal power is very small, the PIN diode is in zero bias state, equivalent to a very large resistor $R_p$ in parallel with a capacitor. They are joined with an equivalent inductor to create a low-pass filter. If the circuit is well designed, the loss in the passband will be minimal. When the microwave signal power exceeds the threshold voltage, the resistance in the PIN diode $R_f$ decreases as microwave power increases. A small part of the microwave power is absorbed by the PIN diode, and most of it is reflected. Therefore, the limiter offers high VSWR in the limit state.

Both the PIN switch and PIN limiter are controlling devices based on the PIN diode. They are the same in principle but different in selection of the PIN diode. For PIN diodes used as limiters, it is required that their impedance is sensitive to signal power, and impedance decreases rapidly with the increase of the input power. Therefore, it is required that the I layer between the P zone and the N zone be very thin, usually around several micrometers, whereas for the PIN diodes used as switches, the I layer is relatively thick, usually 14~20 μm.

#### 4.4.2.5 Circulator/Isolator

The circulator and isolator are important devices of T/R modules. Both are developed for modularization and miniaturization. The study of miniaturized circulators has made considerable progress in recent years, with practical application in many fields; however, until recently, most circulators and isolators have been discrete devices. Although they can be integrated with other components in a T/R module, the size after integration is still larger. Because the circulator and the isolator generally have strong magnetism and easily form interference, it is very important to calculate and analyze the EMC, intermodulation, harmonic suppression, and other parameters of other circuits in the T/R module [8]. Conventional circulators and isolators include drop-in type and microstrip type.

Figure 4.20 shows drop-in-type circulators/isolators. Their input and output are in stripline structure, and their package structures employ types suitable for reflow welding, instead of the traditional method. No screw is used to fix the whole device. Instead, mutual welding is used to package the devices into one piece, so as to reduce size and weight.

Figure 4.21 shows microstrip-type circulators/isolators. The input and output with microstrip lines can realize a planar structure for circulators/isolators, making them convenient to be integrated with other microstrip circuits in a T/R module. And they are superior to the drop-in type in terms of size and weight.

**Figure 4.20** Structures of drop-in-type circulators/isolators.

**Figure 4.21** Structures of microstrip-type circulators/isolators.

These two types of circulators and isolators are mature in technology. Various types of products have been widely used. Thin-film ferrite circulators, low temperature co-firing ferrite (LTCF) circulators, and microelectromechanical systems (MEMS) circulators are currently under research and development. These kinds of circulators have obvious advantages in miniaturization and integration, but their products are still at the laboratory level. The film thickness of thin-film ferrite circulators is less than 100 μm, with small size and light weight. However, they are still in the development stage, and their loss, isolation, and other characteristics have not yet reached the level for practical use. For LTCF circulators, on one hand, low-temperature cofired spin ferrite is used to realize a three-dimensional (3D) graphical structure through low temperature co-fired ceramic (LTCC) technology, produced by silver paste. On the other hand, the cofired spin ferrite substrate is buried into a multilayer ceramic substrate based on the existing device structure. The multilayer ceramic substrate uses silver paste to realize interconnections between different layers and microstrip lines with different characteristic impedances. Study on this method is still in the initial stage. MEMS circulators combine ferrite technology with semiconductor process-based MEMS technology to further reduce size and weight. Mass production is possible with the aid of a semiconductor process, and it is more suitable for high-frequency applications. Currently, samples of MEMS circulators operating above Ku band are available for use.

In a T/R module, the circulator and isolator are usually designed together and are packaged into a circular isolation module, which is an effective method to reduce the size and weight of the T/R module.

### 4.4.3 Wave and Time Sequential Control Device

#### 4.4.3.1 Serial-to-Parallel Converter

Figure 4.22 shows the block diagram of a typical internal wave controller in a T/R module, which is realized by using serial-to-parallel conversion and data latch. The serial-to-parallel conversion begins when the series clock signal arrives, and the serial-to-parallel conversion is fulfilled by reading control information one by one.

When the rising edge in the first-stage latch signal arrives, the parallel control information that has been converted is saved in the first-stage latch, while the second-stage latch still maintains its original data, and output signal of the driver remains constant.

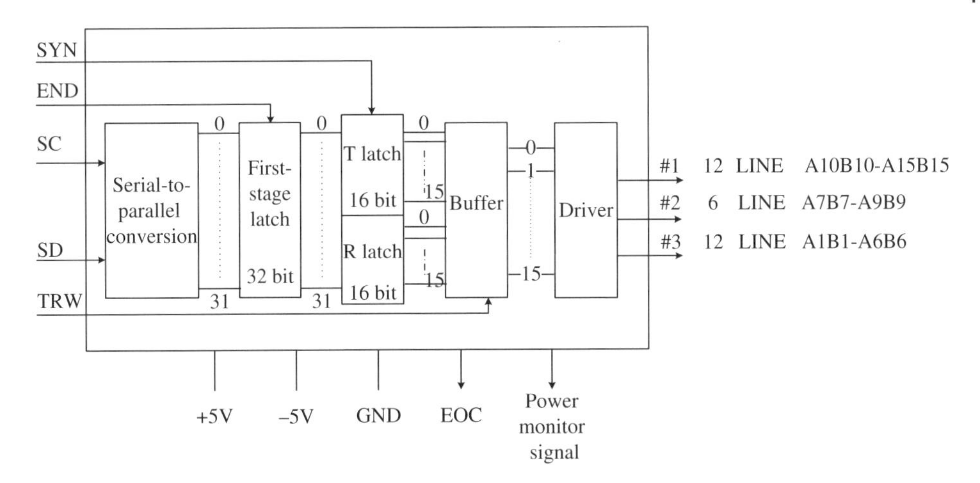


**Figure 4.22** Block diagram of a typical wave controller in a T/R module.

When the rising edge in the second-stage latch signal arrives, the data in the first-stage latch is driven into the corresponding position in the second-stage latch, and the corresponding data is output under the control of a transceiver conversion signal.

The data in the second-stage latch remains the same until the rising edge in the next second-stage latch signal arrives. The driver chip selects the data in a T latch to be output based on high-voltage level "1" of the TRW transceiver conversion signal and selects the data in an R latch to be output based on low-voltage level "0" of the TRW signal.

#### 4.4.3.2 Power Supply Modulator

Because the transmitter and receiver in a T/R module can physically form a loop, if the isolation between the transmitter and receiver is limited and the gain of the transmitter and receiver is high enough, in case of an unreasonable time sequence of the power supply, positive feedback can be generated when the transmitter and receiver operate at the same time, causing oscillation in the module.

Therefore, the time sequence of the power supply needs to ensure that the transmitter and receiver operate in absolute time division. That is to say, the relationship of the rising and falling edges of various power supplies in the T/R module need to be considered, and a corresponding delay needs to be set. In general, a time delay needs to be added between the transmit power pulse and the received power pulse both at the front and at the end, as shown in Figure 4.23.

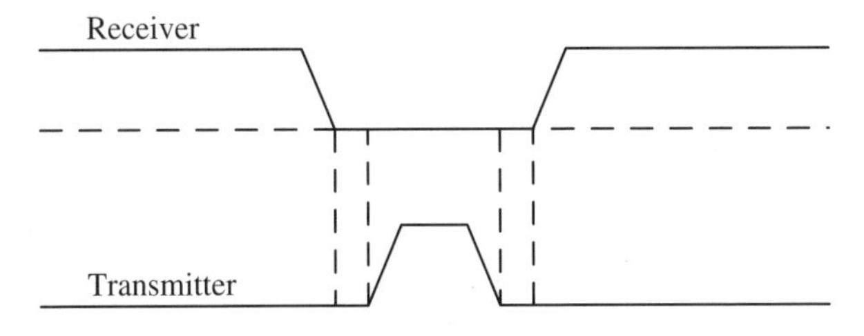


**Figure 4.23** Time sequence of transceiver power supply.

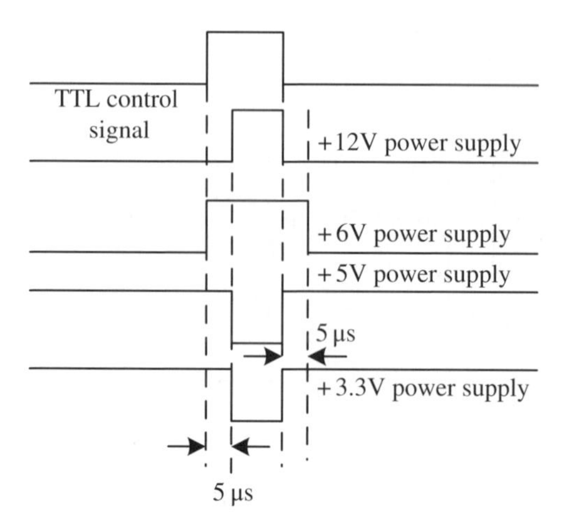


**Figure 4.24** Time sequence of multiple power supplies.

In a T/R module, a specific logic relationship also exists between on/off states of each power supply group and control signals, as shown in Figure 4.24.

## 4.5 T/R Module Manufacture

A T/R module is a complex system made of a variety of circuits, such as microwave circuits, analog circuits, digital circuits, etc. Its material and device characteristics are quite different. It is an important manufacturing technology. The internal layout in a highly integrated T/R module is very compact, which requires highly accurate alignment both in front and rear, left and right, up and down. The narrow distance between solder joint and bonding point and a large number of gold wire bonding points require high assembly precision.

A micro-assembly technique is used in T/R modules to mount components. Many kinds of micro-components and MMIC chips are integrated on one substrate, creating high density and reliable integration, which requires welding, gluing, bonding, and other procedures, in which the welding is performed using different temperatures, different solders, and different processes. After debugging and sealing, the module is still required to undergo multiple tests, such as temperature shock, vibration, burn-in, high–low-temperature experiment, hermeticity detection, particle radiation, noise detection, etc. Generally speaking, micro-assembly technology in the T/R module involves substrate brazing, connector bonding, chip eutectic welding, attachment of bare chips and surface mount devices, wire bonding, welding of isolator and other discrete devices, sealing, etc.

A multichip integration-based T/R module is characterized by high integration, small size, and light weight. For example, the LTCC process can reduce the size of by in one order of magnitude; semiconductor processing can reduce the size of the components by two orders of magnitudes and can realize a streamlined production; and function integration can further reduce the cost and improve the performance and the uniformity.

### 4.5.1 Package

The package of the module is usually made of a baseplate, a frame, a cover, RF connectors, and low-frequency connectors soldered together. Under normal circumstances, brazing is used to weld the baseplate to the frame and to weld RF/low-frequency connectors to the frame. To meet hermetic requirements, a short welding gap should be required, and the tolerances relevant to the welding gap need to be coordinated.

To reduce the size and weight of the T/R module, composite materials with light weight and high thermal conductivity are usually used as package material. They are low in density, high in thermal conductivity, and easy to process. Table 4.4 lists parameters of conventional package materials for weight and thermal dissipation.

In designing a light and miniature T/R module, the key to package design is material selection, including selection of package material, substrate material, and various solders. The matching of heat and stress between these materials should be taken into account. For example, when the difference of the thermal expansion coefficient between two contacted materials reaches 12 ppm $K^{-1}$, thermal fatigue failure may appear after only 100 thermal cycles, which should be avoided.

The comprehensive properties of aluminum silicon carbide materials are good, but their processing and surface modification are relatively challenging. As an electronic package material, the thermal expansion coefficient of titanium alloy is similar to that of chips, LTCC substrate, package, and sealing connectors. Its thermal matching with welding is acceptable with high environment adaptability and reliability. At the same time, the laser welding technology of titanium alloy is mature. The density of titanium alloy is smaller than that of traditional package materials, such as kovar and molybdenum-copper, which is an ideal choice in space-phased array antennas.

The package can be machined through precision numerical control machines, injection molding, precise compression casting, etc. Generally, the surface of the package needs to be plated to meet the assembly requirements of circuit boards and connectors. The plated metal layer not only should be bonded with the composite materials but also should be thermal resistant, antioxidation, and solderable. Common methods to plate the package are magnetron sputtering and chemical coating.

### 4.5.2 Substrate

In a T/R module, conventional substrates mainly include composite dielectric microstrip, ceramic microstrip, LTCC, aluminum nitride, composite laminated

**Table 4.4** Parameters comparison of conventional package materials.

| Product type | Silicon aluminum | Kovar | Aluminum | Aluminum silicon carbide | Titanium alloy |
|---|---|---|---|---|---|
| Composition | Si-Al | Fe/Co/Ni | Al | Al/Si/C | TC4 |
| Thermal expansion coefficient (ppm/°C) | 11 | 5.9 | 23.6 | 6.5–9.5 | 8–10 |
| Density ($g\,cm^{-3}$) | 2.5 | 8 | 2.7 | 3 | 4.5 |
| Thermal conductivity ($W\,m^{-1}\,K$) | 149 | 17 | 238 | 170–220 | 16 |

multilayer microstrip, etc. For a T/R module with relatively simple trace layout, single-layer microstrip substrate is usually used, such as composite dielectric microstrip substrate and ceramic microstrip substrate. As the wiring density of the T/R module increases, single-layer circuit board normally cannot satisfy design requirements, and multilayer microwave circuit board begins to be used. Currently, the most widely used multilayer microwave substrates are LTCC, aluminum nitride, and composite laminated multilayer microstrip board.

#### 4.5.2.1 Composite Dielectric Microstrip Substrate

Rogers, Arlon, and TACONIC are representative manufacturers of composite dielectric microstrip substrate. The circuit pattern is produced by etching a copper bonded dielectric plate, with the characteristics of matured process, relatively good process consistency, relatively low cost in volume production, etc. However, the wiring layout in single-layer circuit greatly restricts the design of the T/R module, and the photolithography process cannot produce a high-definition pattern. Therefore, single-layer microstrip substrate is normally used in T/R modules with relatively low operation frequency and simple structure.

#### 4.5.2.2 Ceramic Microstrip Substrate

Ceramic microstrip substrate generally refers to microstrip circuit substrate that is processed on aluminum oxide ceramic substrate with thick-film or thin-film technology. The dielectric constant of aluminum oxide ceramic substrate may vary from 9.6 to 9.9. After polishing the substrate, a high-precision pattern can be produced with thin-film technology. Due to the relatively high dielectric, constant characteristic dimensions of microwave circuit board can be made even smaller, suitable for high-precision single-layer circuit application. They are commonly used in T/R modules required to have relatively high operation frequency and simple structure.

#### 4.5.2.3 LTCC

LTCC is a multilayer substrate based on thick-film technology. Its advantages include the following: integrated with passive components; 3D design can be performed, which substantially reduces size; chips can be directly mounted on it; and the pattern can be processed with high precision. A common LTCC integration application is shown in Figure 4.25. Due to the relatively high cost of LTCC, it is generally used in T/R modules for satellite platforms or with complex structure requirements.

A typical LTCC fabrication process flowchart is shown in Figure 4.26.

#### 4.5.2.4 Aluminum Nitride Ceramic

Aluminum nitride is a new type of substrate and packaging material with high thermal conductivity, low thermal expansion coefficient, low dielectric constant, low dielectric loss, high mechanical strength, nontoxicity, etc. That is widely used in the field of high-power electronics. Aluminum nitride multilayer ceramic technology facilitates the flexible design of aluminum nitride ceramic products, capable of realizing cavity, multilayer trace, through via interconnection, and hermetic sealing. It also can meet the different requirements of devices, modules, and components to achieve high power, high density, small size, and high-reliability devices, modules, and components. Similar to LTCC substrate, aluminum nitride substrate also uses thick-film technology. The

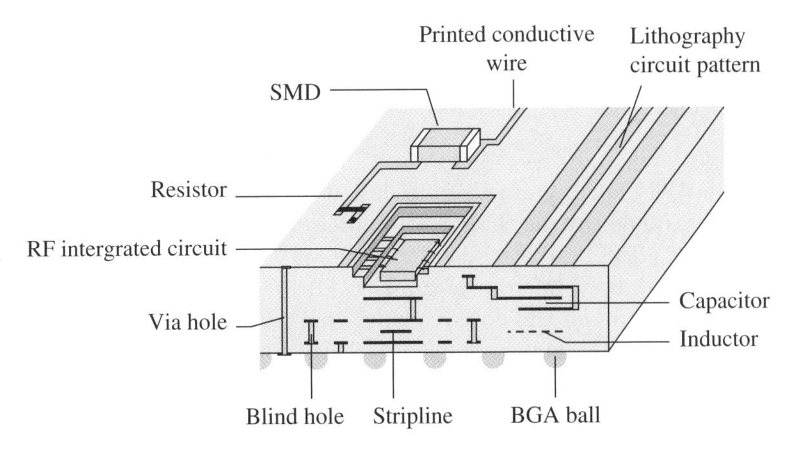


**Figure 4.25** LTCC integration application. BGA: ball-grid array; SMD: surface-mount device.

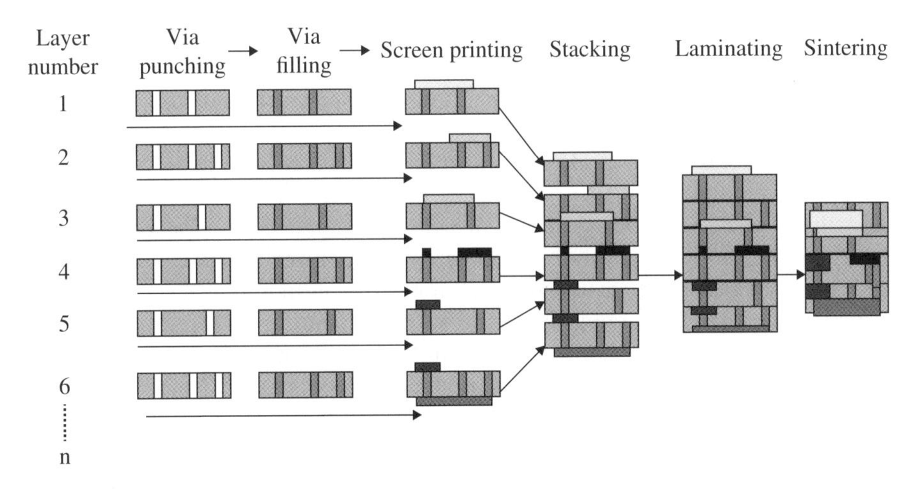


**Figure 4.26** A typical LTCC fabrication process flowchart.

major difference is that the material difference leads to different sintering temperature, as the sintering temperature of aluminum nitride is higher. This in turn leads to different pastes on their printed circuits. LTCC usually uses gold or silver-based paste for low-temperature sintering, whereas aluminum nitride usually uses highly thermal pastes like tungsten and manganese-based for high-temperature sintering. Currently, the multilayer aluminum nitride process is not as mature as LTCC and therefore is not as popular.

#### 4.5.2.5 Composite Laminated Multilayer Microstrip Substrate

Composite laminated multilayer microstrip substrate is based on single-layer composite dielectric microstrip substrate. It is a new type of multilayer circuit board produced by stacking several single-layer microstrip boards with prepreg and interconnecting the multilayers by a vias process. Nowadays, the integration design of buried resistors can be realized. Because the process of composite laminated multilayer microstrip substrate is an extension of that of a single-layer microstrip substrate, the cost of substrate manufacture and material is lower than that of LTCC and aluminum nitride. Therefore, it has obvious cost advantages in T/R modules for large-scale engineering applications.

Generally speaking, multilayer microwave substrate is the carrier of all components in T/R modules. The quality of substrate design directly affects the performance of the assembled module. In the design of multilayer microwave substrate, the most important factors affecting the module stability are grounding and feeding circuits. In the design of multilayer microwave substrate wiring, a distinction should be made between microwave ground and power supply ground, and EMI between them should be minimized. The distance between two microwave ground vias should be smaller than 1/8th of the wavelength. When possible, relative dense grounding vias should be used in the floating ground of a microwave transmission line. In practical wiring, the filled grounds should be combined with the grid grounds. Under the premise of meeting the microwave grounding requirements, bonding strength between the layers needs to be improved to make the inner metal distribution more uniform and to lower the stress of the substrate; consequently, inhomogeneity of the equivalent dielectric caused by inner conductors can be reduced. In addition, using multiple grounding layers inside multiple microwave substrates, and separating high-power modulated power supply signals from RF grounds and digital signals, can not only shield interference but also make the grounding of the module uniform and reduce equivalent impedance of the ground plane.

To facilitate device mounting, microwave circuits are generally placed in upper layers. The grounding of microwave circuits is very important because instability of the module easily happens. There are a number of wire traces in middle layers, with a complicated layout. The signal integrity should be of utmost importance. The controlling signal may be easily disturbed by high-current line in the power supply layer for an RF power amplifier, thus special treatments are normally required.

### 4.5.3 Micro-Assembly Technology

#### 4.5.3.1 Eutectic Bonding

In addition to providing mechanical or electrical connections for devices, eutectic bonding should also provide a good heat sink for devices. The alloy solder of eutectic bonding has the advantages of high mechanical strength, low thermal resistance, good stability, high reliability, etc. Alloy solder is normally used to bond chips with high-power density and carriers (such as package, substrate, etc.) to obtain thermal fatigue resistance, low thermal resistance, and small contact. The quality of eutectic welding affects the grounding and heat dissipation of the chips, which consequently affects stability and lifetime of the chips. According to the bonding process, eutectic bonding can be divided into gold–silicon eutectic friction bonding, gold–tin vacuum sintering, gold–tin sintering in static protective atmosphere, etc.

The process flow of gold–silicon eutectic friction bonding is as follows: under inert gas or reducing gas, the temperature of the bond area is increased to 30~40 °C above the eutectic point of gold–silicon (363 °C). A certain pressure is applied on the chip, and a relative displacement between chip and gold-plated package or substrate is produced to make eutectic liquid infiltrate the bonding surface. The bonding is completed by cooling and solidification. The process of gold–tin sintering in a static protective atmosphere is as follows: first, a piece of gold–tin solder is placed between chip and gold-plated package or substrate. Then a certain pressure is applied under protection of vacuum, nitrogen, or other inert gas, and it is heated up to 30~40 °C above the eutectic point of gold–silicon and for 3~15 s. Finally, the bonding interface is infiltrated by soaked solder and a stable bonding is produced after cooling down.

#### 4.5.3.2 Large-Area Substrate Bonding

In large-area substrate welding, the grounding of substrate significantly influences electrical and mechanical properties. The substrate bonding method includes hot plate bonding and eutectic furnace bonding. The hot plate bonding is performed manually; its advantages are that temperature of the hot plate can be adjusted in real time, and the spreading and the melting in the solder can be visually seen. The eutectic furnace bonding is realized by heating the bonded component as per the preset time-temperature curve. The bondability and bonding resistance should to be considered in substrate bonding measured by brazing rate.

#### 4.5.3.3 Glue and Attachment

Because the GaAs MMIC chip is very thin, brittle and fragile, it greatly increases the difficulty of chip assembly. The attaching methods include conductive adhesive bonding and solder bonding. The conductive adhesive bonding process is simple, easy to operate, and does not introduce foreign substances, such as flux, which may contaminate the chip. For example, the control chip (such as a beam-steering chip) in the component has no metallization layer on the back but can be bonded with conductive adhesive. These chips generally have a small amount of heat. Due to high dissipation in T/R modules, it is required to remove the corresponding substrate, and such a technique as molybdenum-copper heat sink combined with solder bonding is adopted. Chips are usually eutectic bonded molybdenum-copper heat sink using eutectic solder, and the whole piece is bonded to the corresponding bottom position of the package.

#### 4.5.3.4 Wire Bonding

There are usually hundreds of gold wires or gold ribbons in conventional T/R modules. Almost each wire directly affects electrical performance and reliability. It is not possible to perform a destructive pull test on each wire, due to the special demands of microwave circuits and the high-density characteristics in the module. Therefore, it is necessary to carry out a process test and quality evaluation for gold ribbon and gold wire bonding, including factors affecting the bonding strength of gold ribbon and gold wire, damage assessment of bonding pressure on the chip, and environmental test evaluation, etc. Through the optimization of the bonding process, the system analysis and research of the bonding process is carried out to find a reliable control program and method for bonding quality. In general, the load strength, roundness, scratch, deformation, and fracture of gold wires are checked through a high-power microscope. The same specification is used to check the technical status of the bonding equipment for each shift to ensure the stability of the bonding equipment. The parameters of the bonding process are designed for different substrate types, such as silicon-based chips, GaAs-based chips, ceramic chips, LTCC, etc., and the bonding parameters of different devices and materials are optimized. An effective and feasible test method will be designed for a nondestructive pull sample test to replace the existing 100% nondestructive pull test.

When interconnecting microwave components and circuit boards, the impedance between soldering pads and wires and between wires and circuit boards should be matched as well as possible, and the technology parameters, such as wire position, bond pressure, and the inductance produced by bond wires, should be strictly controlled. In general, the wire should be placed as close as possible to chips and circuit substrates, and the wire length should be kept as short as possible.

#### 4.5.3.5 Micro-Assembly Procedure

Micro-assembly technology is an advanced electronic assembly technology that puts together a variety of micro-components (such as integrated circuit chips and chip components) to produce micro-circuit products (including components, parts, subsystems, and systems) with high density, high performance, high reliability, small size, and modularized functions and with such critical technologies as high-density multilayer interconnected substrate, multichip modules, and 3D assembly [9]. Micro-assembly of T/R module includes two parts. One is micro-assembly technology of the module, such as eutectic bonding and gold wire bonding. The other is the board-level assembly and interconnecting technologies, such as low-temperature brazing (vacuum brazing, high-frequency induction brazing, hot plate welding, or hydrogen-protected brazing) and cleaning. A typical micro-assembly procedure is shown in Figure 4.27.

### 4.5.4 Hermetic Package

Because a T/R module is assembled with a large number of bare chips, it is very important to maintain a long-term internal nitrogen environment to reduce failure rate of the components. For the same reason, the hermetic package requirements are relatively high to ensure to long-term reliable operation of the T/R module. Therefore, the module assembly should provide a reasonable protection to components inside the module to guarantee the specified performance of the module under normal storage,

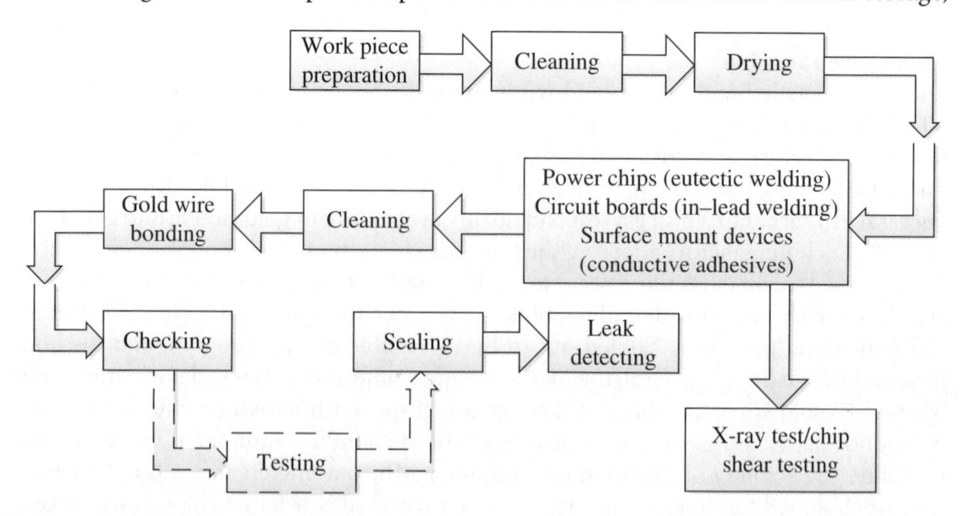


Figure 4.27 Typical micro-assembly procedure.

transportation, and usage environments. In general, it is required that the leakage rate in a module be less than $10^{-3}$ Pa cm$^3$ s$^{-1}$. The module hermeticity is a complicated process, which is usually achieved through three types of sealing technologies, including package substrate integrated bonding, connector bonding, and laser sealing of the cover.

For package substrate integrated bonding and connector bonding, attention should be paid to the structure size relevant to the bonding to produce a good hermetic package bonding. For example, during the machining and coating of the package, the processing accuracy of the package and the thickness of the plated gold should be of concern; during the bonding, attention should be paid to the quantity of solder, temperature curve of bonding, and the fixture so as to ensure the bonding quality. There are many methods to seal the cover of the module, including adhesive seal, gasket seal, soft brazing sealing, parallel seam welding, laser sealing, etc., and the choice of the sealing method is dependent on the materials used in the T/R module.

### 4.5.5 Testing and Debugging

The test procedure examines and verifies electrical properties, mechanical properties, and environment adaptability in a T/R module and evaluates whether the module is qualified or not. When test results do not meet the demands, it is required to debug the module, including rework of the assembly, to ensure that the T/R module satisfies all specifications. Extensive electrical testing is especially important in the research and development stage, which helps to detect problems and risks in the design. In volume production, special attention should be paid to the uniformity, repeatability, and reliability of the module.

Due to the large quantity, multiple tests, huge amount of data to be processed, and complexity of the control signals in mass production, an automatic test system that can

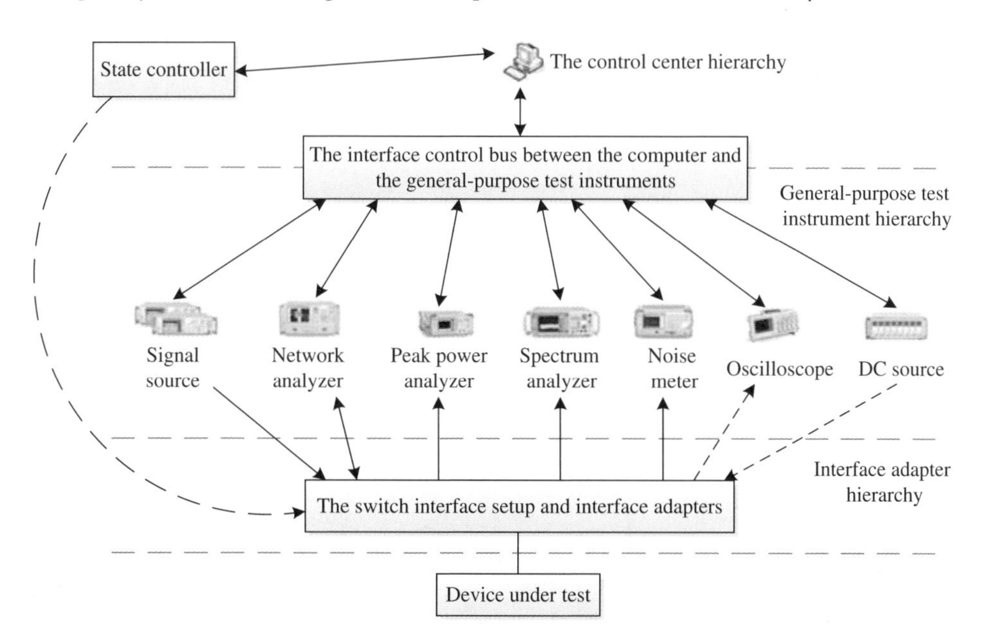


**Figure 4.28** Schematic of the principle of an automatic test system of a T/R module.

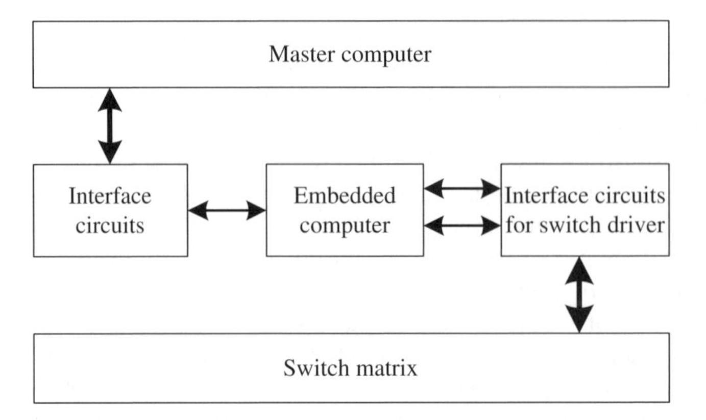


**Figure 4.29** Block diagram of switch interface.

evaluate quality control of T/R modules needs to be established to reduce the labor intensity of test personnel, improve measurement accuracy, shorten the debugging and testing period, improve the speed and accuracy of data processing, generate graphical test results, etc.

A complete automatic test system for T/R modules can be constructed in three hierarchies: the control center, general-purpose test instrument, and interface adapter. The control center is a master computer in which test software is installed. The software controls general-purpose test instruments through control buses like general-purpose interface buses/universal serial buses/local area networks to complete comprehensive

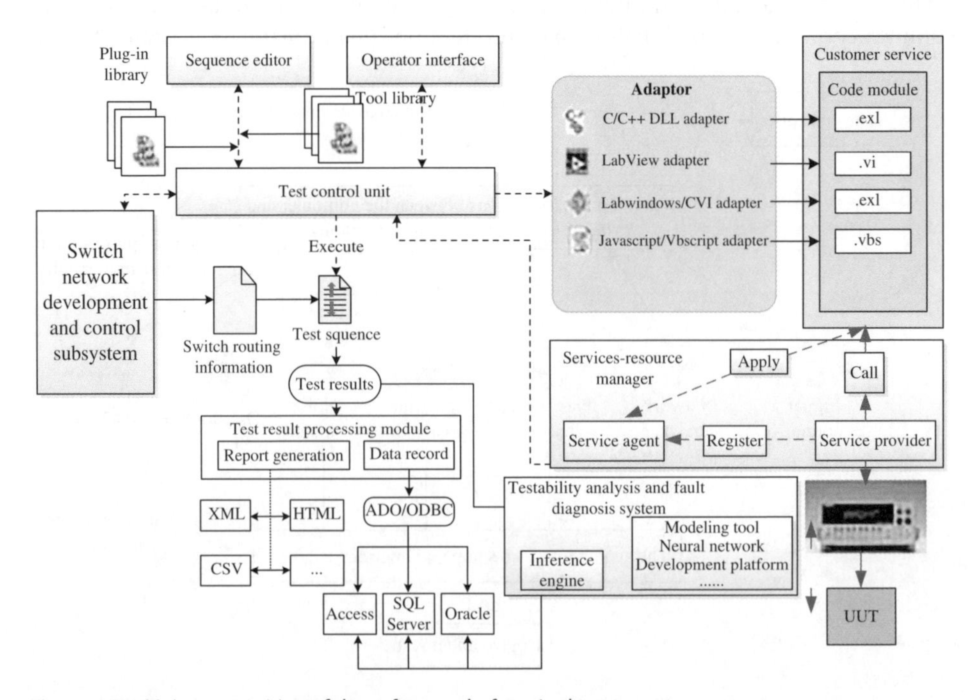


**Figure 4.30** Main composition of the software platform in the test system.

automatic tests on multiple parameters and to complete acquisition, processing, and analysis of the test data. According to the functions and different measured parameters, general-purpose test instruments consist of signal excitation, power parameter test, network analyzer, spectrum analyzer, waveform analyzer, and noise figure meter as well as program-controlled DC power source to supply power to the T/R module. The switch interface setup, computer adapter card, and state controller are hardware at the interface adapter hierarchy that realizes the physical connections of the measured T/R module and provides test channels for automatic tests on multiple parameters and automatic switches of corresponding status. A schematic of the principle of an automatic test system is shown in Figure 4.28.

It should be noted that the core at the interface adapter level is the switch interface setup, which is critical to realize automatic connection and test between the T/R module and general-purpose test instruments and is an important part of an automatic test system. The hardware of interface adapters mainly includes the switch matrix, interface circuit, master control circuit of embedded computer, interface circuit for switch driver, etc. A diagram of principle of switch interface is shown in Figure 4.29.

The test software is used to realize resource management, program maintenance, system calibration, measurements, comprehensive information processes, and other extended functions, providing a variety of operation modes for users, such as automatic test, manual test, independent usage of instruments, etc. The platform of test software is shown in Figure 4.30.

## References


1 H. Schippers, J. Verpoorte, A. Hulzinga (2013). Towards structural integration of airborne Ku-band satcom antenna. *Proceedings of the 2013 7th European Conference on Antennas and Propagation*, Gothenburg, Sweden (April 8–12, 2013). Brussels, Belgium: European Association on Antennas and Propagation.
2 Shanwei, L. (1995). *Fundamental of Microwave Engineering*. Beijing University of Aeronautics and Astronautics Press.
3 Nianxin, H. (2007). Temperature compensation design of GaAs power amplifier. *Cryogenics and Superconductivity* 35 (4): 352–354.
4 M.I. Montrose (1999). Analysis on loop area trace radiated emissions from decoupling capacitor placement on printed circuit board. *Proceedings of the IEEE International Symposium on Electromagnetic Compatibility*, Seattle, WA (August 2–6, 1999). Piscataway, NJ: IEEE.
5 Taosheng, L., Jun, C., and Zhiqiang, W. (2008). Overview of the space radiation environment. *Radiation Protection Communication* 28 (2): 1–9.
6 Collinge, J.P. (1991). *SOI Technology Material to VISL*, 1. Boston: England Kluwer Academic Pub.
7 E.C. Niehenke (2012). The evolution of low noise devices and amplifiers. *Proceedings of the 2012 IEEE/MTT-S International Microwave Symposium*, Montreal, Canada (June 17–22, 2012). Piscataway, NJ: IEEE.
8 Lahey, J. (1989). Junction circulator design. *Microwave Journal* 32 (11): 26–34.
9 Shao, Y. and Wei, W. (2012). T/R module micro-assembly technology. *Vessel Electronic Warfare* 35 (2): 103–107.

# 5

# Receiver Technology

## 5.1 Overview

The main function of the receiver in a radar system is to amplify and process the radar echo, to remove out-of-band space clutter and out-of-band noise in the receiver itself as much as possible by filtering the echo, and, at the same time, to provide high-quality transmitting waveform as well as to provide time and phase reference to the radar system.

The design of a radar receiver is closely related to radar application, waveform, antijamming mode, and signal processing method. It is relative whether an echo is a clutter or a useful signal. For conventional radar, echoes reflected from planes, boats, ground vehicles, and passengers are useful, whereas all echoes reflected from marine and fixed ground objects are clutter. For synthetic aperture radar (SAR), almost all echoes reflected from marine and ground surfaces are useful signals. Generally, it is required that a SAR receiver system offers wideband, low amplitude distortion, wide dynamic range, high sensitivity, high phase stability, etc. For multichannel SAR, phase consistency and stability among channels should be considered. For SAR with moving target detection capability, high-frequency stability and low-phase noise of the frequency synthesizer should be considered as well.

The main function of a SAR receiver includes receiving echoes, generating excitations, synthesizing local oscillators and clocks, etc. SAR not only has the function of high-resolution imaging but is also often combined with other functions, such as moving target detection, target location, and antijamming. Typically, to solve the contradiction between wide swath and high resolution, multichannel sampling is adopted in the azimuth. *One-wideband multiple-narrowband* receiving mode is a commonly used architecture in multifunction SAR. The two main research directions for the receiver are digitization and microelectronics.

### 5.1.1 Digitization

The rapid progress of digital circuit technology makes some receivers able to digitalize radio-frequency (RF) signals directly, thus intermediate-frequency (IF) receivers are no longer needed. Digital receiver technology has developed rapidly in recent years. With the rapid progress in ultra-high-speed digital circuit technology, the digitalization level of radar receivers has become higher and higher. In particular, the development of high-speed multibit analog-to-digital converter (ADC) and direct

*Design Technology of Synthetic Aperture Radar*, First Edition. Jiaguo Lu.

digital synthesis/synthesizer (DDS) technology as well as widespread use of high-speed digital signal processors (DSPs) will provide a good hardware platform for digital radar receivers.

A digital receiver has advantages that cannot be replaced by an analog receiver. For example, instability in analog circuits, such as temperature drift, DC offset, and gain variation, is avoided by replacing traditional analog quadrature demodulation with digital quadrature demodulation. In addition, it can obtain amplitude balance and phase quadrature, which cannot be achieved by analog demodulation. The application of wideband digital receivers in SAR can improve amplitude and phase distortion performance, reduce the amount of hardware, and help to achieve different imaging modes with online programming. As the ADC conversion rate rapidly increases, the dynamic range is becoming larger, significant progress is being made in high-speed digital circuits to process the output of high-speed ADCs, and software algorithms used to process digital output are becoming more and more mature. All of these greatly contribute to rapid development and wide application in the SAR field.

### 5.1.2 Microelectronics

On one hand, due to demands for SAR development, the small platform and multifunction require the receiver to be small in size, light in weight, and highly reliable to improve the radar performance and integration. On the other hand, the development is pushed by monolithic receiver technology. With the rapid development of the microwave monolithic integrated circuit (MMIC) technique, MMIC amplifiers, mixers, switches, digital attenuators, digital phase shifters, and the relative monolithic control circuits in various frequency bands have come into practical application, which promotes development of monolithic receivers.

The development and application of MMICs and ultra-large-scale high-speed digital integrated circuits, multilayer three-dimensional (3D) microwave circuits, and multichip modules (MCMs) all can guarantee the miniaturization of the receiver.

### 5.1.3 Receiver Classification

In general, a SAR receiver needs to digitalize echoes regardless of the means used. Therefore, in a broad sense, SAR receivers are all digital receivers. According to the locations of ADCs and the digitization level, SAR receivers can be classified as baseband digital receivers, IF direct sampling digital receivers, and RF direct sampling digital receivers, as shown in Figure 5.1.

Figure 5.1 shows the basic operation block diagram of SAR digital receivers. Figure 5.1a shows the baseband digital receiver. After being amplified by a low noise amplifier (LNA) and being mixed, the echo is analog quadrature demodulated by a coherent oscillator, producing a quadrature analog baseband signal, also called a *zero IF signal*, then the ADC is sampled to produce a baseband digital signal. The advantage of this receiver is that it can reduce the bandwidth of the radar baseband signal relative to signal bandwidth by 50%. Consequently, the ADC sample rate can be effectively reduced without additional digital filtering. This type of receiver is commonly used in larger wideband systems. Due to the use of analog quadrature demodulation, the consistency of both amplitude and phase in the wideband demodulator is the key

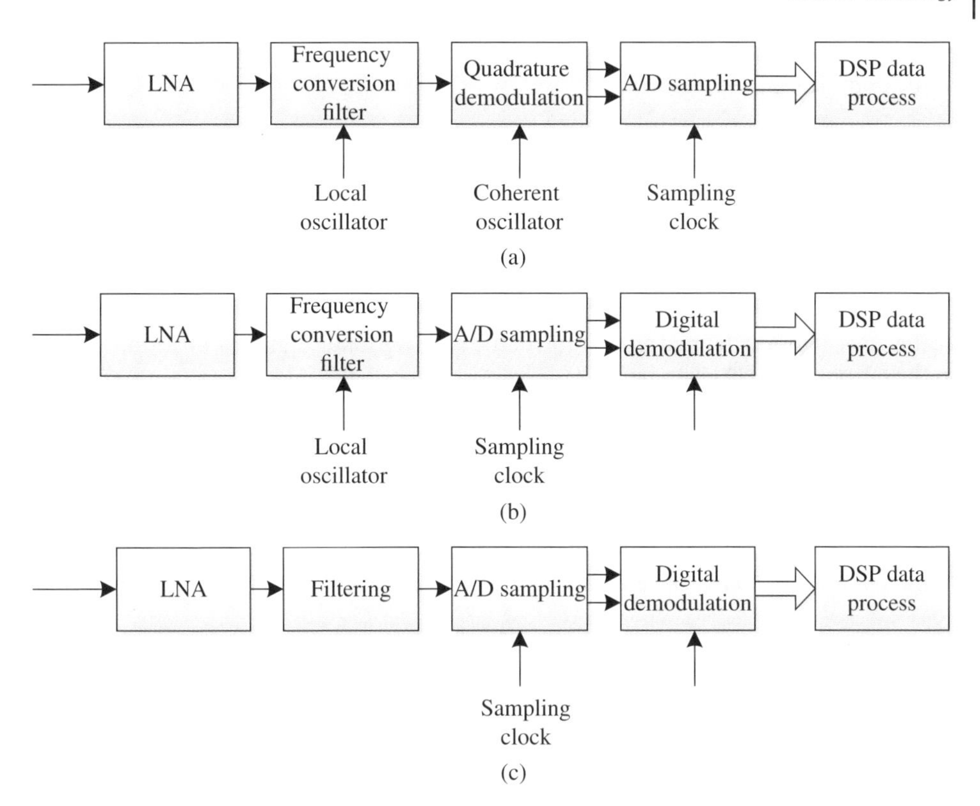


**Figure 5.1** Digital receivers. (a) Baseband digital receiver; (b) IF digital receiver; (c) RF digital receiver. A/D: analog-to-digital.

for practical implementation. Figure 5.1b shows the IF digital receiver. The IF signal is sampled through the ADC directly, after being amplified by an LNA and mixed, followed by digital quadrature demodulation and digital filtering. Then, the obtained digital I/Q baseband signal is sent to the DSP for signal processing. The advantages of digital demodulation are high consistency of amplitude and phase and high integration. Figure 5.1c shows the RF digital receiver. The RF signal is sampled through the ADC directly, after being amplified by the LNA and filtering, and is digital quadrature demodulated. Then, the obtained digital I/Q baseband signal is sent to a DSP for signal processing. This receiver is highly digitalized with high amplitude and phase consistency and high stability. Its disadvantage is overreliance on ADC performance. Currently, it can be realized in some low-operation-frequency radars.

### 5.1.4 Basic Parameters

The basic parameters of the SAR receiver are as follows in the next six sections.

#### 5.1.4.1 Signal Bandwidth

The instantaneous signal bandwidth is one of the most important parameters in high-resolution radar. The signal bandwidth is determined by range resolution in the radar. The bandwidth of the receiver is normally a bit wider than that of the signal, due to the nonlinear phase distortion.

#### 5.1.4.2 Sensitivity and Noise Figure

The sensitivity represents the capability of receiving a weak signal. The higher sensitivity a receiver has, the more weak signal can be received, which means the radar detecting range is much more further, or clearer image is obtained for SAR. One effective method of improving the sensitivity is to decrease the noise figure. Noise is composed of external interference (noise) and receiver internal noise.

The noise figure is defined as the ratio of the receiver input signal-to-noise ratio (SNR) to the output SNR [1] given by

$$F = \frac{S_i/N_i}{S_0/N_0} \tag{5.1}$$

where $S_i/N_i$ is the input SNR, $S_0/N_0$ is the output SNR, and noise figure $F$ represents the noise density with the receiver. For example, $F = 1$ means there is no internal noise in the receiver, which cannot be achieved in practice.

The relationship between sensitivity and noise figure is

$$S_{min} = kT_0B_0FM \tag{5.2}$$

where $k$ is the Boltzmann constant, $k = 1.38 \times 10^{-23}$J/K, $T_0$ is thermodynamic ambient temperature, $T_0 = 290K, B_0$ is bandwidth of the system noise, and $M$ is identification factor, whose value depends on the different radar systems. For SAR, $M = 1$.

It should be noted that this refers to the noise sensitivity, which is the corresponding minimum input power level when the SNR of the receiver is equal to 1, instead of actual receiving sensitivity of the radar, because the final SNR can be improved by coherent accumulation (noncoherent accumulation can be performed on some radars). The added gain is 10log (B$\tau$), where B is signal bandwidth and $\tau$ is signal time-width.

#### 5.1.4.3 Gain and Dynamic Range

The gain indicates amplification capability of a receiver, which is defined as the ratio of output power to input power, i.e. $G = S_0/S_i$. The dynamic range indicates the power variation range of the input signal allowed by the receiver under normal operation. When the input signal level is too high, the receiver will be saturated or overloaded, causing the targets with smaller echo to be lost. Higher receiver gain does not mean better. It is commonly determined by receiver dynamic range, sensitivity, and ADC quantization performance. The receiver gain identifies the amplitude of the receiver output signal. Generally, the gain distribution in the receiver is closely related to noise figure and dynamic range.

The maximum input power is the input power point at which the receiver gain drops 1 dB, and the minimum input power generally refers to the receiver sensitivity. The dynamic range can be usually described in three ways: spurious free dynamic range, instantaneous dynamic range, and total dynamic range. Spurious free dynamic range refers to the power difference between the maximum input signal and simultaneously detectable minimum input signal. It is the most valuable dynamic range. Instantaneous dynamic range is allowable input power range without any radar gain adjustment. Total dynamic range is allowable input power range with radar gain adjustment.

Similar to the sensitivity, dynamic range here refers to the noise sensitivity, not that of the actual radar, because the radar process system can amplify the signal, and the gain is related to radar signal time-width, bandwidth, accumulation time, etc.

#### 5.1.4.4 Amplitude and Phase Distortion

The amplitude and phase distortion is a more important parameter for SAR than for conventional radar. It is directly related to the quality of a radar image. Wideband pulse compression is a major means to obtain high-range resolution in SAR. The existence of amplitude and phase distortion causes the main lobe of the compressed pulse and deteriorates the main/side lobe ratio. High-order amplitude and phase distortion will produce paired echoes, where the influence and control difficulty of phase nonlinearity for the system is more important than amplitude distortion.

#### 5.1.4.5 Multichannel Amplitude and Phase Stability and Consistency

For the SAR with multi-receiving channels, the functions of moving target detection, polarization characteristic analysis, wide-swath coverage, etc. are realized by detecting the amplitude and phase characteristics between multiple signals. Therefore, the consistency of amplitude and phase between multiple receiving channels has an important impact on radar performance. In practical applications, it is extremely difficult to achieve high consistency in amplitude and phase by hardware. The amplitude and phase should be calibrated instead. For amplitude and phase stability in the receiver, due to the fact that the amplitude and phase in the radar system can be calibrated, more attention should be paid to short-term stability. Short term normally means one cycle of spaceborne SAR (usually about 15 min). The amplitude and phase stability within one aperture time should be specially emphasized. Amplitude and phase stability in multichannel receivers are more complicated, including room temperature stability and wide-range temperature stability. Relative high amplitude and phase stability is desired to avoid frequent calibration.

#### 5.1.4.6 Frequency Stability

The frequency stability mainly refers to the frequency stability in the radar frequency source. The frequency source provides transmitter carrier frequency and system reference frequency for the radar, which is directly related to radar image quality. SAR is more concerned with short-term frequency stability (usually in milliseconds), which is usually described by single sideband phase noise power density spectrum. Random vibration has a significant influence on the frequency stability and the phase noise. In some cases, it can seriously deteriorate the phase noise. For SAR operating on a moving platform, random vibration is widespread. On certain special platforms (such as helicopters and missiles), the influence of random vibration is very critical. Therefore, for SAR, stability of the frequency source on a vibration platform is more important than in conventional static state.

## 5.2 Receiver Technology

Compared with a conventional radar receiver, SAR has special requirements on the receiver, mainly in signal bandwidth and phase fidelity. The signal bandwidth in a high-resolution SAR reaches several gigahertz, and sometimse the relative bandwidth is larger than 50%. There are mainly two types of SAR receivers: direct quadrature demodulation receivers and dechirp receivers. Direct quadrature demodulation receivers can be divided into analog type and digital type. A direct quadrature demodulation receiver is suitable for large swath, high-quality imaging, and relatively easy wideband distortion compensation. Its disadvantages are relatively high requirements on sampling rate and data process, especially with high resolution. A dechirp receiver has great advantages in high resolution and small swath systems, and the sampling rate and the complexities in the signal process design can be effectively reduced. Its disadvantage is that compensation of the wideband distortion can be relatively complicated.

### 5.2.1 Analog Demodulation Receiver

The analog quadrature demodulator in a wideband direct demodulation receiver can reduce bandwidth of the system to half while retaining the amplitude and phase information. At the same time, problems of blind phase in single-phase detector and inability to identify Doppler speed direction of the target can be solved. A common method of realizing quadrature demodulation is to directly sample the digital IF signal and to isolate the I/Q signal by using a digital filter. However, current devices and processes have limited the improvement of the sampling rate. Consequently, an analog quadrature demodulator in high-resolution SAR is preferred, and it is assisted by digital amplitude and phase correction if necessary.

The operation principle of an analog quadrature demodulator is that the phase in the input signal is detected by two coherent local oscillator signals after the input signal is divided by a 90° power divider, then two mutually orthogonal zero IF signals, that is, the in-phase I signal and a quadrature phase Q signal, are generated through low-pass filtering, amplification, leakage suppression, and high-frequency intermodulation rejection. Due to the limitation of phase shifting precision in the power divider network and the asymmetry of circuits, the amplitude and phase of an I/Q signal is unbalanced, which generates a false image signal. Figure 5.2 shows the block diagram of a typical wideband demodulator circuit.

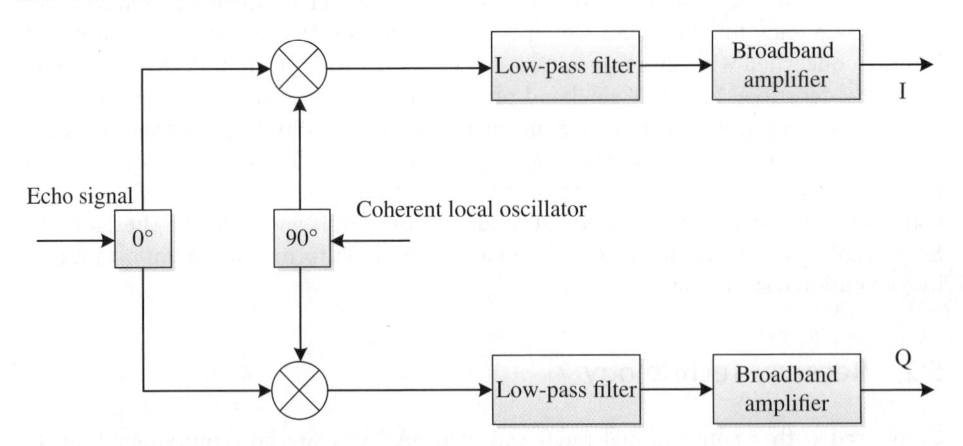


**Figure 5.2** Broadband quadrature demodulator.

Conventional discrete devices usually cannot meet specifications of wideband, high-performance demodulators. Generally, wideband I/Q processing is achieved by a high-performance integrated broadband demodulator combined with highly consistent ultrawideband zero IF processing technology. A Chebyshev low-pass filter (LPF) is usually used to realize good in-band flatness and high out-band suppression in the demodulator. Due to the large overshoot distortion at the turning point of the amplitude-frequency characteristics of the group delay in the Chebyshev LPF, cutoff frequency of a two-way LPF is normally 1.5 times its bandwidth, and group delay needs to be strictly matched. The wideband quadrature demodulator requires a large dynamic range. Therefore, wideband operational amplifiers with big signal output capability should be selected. Experiments show that when there is high gain in the wideband operational amplifier, the in-band flatness in the gain is significantly deteriorated. Therefore, an effective method of realizing wideband flatness is to cascade two stages of broadband operational amplifiers and insert a resistor matching network between them for isolation.

For wideband quadrature demodulators, measurement is one of the most important factors that guarantees high performance [2]. The bandwidth and precision of a conventional amplitude and phase imbalance test instrument should be able to satisfy measurement requirements, and a vector network analyzer, with frequency scanning and vector analysis, can provide fast S-parameter measurement for two-port networks and also has an error correction function and high measurement precision. A new type of network analyzer with frequency bias mode is an ideal test instrument for quadrature demodulators. Because of inconsistency between test channel 1 and test channel 2 (including instruments and the connecting cables between I and Q), a relatively large measurement error may occur in I/Q amplitude imbalance, especially for a wideband system, thus error correction is required during testing. For wideband quadrature demodulators, cables connecting I/Q and ADC should be measured as part of the quadrature demodulator. Otherwise, there will be great difference between parameters in actual use and the measured results in the quadrature demodulator.

### 5.2.2 Digital Demodulation Receiver

As mentioned before, there are many ways to realize a digital receiver. Taking into account operation frequency of an ADC, sampling clock, and amplitude-phase characteristic in the demodulator, high-resolution SAR usually uses a conventional baseband digital receiver, whereas middle- or low-resolution SAR usually uses an IF digital receiver, i.e. digital demodulation. The implementation of a SAR IF digital receiver is discussed below. Figure 5.3 shows the block diagram of a typical receiver.

An LNA is used to amplify weak echoes received from the antenna to ensure the receiver sensitivity. The RF filter, sometimes called the *image reject filter*, is mainly used to reject external RF interference and to suppress image noise. To improve the anti-interference capability, a preselected filter is added in front of the LNA. An IF bandpass filter is used to restrain bandwidth of the signal and noise from entering the ADC, which is instrumental to facilitating selection of nonaliasing sampling frequency. The ADC converts input analog signal into digital signal and provides digital input signal to the digital processor. The digital processor is used to downconvert and to filter input high-speed digital signal, i.e. digital demodulation filter. A digital demodulation

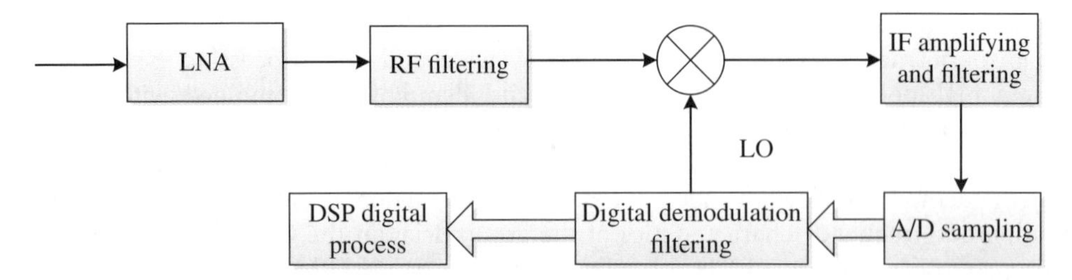


**Figure 5.3** Block diagram of typical IF digital receiver. LO: local oscillator.

filter usually consists of numerically controlled oscillator (NCO), digital mixer, and digital filter.

The interface design of the front amplifier and filter circuit of the ADC is of great importance in practice. The matching between them will directly affect sensitivity and dynamic range in the receiver. Therefore, in practical design, the parameters of the ADC front amplifier and bandpass filter should be determined based on ADC characteristics (such as bit number, maximum sample rate, input power, etc.).

The sample rate of a digital receiver is mainly dependent upon the signal bandwidth, the ADC characteristics, the sampling method, etc. The sampling method greatly influences the sample rate. Commonly used sampling methods are described below.

#### 5.2.2.1 Oversampling Technology

Oversampling samples the signal with a much higher sample rate. Its advantages are that the SNR and the dynamic range are improved and performance demands on the front nonaliasing filter are reduced. However, the computation amount in the system is increased, which influences real-time information processing.

#### 5.2.2.2 Quadrature Sampling Technology

Quadrature sampling is widely used in baseband digital receivers and is actually an undersampling technology. The sample rate can be reduced to a half as two orthogonal components (I and Q) are sampled separately.

As the bandwidth of a SAR radar signal is wide, bandpass sampling is usually adopted, instead of conventional low-pass sampling. The bandpass sampling should satisfy the following formula:

$$\frac{2f_h}{k} \le f_s \le \frac{2f_l}{k-1} \tag{5.3}$$

$$2 \le k \le \frac{f_h}{f_h - f_l} f_s \ge 2B \tag{5.4}$$

where $f_h$ is upper sideband frequency, $f_l$ is lower sideband frequency, $f_s$ is sample frequency, and $k$ is a positive integer larger than 2.

The digital receiver processor is embedded, where decimation ratio of the decimation filter, filter coefficient, and output data format are all set by built-in control register and random access memory (RAM).

In a digital receiver, because there is no temperature drift, gain variation, or DC voltage drift, commonly seen in analog receivers, its orthogonal characteristic cannot be compared to an analog receiver. In a digital receiver, either IF or RF signals are firstly

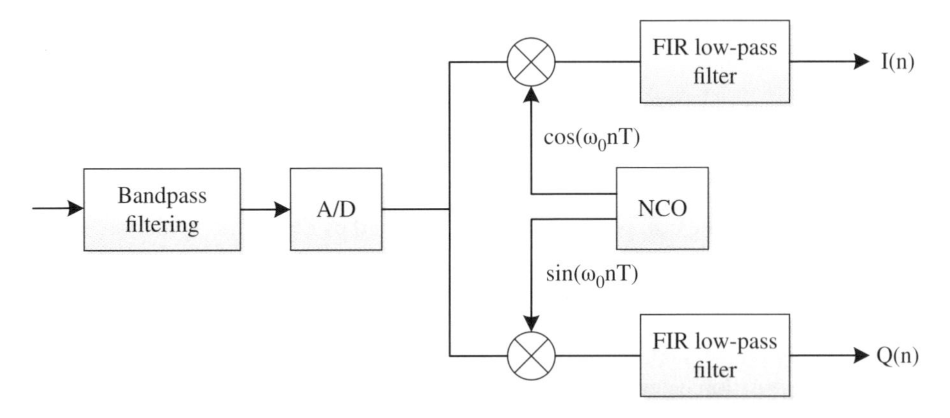


**Figure 5.4** I/Q isolation with digital mixing and low-pass filtering.

performed by ADC sampling, followed by digital quadrature phase demodulation and digital filtering [3, 4]. The biggest advantage in digital quadrature phase demodulation is that high precision and stable I/Q can be achieved. There are many methods to realize digital I/Q. Here, three methods are briefly introduced: digital mixing and low-pass filtering, interpolation, and Hilbert transform.

#### 5.2.2.3 Digital Mixing and Low-Pass Filtering

The principle is shown in Figure 5.4, similar to the analog quadrature phase demodulation, where the mixer, LPF, and coherent oscillator are digitally implemented. The coherent oscillator is realized by NCO, which outputs two orthogonal signals of sine and cosine. Because formation and multiplicity of these two orthogonal oscillation signals are results of digital calculation, their orthogonality can be fully guaranteed; only the calculation precision needs to be guaranteed.

#### 5.2.2.4 Digital Interpolation

Alternate I(n) and Q(n) are firstly obtained by an ADC converting IF signal through a proper sample frequency, then two integral I and Q signals are obtained by interpolation with an interpolation filter.

Assume the signal is $S(t) = A(t)\cos[\omega_0 t + \phi(t)]$ and the sample frequency is

$$f_s = \frac{1}{T} = \frac{4}{M+1} f_0 \tag{5.5}$$

where $A(t)$ is the time variable function of signal amplitude, $\phi(t)$ is the time variable function of the phase added in the signal, $\omega_0$ is the angle frequency of the signal, $f_0$ is the signal frequency, and $T$ is the sample period.

Equation (5.5) should meet the sampling theorem requirements, that is $f_s \geq 2B$, then

$$\omega_0 T = 2\pi f_0 \frac{M+1}{4}\frac{1}{f_0} = \frac{\pi}{2}(M+1) \tag{5.6}$$

Let $M+1 = n$, $M = 0, 1, 2, 3, 4\ldots\ldots$

$$\begin{aligned} S(nT) &= A(nT)\cos[n\omega_0 T + \phi(nT)] \\ &= A(nT)\cos\phi(nT)\cos\left(n\frac{\pi}{2}\right) - A(nT)\sin(nT)\sin\left(n\frac{\pi}{2}\right) \end{aligned}$$

$$= I(n)\cos\left(\frac{n\pi}{2}\right) - Q_n \sin\left(\frac{n\pi}{2}\right)$$

$$= \begin{cases} (-1)^{\frac{n}{2}} I(n) & n \text{ is even} \\ (-1)^{\frac{n-i}{2}} Q(n) & n \text{ is odd} \end{cases} \tag{5.7}$$

Obviously, $I(n)$ is obtained when $n$ is even, and $Q(n)$ is obtained when $n$ is odd. The even point $Q(n)$ and the odd point $I(n)$ can be obtained by digital interpolation. Finally, two complete I and Q baseband digital signals can be obtained. Here a relatively simple Bessel interpolation method is introduced. When the relationship between sampling frequency $f_s$ and signal frequency $f_0$ is $f_0 = \frac{M+1}{4} f_s$, the output of ADC sampling is an I/Q interval digital signal. Quadrature digital signals, I and Q, can be obtained by the following interpolation calculation:

$$I(n) = \frac{1}{2}[I(n-1) + I(n+1)] + \frac{1}{8}\left[\frac{1}{2}(I(n-1) + I(n+1))\right] - \frac{1}{16}[I(n-3) + I(n+3)]$$
$$n = 2\,m + 1,\ m = 0, 1, 2\ldots\ldots \tag{5.8}$$

$$Q(n) = \frac{1}{2}[Q(n-1) + Q(n+1)] + \frac{1}{8}\left[\frac{1}{2}(Q(n-1) + Q(n+1))\right] - \frac{1}{16}[Q(n-3) + Q(n+3)]$$
$$n = 2\,m,\ m = 0, 1, 2\ldots\ldots \tag{5.9}$$

#### 5.2.2.5 Hilbert Transform

Hilbert transform of function $x(t)$ is defined as convolution between $x(t)$ and $h(t)$, given by

$$H[x(t)] = x^h(t) = x(t) * h(t) = x(t) * \frac{1}{\pi t} = \frac{1}{\pi}\int_{-\infty}^{\infty} \frac{x(\tau)}{t-\tau} d\tau \tag{5.10}$$

where $h(t) = \frac{1}{\pi t}$. In frequency domain, the Hilbert transform of $X(f)$, which corresponds to $x(t)$, is given by

$$X^h(f) = X(f)H(f) \tag{5.11}$$

from Fourier transform:

$$F[h(t)] = H(f) = jsgn(f) = -j\begin{cases} 1 & f > 0 \\ -1 & f < 0 \end{cases} \tag{5.12}$$

where $sgn(\ )$ is a sign function. From Eq. (5.12), it can be seen that frequency domain Hilbert transform is realized by multiplying $X(t)$ by $j$ at negative frequencies and multiplying $X(t)$ by $-j$ at positive frequencies. The Fourier transform of $\sin(2\pi f_i t)$ and $\cos(2\pi f_i t)$ are given by:

$$H[\sin(2\pi f_i t)] = -\cos(2\pi f_i t)$$

$$H[\cos(2\pi f_i t)] = \sin(2\pi f_i t) \tag{5.13}$$

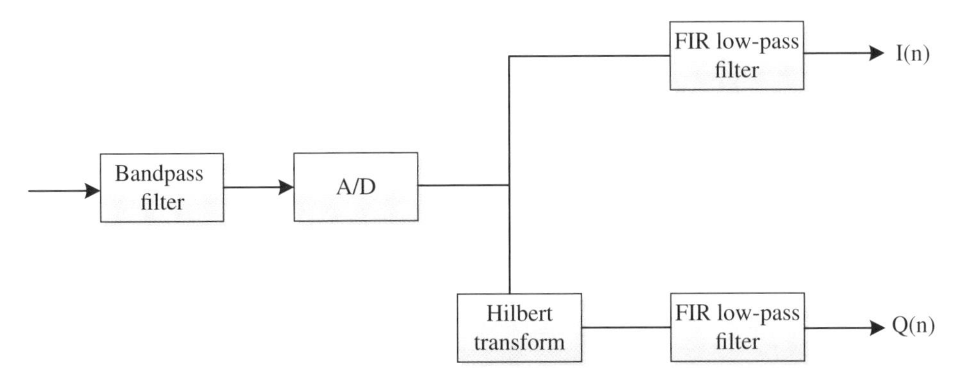


**Figure 5.5** Block diagram of I/Q separation with Hilbert transform.

Equation (5.13) shows that Hilbert transform introduces a 90° phase shift, without influence on amplitude of the spectral component. In digital I/Q quadrature phase discrimination, the input $x(t)$ is firstly digitalized by ADC, then the quadrature components are calculated by discrete Hilbert transform, with a finite impulse response (FIR) filter. For example, a FIR filter with either rectangle window or Hamming weighted window can realize discrete Hilbert transform, as shown in Figure 5.5.

### 5.2.3 Dechirp Receiver

The above-described receiver is one of the main types of SAR receivers that belongs to the category of direct demodulation receiver. Its amplitude and phase distortion mainly depends on the phase nonlinearity of the receiver filter and the amplitude and phase characteristics of the quadrature demodulator. It has high amplitude and phase fidelity. When the radar bandwidth is wide, high requirements for the demodulator and the ADC are put forward, especially under wideband operation. In the meantime, the platform (such as drone and missile) requires that the system be small in size and low in power consumption and that the image is processed in real time. In this case, a dechirp receiver is an ideal choice. Dechirp receiving is an effective way to compress the receiving bandwidth, to reduce ADC sample rate, to reduce the amount of stored data, and to reduce the speed of the imaging processor [5]. Therefore, a dechirp receiver is commonly used in wideband SAR; its working principle is shown in Figure 5.6.

The dechirp receiver has the advantages of low receiver bandwidth compression. In addition, the distortion of a radar system with a dechirp receiver changes according to the time variation of echo, known as *space variant distortion*. Therefore, conventional closed-loop transceiver compensation in a radar system alone does not work. It needs to be combined with predistortion, which increases the complexity of the system. Because of this, a dechirp receiver is normally only used in a high-resolution and small-size SAR system.

### 5.2.4 Multiband Receiver

Multiband SAR illuminates the same area with multiband RF signals and receives the echoes of multiband RF signals. The number of receiver bands is determined during

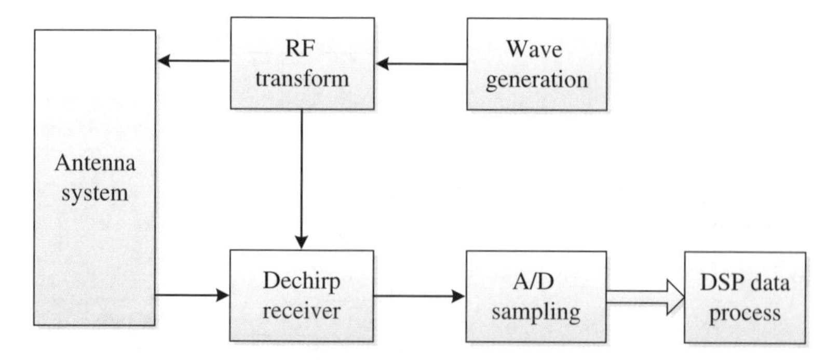


**Figure 5.6** Architecture of a dechirp receiver.

SAR system design. As an example, the design of a four-band airborne SAR receiver is described below.

As mentioned before, there are two main types of high-resolution SAR receivers: dechirp quadrature and wideband direct quadrature demodulation. The former is superior in systems with high resolution and small swath. It can effectively reduce sample rate and design difficulty in the video frequency process, but difficult to compensate wideband distortion. A direct quadrature demodulation receiver is suitable for systems with relatively low resolution, large swath, and high quality of imaging. It is relatively easy to compensate wideband distortion. For a multiband airborne SAR, assuming its maximum range resolution is 1 m, and the corresponding maximum bandwidth is 200 MHz, a direct quadrature demodulation receiver can be adopted. The receiver usually consists of four-band receiving channels, four-band excitation sources, a frequency synthesizer, and a power source. The system working principle is that a relatively narrow-band linear frequency modulation (LFM) signal is generated based on DDS, and then the excitation signals at P, L, X, and Ku bands are generated by expanding and shifting LFM signals. After LNA amplification and one-time frequency conversion, the echo is wideband quadrature demodulated and then sent to the data collector. The block diagram of the receiver is shown is Figure 5.7.

The spectrum of multiband RF signals is shown in Figure 5.8.

### 5.2.5 Multichannel Receiver

To realize high resolution and large swath, usually single-output, multiple-input or multiple-input, multiple-output in the azimuth is used in SAR, which requires SAR to be capable of receiving in multiple channels. Amplitude and phase consistency is a key characteristic of multichannel receivers. Although relatively good multichannel amplitude and phase consistency can be obtained by strict hardware matching [6], the scattering of the device parameters, temperature variations, and noise interference make it difficult to meet the requirements for consistency. Calibrating amplitude and phase of the channels is an effective method to obtain multichannel amplitude and phase consistency, as shown in Figure 5.9.

The transmitting signal in calibration mode is shut down, and the calibration signal is sent into multichannel receiver. The amplitude-phase characteristic of one receiving channel is used as reference, and the rest of the channels are then calibrated.

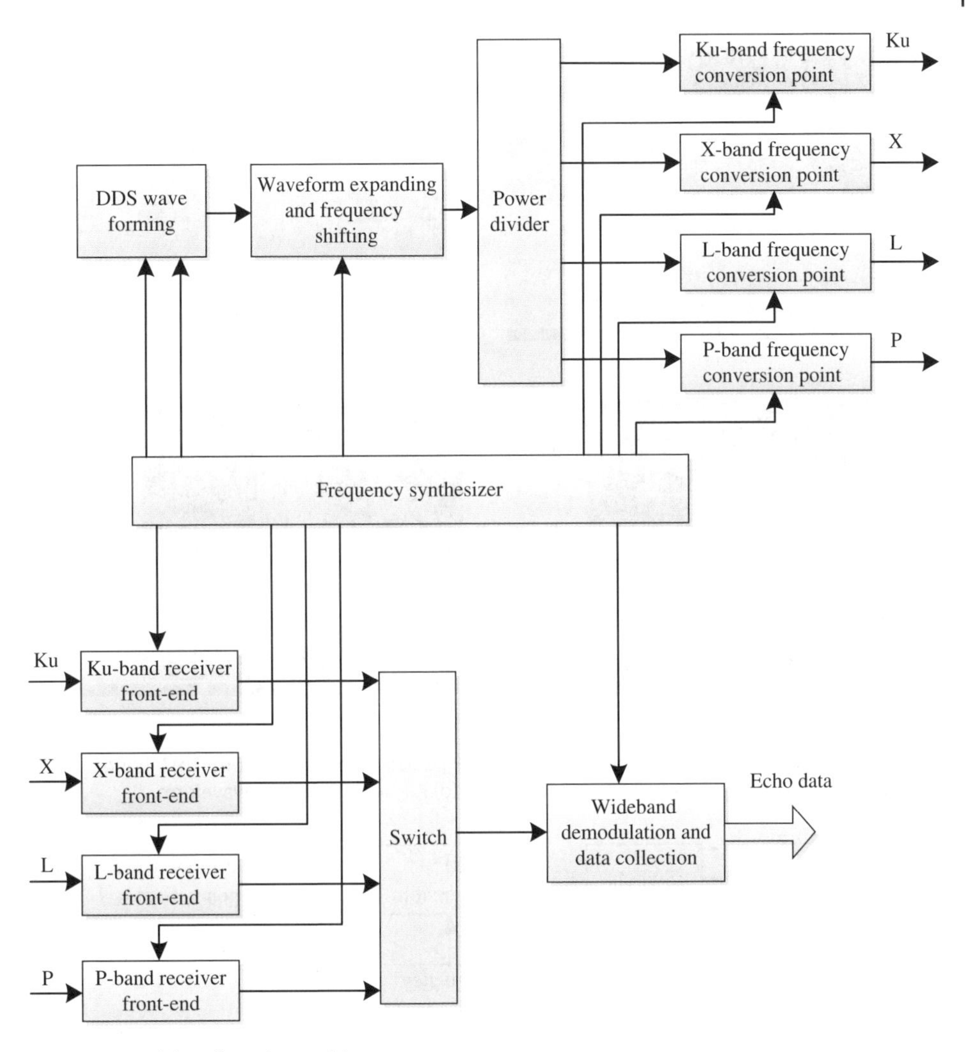


**Figure 5.7** Multiband receiver architecture.

### 5.2.6 Monolithic Receiver

The large size and heavy weight of a conventional microwave analog frequency conversion receiver are unacceptable for SARs with strict weight and size limitations, especially for such platforms as drones and missiles, etc. Microwave monolithic receivers are an important development trend. They are characterized with lightweight and miniature MMIC lightweight and miniature MMIC receiver. With the development of MMIC, the chip dimensions of amplifiers, mixers, switches, digital attenuators, digital phase shifters, and their corresponding driving circuits are all on the order of square millimeter at multiple frequency bands. A monolithic receiver integrates active components and passive filters (such as microstrip, dielectric stripline, thin-film LC filter, etc.) in a minimized-space MCM technology. However, monolithic filters, including active filters

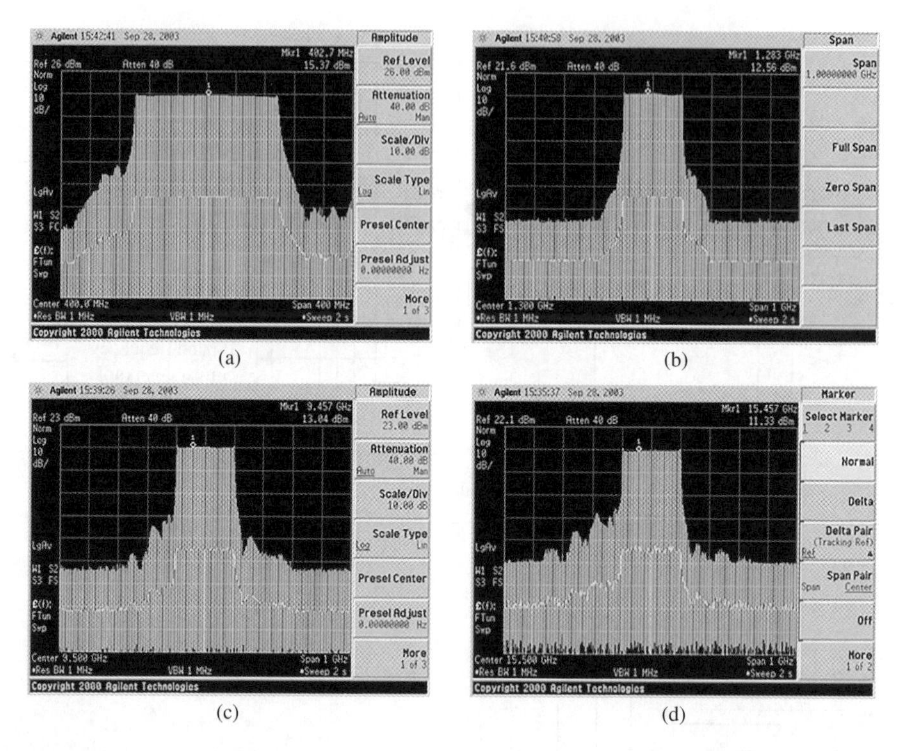


**Figure 5.8** The spectrum of multiband waveform. (a) P-band wideband waveform; (b) L-band wideband waveform; (c) X-band wideband waveform; (d) Ku-band wideband waveform.

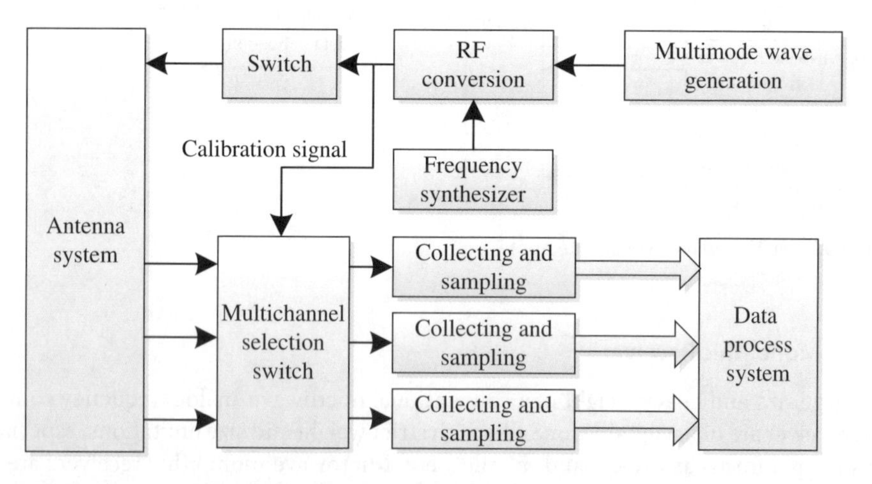


**Figure 5.9** Architecture of a multichannel receiver.

and passive filters (lumped parameter or distributed parameter), are far less mature than monolithic amplifiers, mixers, and oscillators.

The design difficulty of the monolithic receiver is in overall layout, unit circuit electric performance, space distribution, electromagnetic compatibility (EMC), thermal dissipation, etc. The system design level determines whether performance of the monolithic receiver is good or not [7]. Mutual interference can be easily generated in the whole circuit design, from RF input to IF output, between RF and IF signals, between large and small signals, between high gain and dense components. In circuit design, the power line that provides DC power to the LNA is usually isolated from the coming circuits. And a filtering circuit is added at every bias network to avoid interference between power supply and microwave signal. The layout should be distributed properly, separating large signal from small signal and increasing the distance between them. High-frequency signals should be isolated from low-frequency signals by grounding. The crossover between signal line and power line should be avoided by using an intermediate transition layer where transmission lines and power lines intersect. Two ground layers, above and below the intermediate layer, are added to increase isolation as well as to reduce mutual interference dramatically and to ease the contradiction between miniaturization and interference.

#### 5.2.6.1 MCM Design

In electrical design of wideband MCMs, the graphic parameters should be selected properly, such as the width of the signal line, the thickness of the signal line, the space between signal lines, the dielectric thickness, and material properties (such as conductivity, dielectric constant, etc.). Once these parameters are defined, an electromagnetic model is used to transfer these parameters into an equivalent circuit. The mathematical model of the MCM is transformed from the electromagnetic model, which is valid in certain frequency domains. This model includes the full-wave solution of Maxwell's equations in this frequency domain. Generally, this model is fully capable of handling electronic packaging structure with non-transverse electromagnetic wave transmission. It also includes inhomogeneous current produced in the conductor due to skin effect at high frequencies.

#### 5.2.6.2 Multilayer Substrate

Multilayer substrate is a crucial technology of MCM manufacture. The basic classifications of MCM are laminated MCM, ceramic thick-film MCM (MCM-C), and deposited thin-film MCM, etc. As an example, the design of MCM-C is described below.

MCM-C assembles chip components and chips on high-density thick-film multilayer substrate or cofired ceramic multilayer interconnection substrate. Its advantages are multiple wiring layers, high density, high package efficiency, and high performance, and it is applicable to higher frequencies. MCM-C usually uses ceramic multilayer substrate, in which there are two types available: thick-film multilayer substrate and cofired ceramic multilayer substrate.

Cofired ceramic multilayer substrates can be divided into two types: high temperature co-fired ceramic (HTCC) multilayer substrate and low temperature cofired ceramic (LTCC) multilayer substrate. The foundation of ceramic multilayer substrate is thick-film and ceramic multilayer technology. Ceramic multilayer substrate includes the following parts: component mounting layer (top layer), signal layer, power layer, ground layer, external connection layer (bottom layer), etc. The ceramic dielectric layer is between each conductor layer for electrical isolation.

The top layer offers all kinds of pads used to mount the corresponding electrical components. To improve mounting density, double-sided component mounting on multilayer substrate can be used, that is, both the top layer and bottom layer are used to mount electrical components. The signal layer of multilayer substrate is located below the top layer and used to place interconnection lines between components. The number of the

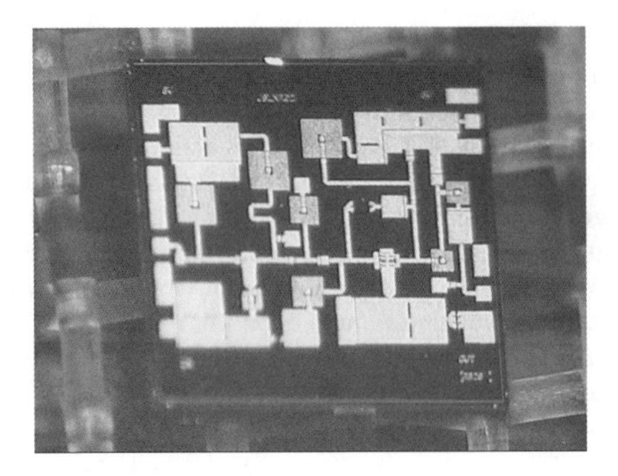

Figure 5.10 A microwave monolithic LNA.

signal layer is defined by the scale of the module and layout density. The power layer and ground layer are usually independently passed and designed according to performance requirements of the module. Vertical metal vias are used to interconnect layers above the ceramic substrate.

HTCC is used in common MMIC receivers as multilayer substrate, with four layers in total. The signal layer is at the top. The power layer is in the middle and connected to the top layer through vertical via holes, which reduces the amount of on-chip interconnects. The RF input and output signals pass through the vertical vias and are led out by 50 Ω insulators, which greatly reduce the size and weight of the receiver.

#### 5.2.6.3 Design of MCM Mounting

With micro-assembly, chips are epoxy attached or soldered to multilayer substrate, and input/output port is connected by gold wire bonding. A microwave monolithic LNA is shown in Figure 5.10, and a microwave monolithic mixer amplifier is shown in Figure 5.11.

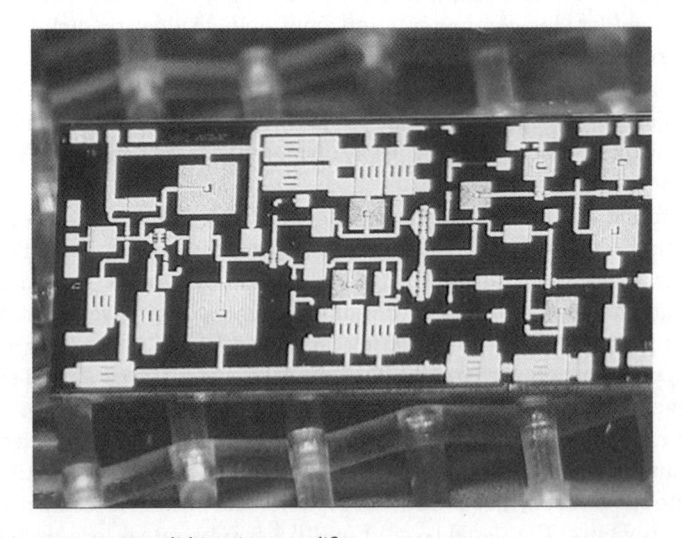

Figure 5.11 A microwave monolithic mixer amplifier.

## 5.3 Frequency Synthesizer Source

The stability of frequency source directly affects the edge jitter of the pulse signal and phase stability of the system, and consequently affects resolution and image quality of SAR. The frequency source in the receiver provides local oscillation and coherent reference signal as well as generates the clock signal for the radar system.

A highly stable reference source is used in the frequency source as the standard (such as a crystal oscillator or atomic clock), producing a series of predefined stable frequency points by arithmetic operations (add, subtract, multiply, and divide), phase lock, or data collection. The stability of frequency source refers to relative frequency variation within a certain period. When the sampling time is longer, such as an hour, day, year, etc. the random jitter produced in the noise is averaged. The frequency fluctuation is approximately a stationary process. The frequency stability can be approximately measured by standard deviation, i.e. the standard deviation is suitable for the stationary process. When the sampling time is shorter, due to random jitter and drift, with the former dominating, the frequency fluctuation is a nonstationary process. The frequency stability is normally measured by Allan variance. Allan variance is defined by firstly performing multiple gapless and continuous measurements at one frequency, obtaining a group of relative frequency fluctuations, then calculating root mean square between two adjacent measurements, also called *two-sampling variance.* Allan variance is widely used in nonstationary random processes. In practice, short-term stability in a radar frequency source is of more concern. Phase noise is usually used to represent fine time characteristic and frequency domain noise distribution of the stability in the frequency source.

A frequency synthesizer usually consists of reference source, synthesizer circuit, and transmission line. The reference source has an important influence on the synthesizer performance. There are two major types of high-performance radar reference sources: one is oven-controlled crystal oscillator (OCXO), which has a moderate frequency of generally around 100 MHz, with short-term frequency stability and relatively small phase noise. For some SAR systems requiring frequency precision and long-term stability, low-frequency crystal oscillator (for example, at 10 MHz) is usually adopted, which has better near-carrier equivalent noise than high-frequency crystal oscillator (at 100 MHz), i.e. better long-term frequency stability. The other choice is atomic clock, which has a relatively low frequency (generally around 10 MHz). The phase noise in the super near zone is good, the cost is relatively high, and the long-term frequency stability is high, but the phase noise in the equivalent mid-far zone is worse than that of OCXO. For SAR systems requiring high stability in both the long term and short term, atomic clock phase-locked OCXO can be used. With the development of microwave photoelectric technology, the focus has turned to optoelectronic oscillators, with high frequency (usually around 10 GHz), and the mid-far range phase noise is one or two orders of magnitude higher than that of conventional frequency synthesizers.

With continuous improvement of SAR resolution and swath area, more and harsher requirements are put on stability of the frequency source, especially on short-term stability. The phase instability of the frequency source will lead to relative position fuzziness of the target in the range direction and will influence the Doppler effect in the azimuth. Only full-phase-coherent highly stable frequency synthesizers can meet SAR requirements on phase synchronization and frequency stability. The frequency synthesizer in a full-phase-coherent frequency source can be realized by direct synthesis or indirect synthesis (phase locked). In view of the realization method, frequency synthesis can be classified into three major types: conventional direct analog frequency synthesis, phase-locked frequency synthesis, and DDS.

### 5.3.1 Direct Analog Frequency Synthesis

Direct analog frequency synthesis is the most fundamental frequency synthesis method, with a simple principle that is easy to realize and relatively low phase noise. Frequency hopping time is mainly dependent on microwave switching time and filter delay time. Sometimes hopping time of the synthesizer can be made equal to microwave switching time. Consequently, frequency hopping time is extremely short, mainly used in frequency agility radar. In this synthesis method, the required signals are obtained by frequency multiplier, mixer amplifier, and switching filter, thus a lot of equipment is required. More importantly, when the synthesis method is complicated, a large number of IFs are generated, with inferior EMC, and the method cannot be used in highly dense platforms, such as airborne, spaceborne, missile, etc. This synthesis method is seldom used in SAR.

A highly stable crystal oscillator is used in indirect analog frequency synthesis as a reference to obtain a group of highly stable frequency signals through frequency multiplier, mixer amplifier, and switching filter. Figure 5.12 shows the block diagram of conventional direct analog synthesis. The frequency bands of two comb wave generators are dependent on comprehensive consideration of the synthesizer frequency, bandwidth, and required frequency interval. The number of components used in a direct analog frequency synthesizer is large. Generally, the frequency synthesizer is large in size. The usage of a large number of components also affects the reliability. With the development of microelectronic technology, and application of various monolithic amplifiers, mixers, lumped element filters, and monolithic high-speed analog switches, miniaturization of direct analog synthesis is improved to some degree. In direct analog synthesis, the usage of frequency doubling, frequency division, and mixing at many different frequencies makes the requirements for noise suppression very strict. This is difficult, and a series of actions need to be taken. First, high-speed and high-isolation analog switches

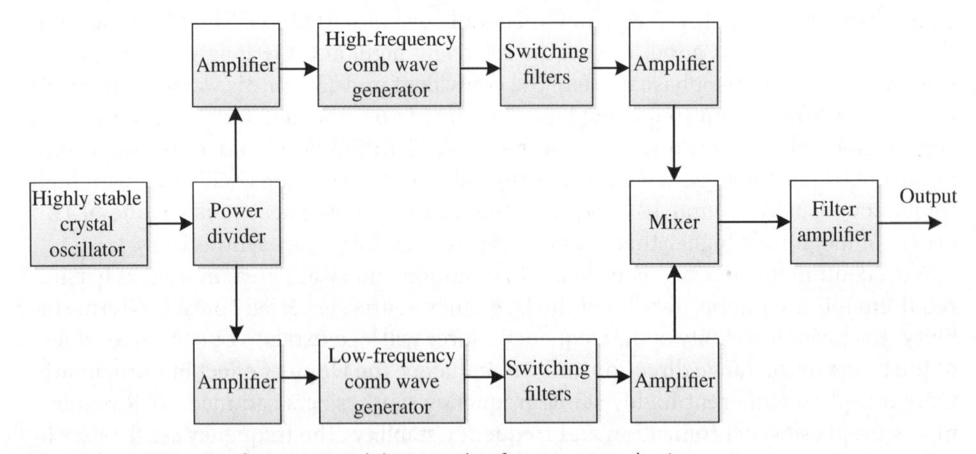


**Figure 5.12** Diagram of conventional direct analog frequency synthesis.

are selected. Second, the design must be optimized, including a reasonable choice of mixing window. In addition, filters with high out-of-band rejection and EMC design are adopted to improve clutter suppression in the output signal of the frequency synthesizer.

### 5.3.2 Phase-Locked Frequency Synthesis

The foundation of phase-locked frequency synthesis is control theory, of which phase-locked feedback loop is critical. Phase-locked loop (PLL) is a closed-loop automatically controlled system, in which phase of the output frequency can track phase of the reference frequency. Phase-locked frequency synthesis locks phase of the microwave oscillator to that of a highly stable reference source. It is simple in circuit structure, a few IF distortion, phase synchronization, easily integrated EMC, are commonly adopted in the SAR.

Phase-locked frequency synthesizer uses phase tracking characteristics of PLL to strictly synchronize phase of the output signal to that of the reference source. It has fewer switching filters than direct analog synthesis. Therefore, it is small in size. In addition, due to use of closed-loop tracking, good synchronization is ensured among signals, which is very important for SAR with high speed and multiple clocks. With the development of electronic technology, high-performance digital integrated phase detectors and digital multimode controllable frequency dividers are applied in microwave frequency synthesis, which dramatically improves the stability and performance of the tracking loop.

The frequency synthesizer is developed to be digitalized. Digital PLL has two major types. One is all-digital PLL, in which phase detector, loop filter, oscillator, and programmable frequency divider are all digital, with high integration and good stability. However, its operation frequency is relatively low. Both spectrum purity and phase noise are much worse than those of a high-performance frequency synthesizer. The other is quasi-digital phase locked, whose phase detector and programmable frequency divider are digital, while the loop filter and oscillator are still analog circuits. It has high operation frequency and good performance. And it is one of the main synthesis methods in the current frequency synthesizer.

If front frequency dividing phase lock is used, the total loop frequency dividing ratio is large, seriously deteriorating phase noise and tracking characteristics of frequency synthesizer. Downshifting the frequency (i.e. frequency shift feedback) in the feedback branch may effectively reduce the loop frequency dividing ratio, which is helpful to improve phase noise and dynamic response of the system. It is commonly used in radar phase-locked synthesis. Its working principle is that the phase comparison of the reference signal and feedback signal is performed. Then the output voltage is used to suppress the noise by loop filter and to control the voltage-controlled oscillator (VCO) by high-frequency components. Finally, frequency agility is realized by controlling the frequency dividing ratio of the feedback branch after frequency shift [8]. When the loop is locked, output frequency is $f_0 = (M + N)f_i$, and minimum frequency interval is $\Delta f = f_i$, where $f_i$ is reference frequency, i.e. phase detection frequency. The working principle of digital phase-locked synthesis is shown in Figure 5.13.

Assume $k_d$ is the sensitivity of the phase detector, $F(s)$ is the transfer function of the loop filter, $k_o$ is the sensitivity of VCO, $N$ is equivalent frequency divide ratio, and $M$ is the multiplication number of the auxiliary multiplier. Suppose $K = \frac{k_0 k_d}{N}$, then the

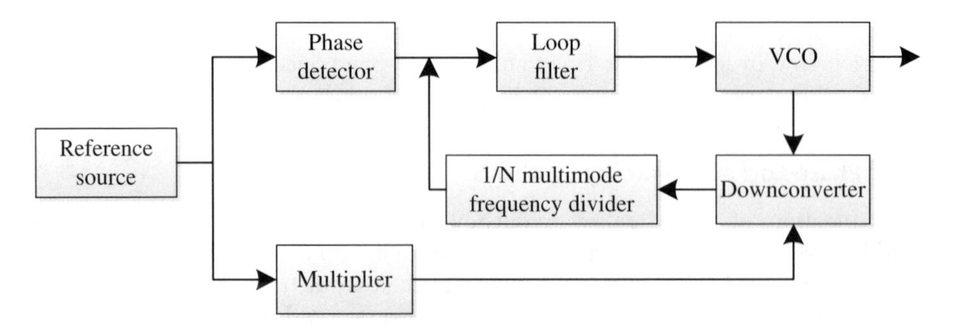


**Figure 5.13** Block diagram of conventional digital phase-locked synthesis.

open-loop transfer function is given by:

$$G(s) = \frac{KF(s)}{s} \tag{5.14}$$

The closed-loop transfer function is given by

$$H(s) = \frac{KF(s)}{s + KF(s)} \tag{5.15}$$

The closed-loop equation of Figure 5.13 is given by

$$\phi_0 = \left[(M+N)\phi_i + \frac{N}{k_d}\phi_{PD}\right] H(s) + [1 - H(s)]\phi_{VCO} \tag{5.16}$$

The phase noise power spectrum density can be expressed as

$$S_{\varphi_O} = |H(S)|^2 \left[(M+N)^2 S_{\varphi_i} + \left(\frac{N}{k_d}\right)^2 S_{\varphi_{PD}}\right] + |1 - H(j\omega)|^2 S_{\varphi_{VCO}} \tag{5.17}$$

where, $\varphi_i$, $\varphi_{\varphi PD}$, $\varphi_{\varphi vco}$, and $\varphi_0$ are the phase jitter of the reference source, phase detector, VCO, and output signal, respectively, and $S_{\varphi i}$, $S_{\varphi PD}$, $S_{\varphi vco}$, and $S_{\varphi o}$ are the phase noise power spectrum density of the reference source, phase detector, VCO, and output signal, respectively.

Loop filter usually uses active proportional integral filter, whose characteristic is very close to an expected integral filter, so as to make the loop realize a steady-state phase error of approximately zero. This type of filter can adjust the damping coefficient and frequency of free oscillations independently, facilitating the design of loop response time and noise. In the meantime, the lag-lead characteristic of active proportional integral filter is beneficial to loop stability. The approximate estimation is

$|H(j\omega)| \approx 1\ \omega \ll \omega_n$, $|H(j\omega)| \ll 1\ \omega \gg \omega_n$

Thus, Eq. (5.17) is equivalent to

$$S_{\varphi_0} = (M+N)^2 S_{\varphi_i} + (N/k_d)^2 S_{\varphi_{PD}} \quad \omega \ll \omega_n \tag{5.18}$$

$$S_{\varphi_0} = S_{\varphi_{VCO}} \quad \omega \gg \omega_n \tag{5.19}$$

It can be seen from Eq. (5.18) that the phase noise reduction can not only improve the phase noise performance of the system but also increase $\omega_n$ of the loop. These are instrumental in improving hopping speed, capture, and tracking performance of the system.

### 5.3.3 DDS

Direct digital synthesis is one of the major means for SAR to generate waveform. The working platforms of SAR are all moving types, and random vibrations of some platforms are very strong, such as missiles and helicopters, seriously deteriorating phase noise of the synthesizer. Therefore, a frequency synthesizer in SAR must have high-frequency stability under strong vibration environments. A SAR frequency synthesizer, by using phase lock as the main technology, is combined with direct digital synthesis and mainly solves the influence of phase synchronization, miniaturized integration, and random vibration on phase noise.

DDS has extremely short frequency agility time (in nanosecond order), quite high frequency resolution (in hertz), and excellent phase noise performance. A variety of modulations can be easily realized. DDS is a fully digital programmable system with high integration.

DDS is a type of frequency synthesis that directly synthesizes the desired waveform based on the phase concept. Generally, a direct digital frequency synthesizer consists of phase accumulator, amplitude and phase converter, digital-to-analog converter (DAC), and low-pass filter (LPF). Its block diagram is shown in Figure 5.14.

The shaded section is the core part of DDS, in which the phase accumulator is a typical feedback accumulator composed by cascading the accumulator and phase register, which completes the phase accumulation function. There are two inputs in this structure: frequency/phase tuning word and clock reference $f_c$. N is the digit number of the phase accumulator. $f_0$ is output frequency. The phase accumulator can be seen as a modulus controllable counter determined by frequency tuning word $K$. Under the action of system clock $f_c$, the phase accumulator cycles up with a mode value of $K$. When the output of the phase accumulator reaches the full scale, overflow emerges. The time for the accumulator reaching overflow is the period of the output signal. Apparently, the phase accumulator overflows one time after every $2^N/K$ clock period. And period of the output signal is $(2^N/K)T_c$. So more frequency tuning word $K$ *is*, less the increment of the phase is, and the longer time for the accumulator to overflow is, the shorter the frequency of the output signal is, and vice versa. Therefore, frequency tuning word $K$ determines sample points in one complete period.

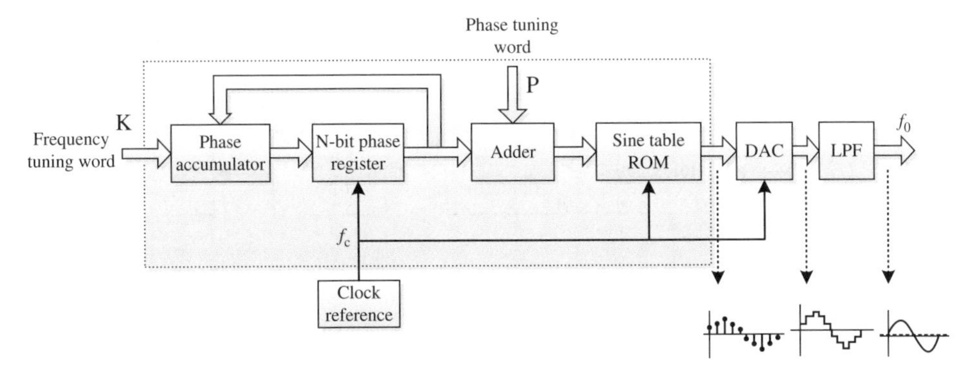


**Figure 5.14** DDS block diagram.

Assuming the frequency of the reference clock is $f_c$, and bit number of the phase accumulator is $N$, output frequency of DDS is given by

$$f_0 = Kf_c/2^N(K = 0, 1, 2 \ldots 2^{N-1}) \tag{5.20}$$

where $K$ is the frequency tuning word, predefined by control circuits. When the clock frequency $f_c$ and digit number of the phase accumulator $N$ are fixed, output frequency $f_o$ is determined by $K$ only:

$$K = 2^N f_0/f_c \tag{5.21}$$

When $K = 1$, minimum frequency of the sine signal generated by DDS, i.e. the frequency resolution is given by

$$\Delta f = f_c/2^N \tag{5.22}$$

Theoretically, the phase noise of DDS is mainly dependent on noise of the reference clock, and DDS is equivalent to a fractional divider. However, the phase truncation of DDS and the definite word size of read-only memory (ROM) have affected the phase noise of DDS. Theoretical analysis indicates that phase noise of DDS is mainly determined by the definite word size of ROM and digit number of DAC.

The major disadvantages of DDS used as frequency synthesizer are as follows: one is that operation frequency is relatively low at the moment. The other is the existence of strong scattering. A DDS-based frequency synthesizer, operating at microwave frequencies, must be combined with a PLL synthesizer or direct analog synthesizer for frequency shifting. The principle of a frequency synthesizer that combines DDS with PLL is shown in Figure 5.15.

Due to the limitations of the oversampling rate, phase truncation, and digit number of DAC, the clutter distribution of the DDS output signal is wide, and the SNR is bad. Although a narrow-band characteristic of the loop can eliminate far-range clutter, the near-range clutter will have 20log $N$dB deterioration at the output port according to frequency division ratio N. Therefore, this method fails to make full use of the DDS advantages and overcomes its disadvantages. If the DDS is inserted in front of the frequency divider as a frequency standard, the loop frequency divider will not cause the noise of the DDS to show the above multiplication relationship at the output port of the frequency synthesizer. The loop division ratio N is equivalently reduced to improve the frequency synthesizer clutter and phase noise performance. Figures 5.16 and 5.17 show typical architectures based on this idea. When the output frequency is relatively low, for instance, below L-band, the architecture in Figure 5.16 is normally used.

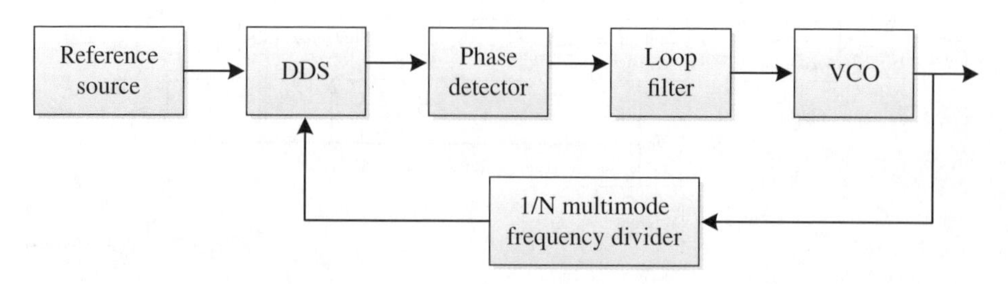


**Figure 5.15** DDS combined with PLL.

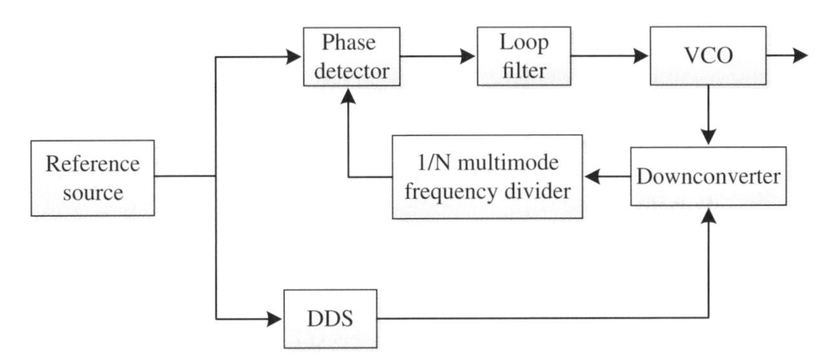


**Figure 5.16** DDS combined with PLL, output frequency relatively low.

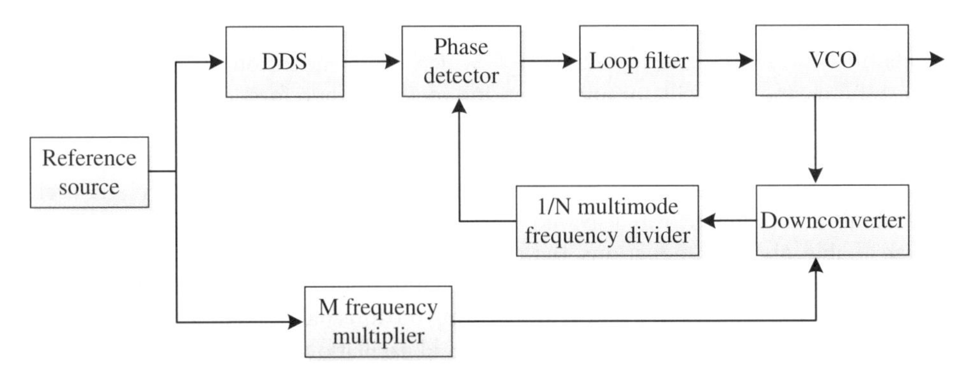


**Figure 5.17** Frequency synthesizer that combines DDS with PLL, output frequency relatively high.

When the output frequency is relatively high, for instance, above C-band, the architecture in Figure 5.17 is normally used.

### 5.3.4 Antivibration Characteristic of Frequency Synthesizer

Unlike general-purpose circuits, random vibration not only affects the reliability of a frequency synthesizer but also deteriorates its electronic characteristics, such as spur suppression and phase noise. The unique pressure-frequency effect of highly stable crystal oscillator makes it the most sensitive component of a frequency synthesizer. For a simple harmonic vibration $\vec{a} = \vec{A}\cos(2\pi f_v t)$, a sideband modulation is produced, in which amplitude is approximated

$$I(f_v) \approx 20lg\frac{(\overrightarrow{\Gamma A})f_0}{2f_v} \tag{5.23}$$

where $\vec{A}$ is the peak acceleration of harmonic oscillation (g), $f_v$ is the frequency of harmonic oscillation (Hz), $\vec{\Gamma}$ is the acceleration sensitivity of harmonic oscillation, and $f_0$ is static frequency of the crystal oscillation (Hz).

As shown in Eq. (5.23), both the acceleration and sensitivity are vectors. They change with the platform variation of the 3D vibration directions, though the difference of their absolute value is small. Normally only the scale value of the acceleration sensitivity in the crystal oscillator is given. For example, for a synthesizer using stress complementary–cut crystal oscillator with a frequency of 100 MHz, $|\vec{\Gamma}|$ is $2\times 10^{-10}$/g. For harmonic oscillation with frequency of 50 Hz and peak acceleration of 5 g, the sideband clutter suppression is −60 dBc. In the condition of random vibration, the phase noise of a crystal oscillator is almost not relevant to its static phase noise, and the power density is given by

$$S_{\varphi REF}(f) = \frac{(|\vec{\Gamma}|f_0)^2 G(f)}{4f^2} \quad (f_A \le f \le f_B) \tag{5.24}$$

where $S_{\varphi REF}(f)$ is phase noise power density of a crystal oscillator at random vibration, $G(f)(g^2/Hz)$ is vibration power density sensed by crystal oscillator, $f_A$ is minimum frequency of random vibration, and $f_B$ is maximum frequency of random vibration.

It can be seen from Eqs. (5.23) and (5.24) that it is very important to choose a crystal oscillator with low $|\vec{\Gamma}|$ and to reduce $G(f)$, which is dependent on the cutting type of the crystal harmonic oscillator and mounting method. Therefore, reducing the equivalent $G(f)$ is one of the key points in designing a SAR frequency synthesizer. It is well known that performing vibration isolation only on a crystal oscillator does not produce good effect. More preferred is firstly to perform vibration isolation on a crystal oscillator alone, and then perform a second vibration isolation on the whole synthesizer to realize double vibration reductions. In the meantime, a variety of auxiliary antivibration methods are combined, for instance, full surface mounting on circuits, a soft shield bar inserted between the lid and the box, and making the internal resonant frequency of the synthesizer much higher than the upper threshold oscillation frequency. The features of second vibration isolation are wideband vibration isolation, good vibration isolation at ultra-low frequency, and low added weight.

## 5.4 Wideband Waveform Generation

A digital method is more popularly seen to generate a waveform. Digital waveform generation can not only realize a variety of waveform agility but also improve waveform quality by amplitude and phase compensation. A digital waveform generator can realize a desired waveform by programming the digital storage with flexibility, repeatability, and consistency, so that more attention is paid to digital waveform generation. Digital waveform generation can be divided into digital IF direct generation and digital baseband generation.

IF direct generation is based on DDS technology. In recent years, with the development of high-speed digital circuits, it is possible to synthesize the required waveform at relatively high IF. Its main advantages are simplicity in circuit structure and multimode in bandwidth and time-width. But its disadvantage is that due to the fact that wideband waveform is generated at IF to enable upconverting and filtering, the bandwidth should not be too high and the IF should not be too low. Therefore, the frequency of the output signal relative to the oversampling rate in the sample clock is relatively low, and signal distortion is relatively high. In addition, the waveform is generated based on phase accumulator and frequency adder inside DDS, which makes it difficult to perform predistortion compensation.

The principle of the baseband generation is to produce I and Q baseband signals directly by using digital storage. After that, the baseband signals are modulated to an IF carrier by an analog quadrature modulator. In this method, the amplitude and phase information are prestored in truth table format. Its advantage is that a variety of flexible waveforms can be generated, and system distortion can be compensated by using the data prestored in memory. As the waveform is generated from the baseband, under the same sampling rate, the oversampling rate is much higher, which is useful for signal fidelity. The disadvantage is that due to the usage of the truth table, the requirement for the storage space is higher under multimode operations. In addition, as the analog quadrature modulator is introduced, it is difficult to obtain the ideal amplitude phase balance. The false image signal and carrier leakage appear in the output waveform, which consequently affects the main side lobe ratio, especially noise floor of the pulse compression. In practical imaging application, the pulse compression noise floor has more influence on the image quality than the main side lobe ratio.

### 5.4.1 DDS-Based Direct Waveform Generation

The basic principle of DDS is that the frequency control is realized by controlling the phase accumulation value of the DDS phase accumulator in unit time (generally referring to the period of DDS sampling clock), and frequency adjustment is realized by varying the accumulation value as per a certain rule. The unique DDS structure makes it flexible to control phase accumulation value and variation law in the phase accumulation, so that the waveform generation method is more flexible and more suitable for radar systems with waveform agility and multimode. DDS is used to generate radar waveform by making full use of the DDS features. The desired waveform can be realized by selecting DDS chips suitable for the waveform synthesizer as well as with complete functions. It should be noted that the selected chip should have not only high sampling frequency but also a high frequency update rate. In the meantime, the quality index of the generated waveform (such as scattering, SNR, etc.) should meet the specifications of the radar system.

The implementation method depends on bandwidth and frequency of desired waveform. If the desired frequency is low with narrow bandwidth, the waveform can be directly generated at IF by DDS. In case of high frequency and large bandwidth, shifting and extension can be used to improve the operating frequency and signal bandwidth [9].

The method of generating waveform directly at IF by DDS is shown in Figure 5.18. In this method, the dominating DDS chip and the additional logic circuits are used to realize different waveforms. In Figure 5.18, $F_{CW}$ is initial frequency tuning word, $K_{CW}$ is modulation chirp rate control word, and $T_{CW}$ is the time tuning word. Three common types of radar waveforms can be generated: LFM, non-LFM, and phase coding.

In SAR, the radar waveform is commonly required to have a large bandwidth and time-width product. In practice, DDS is not directly used to generate wideband or ultra wideband LFM signals to obtain good wideband signal performance. Instead, a relatively narrow bandwidth signal is generated first, then bandwidth extension or frequency shifting is used to generate the desired wideband or ultra wideband waveform.

Frequency doubling and frequency upconversion is an effective approach to realize wideband waveform. Figure 5.19 shows the block diagram of this approach, which requires that the frequency multiplier have good linearity. Both filter and amplifier

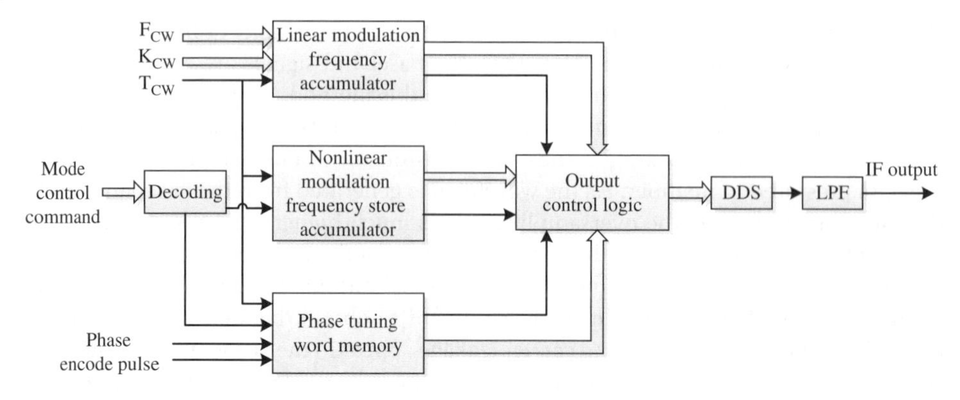


**Figure 5.18** Architecture of DDS-based IF waveform generation.

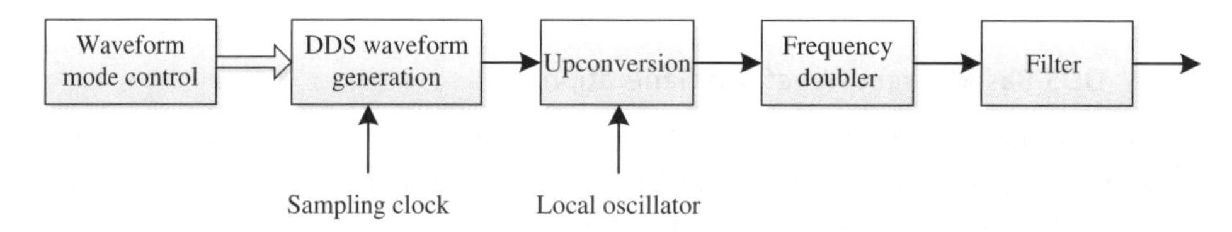


**Figure 5.19** Wideband waveform generation by using upconversion and frequency multiplier.

have better amplitude flatness and smaller nonlinear distortion. In addition, intermodulation of the frequency converter needs to be calculated strictly to prevent intermodulation produced by upconversion from falling into the band, which leads to intermodulation distortion and affects signal quality.

It should be noted that the frequency multiplier will increase the phase nonlinear distortion of the waveform. Therefore, the order of doubling should not be too high; normally, less than four times. It is not advisable to generate wideband waveform by only using frequency doubling. DDS-based direct waveform generation is usually applied in a radar system requiring relatively low bandwidth (for example, below 200 MHz) and multiple operation modes.

### 5.4.2 Parallel DDS IF Waveform Generation

DDS-based direct IF waveform generation eliminates the wideband quadrature modulator, which simplifies design of the waveform generation and avoids image frequency in the modulator and RF carrier leakage at the same time. However, due to the limitation of the working frequency of the general DDS, its working bandwidth and operating frequency are greatly restricted. With the development of field-programmable gate array (FPGA) technology, the programmable function of FPGA is used to realize high-speed DDS function based on multichannel parallel DDS structure, which directly generates large bandwidth signals. In the meantime, high-frequency and IF output can be realized on high-speed DAC by using the high-order Nyquist area. Theoretically, a digital equalization filter can be added behind the DDS to compensate predistortion, as shown in Figure 5.20.

**Figure 5.20** Block diagram of direct IF waveform generation.

With the application of new technology and new material, the frequency of GaAs-based multiplexer (MUX) and DAC can reach 2.4 GHz in practice. After process modification, 3.5 GHz MUX and DAC can be produced. The frequency of InP-based DAC can reach 9.2 GHz. Its development prospects are immeasurable. DAC with 8 GHz conversion rate and 12-bit resolution can be found on the market. With the development of high-speed devices and electronic processing technology, generating 2 GHz bandwidth or an even larger bandwidth digital signal is completely feasible.

### 5.4.3 Digital Baseband Waveform Generation

Baseband signal generation combined with wideband quadrature modulation may double the signal bandwidth as well as shift the RF signal. The amplitude and phase distortion in the entire link looks relatively small, and the DAC digital sample rate is reduced, which is one of the major methods to generate wideband waveforms for high-resolution imaging radar, as shown in Figure 5.21. The difficulty of this approach is how to solve the amplitude and phase inconsistency in the wideband quadrature modulator and leakage in the local oscillator. For the wideband quadrature modulator, the amplitude and phase consistency is limited due to amplitude and phase unbalance of the analog I/Q channels and the temperature dependency. Image clutter is produced. Generally, the image clutter and carrier leakage for high-performance wideband quadrature modulator is around −28 dB.

Due to usage of digital direct memory and readout, distortion compensation can be stored in the memory in advance, to realize predistortion compensation. This function is very important for systems unable to realize closed-loop compensation, such as wideband dechirp SAR. This approach has high requirements for memory space, especially under long time-width and multimode conditions. Therefore, DDS baseband signal generation combined with quadrature modulation is widely used and developed.

With the development of FPGA technology, using FPGA programmable functions to generate digital baseband signals based on parallel DDS structure greatly reduces the requirements of DAC conversion rate while reducing the complexity of implementing

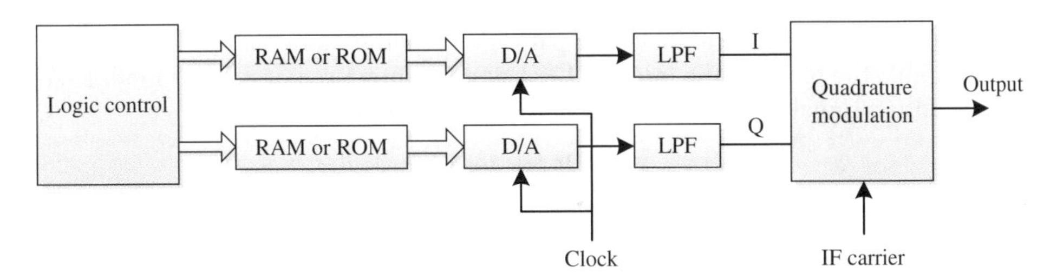


**Figure 5.21** Wideband waveform generation by using quadrature modulation. D/A: digital-to-analog.

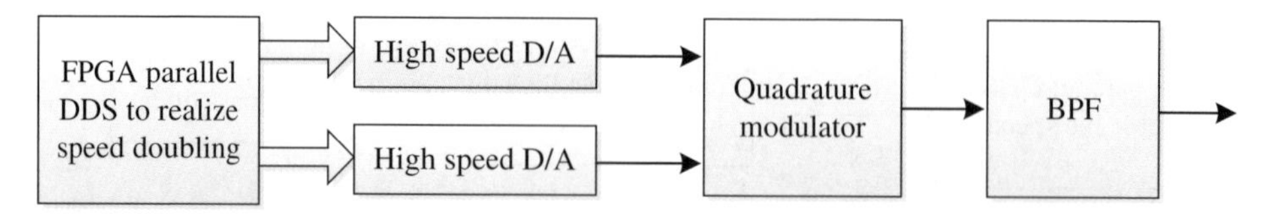


**Figure 5.22** Baseband waveform generation principle. BPF: bandpass filter.

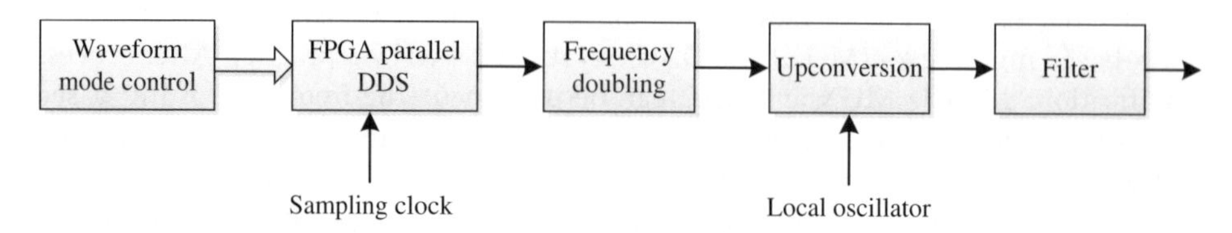


**Figure 5.23** Block diagram of wideband waveform direct generation.

DDS in parallel within FPGA. The advantage of using parallel DDS instead of RAM is that bandwidth and pulse width of the output signal may be changed in real time as required by the system, which substantially reduces the required memory space. It also has no limitation on the type of output signal (without on-chip or off-chip RAM). In the meantime, real-time predistortion (the loop needs to be calibrated) can be realized by adding a digital equalization filter behind the DDS. The principle of baseband signal generation is shown in Figure 5.22.

An example of X-band 2.4 GHz wideband signal generation is introduced below. Parallel DDS is realized by FPGA controlling. A 1.2 GHz wideband signal is directly generated at IF by combining high-speed DAC in the trans-Nyquist area. A wideband 2.4 GHz waveform is generated after wideband frequency doubling. Then an X-band 2.4 GHz LFM signal is generated by frequency shifting, filtering, and amplifying, as shown in Figure 5.23.

Parallel DDS generates 1.2 GHz bandwidth signal directly and then generates 1.2 GHz multiplier to get 2.4 GHz. The frequency multiplier will bring phase nonlinearity and in-band amplitude fluctuation. The amplitude and phase contains first-order, second-order, and high-order errors. The influence on high-order error is neglected and error of frequency doubling is analyzed as follows.

Assume the input signal expression is

$$V_i(t) = \cos[2\pi f_0 t + \pi K t^2 + \theta(t)] \tag{5.25}$$

where $f_0$ is the carrier frequency, $K$ is the linear modulation frequency, $t$ is time variable, and $\theta(t)$ is the parasitic phase modulation, i.e. phase noise.

Without considering the influence of parasitic amplitude modulation, in frequency doubling link, the phase of the input IF LFM signal is interfered or noise is modulated, and the output signal after M-time frequency doubling is given by

$$V_0(t) = \cos[2\pi M f_0 t + \pi M K t^2 + M\theta(t)] \tag{5.26}$$

Namely, the phase noise $\theta(t)$ is amplified by M times. Figure 5.24 shows the simulated spectrum of IF LFM signal before and after frequency doubling, when 5° phase error (around 0.0873 rad) with zero-average white noise is added. The parameters of LFM signal in simulation are bandwidth B = 1200 MHz, time-width T = 20 μs, and carrier

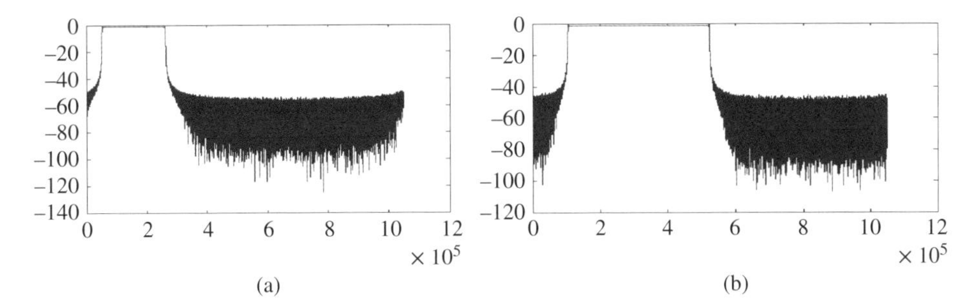


**Figure 5.24** Signal quality before and after frequency doubling. (a) Spectrum before frequency doubling; (b) spectrum after frequency doubling.

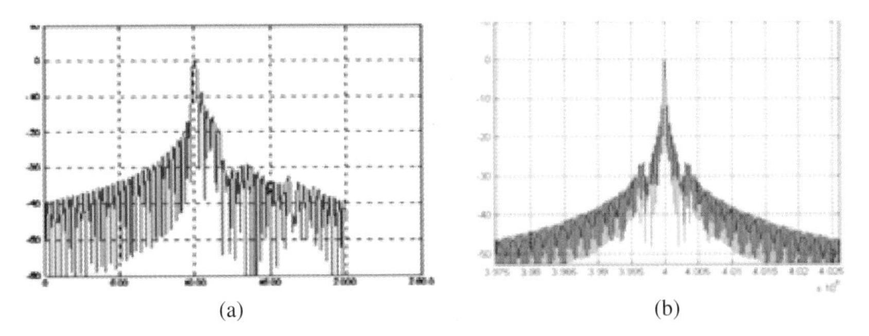


**Figure 5.25** Pulse voltages before and after compensation. (a) Pulse compression before compensation; (b) pulse compression after compensation.

frequency $f_0 = 3000$ MHz. It can be seen from Figure 5.24a and b that frequency doubling makes the out-band noise floor of the phase noise elevated from about −55 dB to about −48 dB. In the meantime, inside the effective band, amplitude fluctuation caused by phase noise is increased due to frequency doubling. On one hand, the frequency doubling may lead to an increase in the phase noise, namely the phase error increases from $\theta(t)$ to $M\theta(t)$ after M times doubling. On the other hand, it may lead to a decrease in the SNR, in other words, the effective output power decreases. For example, the SNR decreases 20logM dB after M times frequency doubling. For frequency multiplier, the phase noise leads to the SNR decrease to 6.02 dB. This affects the pulse compression characteristic of the generated signal. Therefore, it is necessary to compensate the signal in amplitude and phase. Figure 5.25 shows the pulse compression before and after compensation.

### 5.4.4 Multiplex Splicing Waveform Generation

It is difficult to generate such a signal depending only on device technology due to limitations in the development of DAC devices when the bandwidth is wide. An approach of generating wideband waveform is proposed that is based on regular DAC or DDS and multiplex (similar to an ADC, which generates high sampling rate and wideband signal by splicing multiple ADCs). The principle is that multiple signals with head-to-end connected time and frequency are generated by multiple DACs or DDSs, respectively.

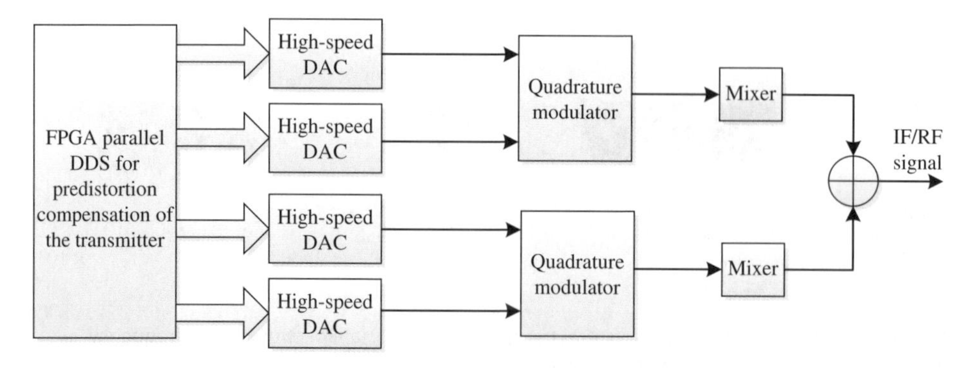


**Figure 5.26** Block diagram of two-channel baseband waveform splicing.

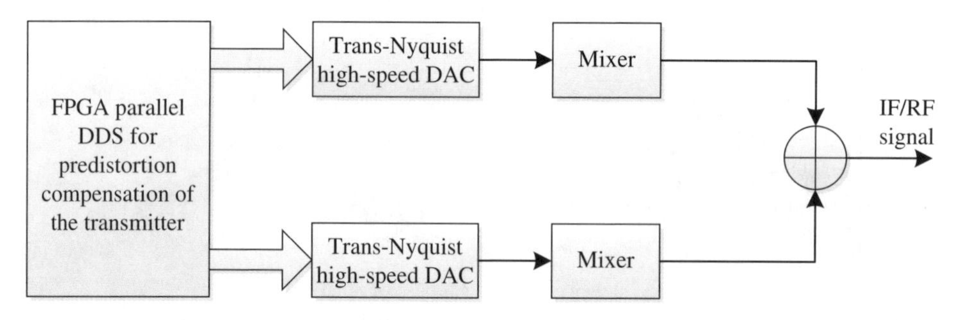


**Figure 5.27** Block diagram of two-channel IF waveform splicing.

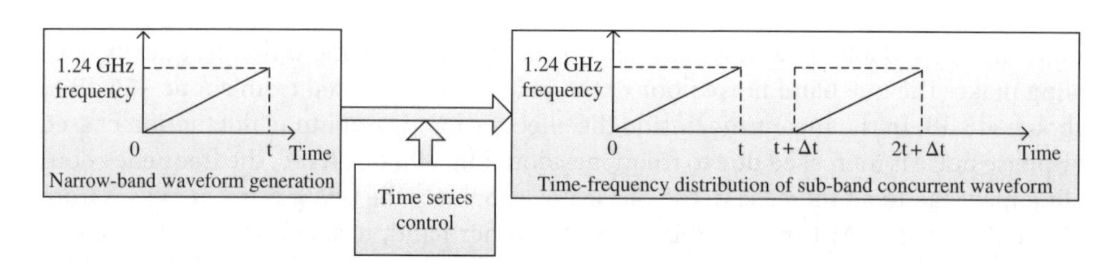


**Figure 5.28** Sub-band concurrent wideband waveform generation.

After being mixed, these signals are synthesized in analog domain to obtain a wideband signal [10]. The synthesis includes bandwidth and time-width. Block diagrams of generating baseband signal and IF signal are shown in Figures 5.26 and 5.27, respectively.

### 5.4.5 Sub-Band Concurrent Wideband Waveform Generation

The principle of sub-band concurrent wideband waveform generation is to divide a wideband signal into several narrow-band signals, generating several narrow-band signals sequentially, under the action of a timing sequence controlling circuit. Frequency splicing is used in receiving to realize super wideband equivalent processing. Therefore, the waveform generation corresponds to time splicing of multiple narrow-band signals, as shown in Figure 5.28.

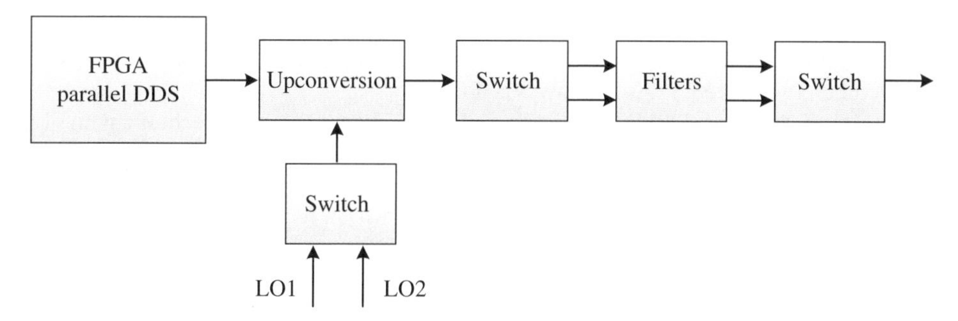


**Figure 5.29** Implementation block diagram of sub-band concurrent wideband waveform generation.

For example, for a 2.4 GHz waveform, two sub-bands are used for splicing; each sub-band has a bandwidth of 1.22 GHz. Two different local oscillation signals are selected during band shifting, so that the output frequency band extends to 2.4 GHz. The implementation block diagram is shown in Figure 5.29.

Sub-band concurrent wideband waveform generation reduces the design difficulty in digital waveform, receiver, and digital acquisition, at the cost of increasing the amount of hardware and increasing the complexity of time series and digital processing.

The reduction of bandwidth reduces the difficultly in designing digital waveform and reduces the order of frequency doubling, and it even eliminates the frequency doubling directly. If the subsequent processing is not considered, the waveform generated by sub-band concurrently is superior in quality to that generated by wideband waveform directly.

## References

1 Bahl, I. and Bhartia, P. (1988). *Microwave Solid State Circuit Design*. Wiley: Hoboken New Jersey.

2 Li, X. and Liu, G. (2000). UWB quadrature demodulation receiver performance analysis. *Journal of Systems Engineering and Electronics* 22 (5): 58–60.

3 Xie, Y. and Ji, L. (2014). An UWB digital radar receiver based on FPGA. *Modern Radar* 36 (1): 62–65.

4 J.A. Wepman, J.R. Hoffman (1996). RF and IF Digitization in Radio Receivers: Theory, Concepts, and Examples. U.S. Department of Commerce, National Telecommunications and Information Administration. *NTIA Report 96–328*.

5 Fang, L., Ji, Z., Ma, J., and Tan, J. (2010). A high resolution SAR receiver and signal generation base on stepped frequency modulation. *Radar Science and Technology* 8 (4): 335–338.

6 Li, P. (2005). An integrated design and implementation of microwave radar receiver circuits. *Microelectronics* 35 (4): 441–444.

7 Chen, X. (2005). Monolithic LNA design and its application in digital T/R module. *Microelectronics* 6: 326–328.

8 Fang, L. (2004). Study on sampling phase locked frequency synthesizer. *Modern Radar* 26 (8): 49.

9 Postema, G.B. (1987). Generation and performance analysis of wideband radar waveforms. *Proceedings of the IEEE International Radar Conference*, London, UK (October 19–21, 1987). Piscataway, NJ: IEEE.

10 Li, H. and Xiang, R. (2011). Application of direct digital synthesis technology in radar receiver. *Radar Science and Technology* 9 (6): 579–584.

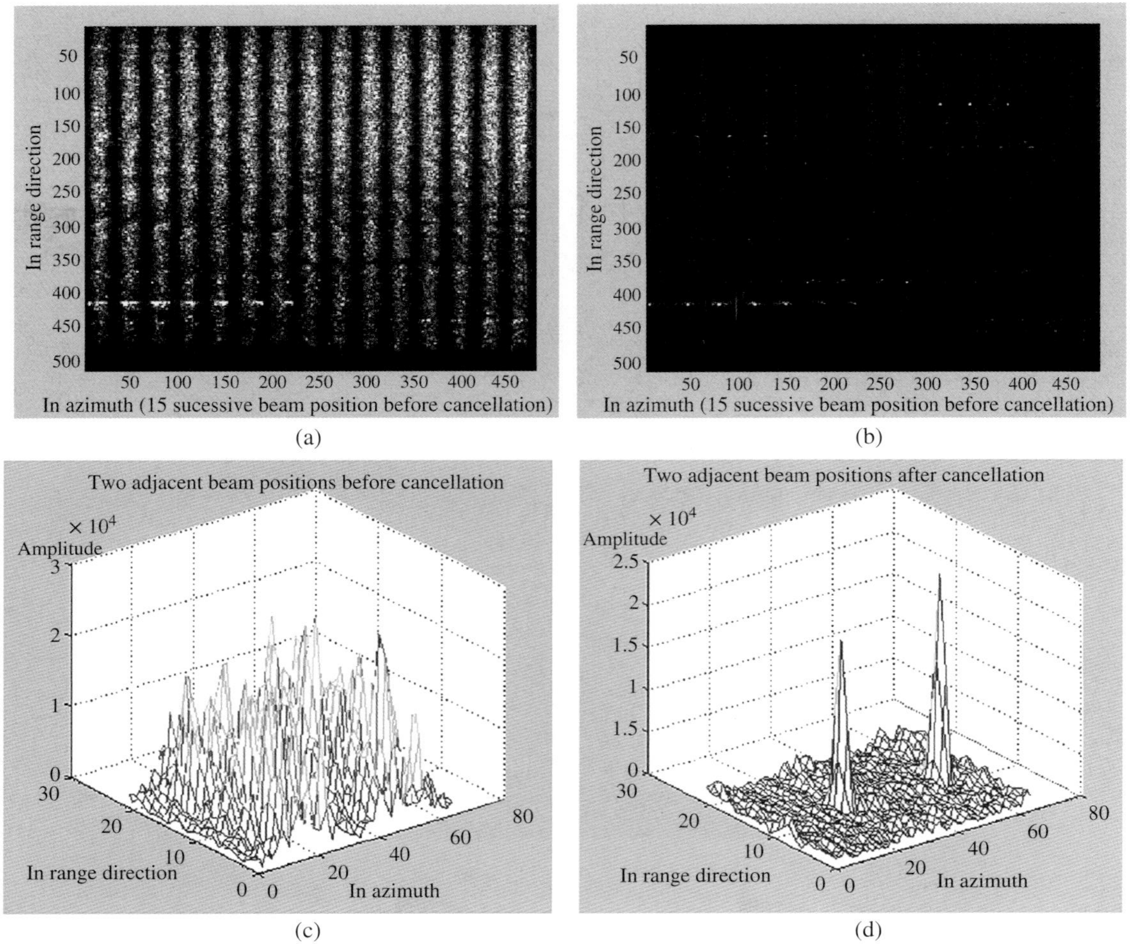

**Figure 6.16** Findings of airborne wide-area MTI. (a) Fifteen successive bursts before cancellation; (b) 15 successive bursts after cancellation; (c) two adjacent bursts before cancellation; (d) two adjacent bursts after cancellation.

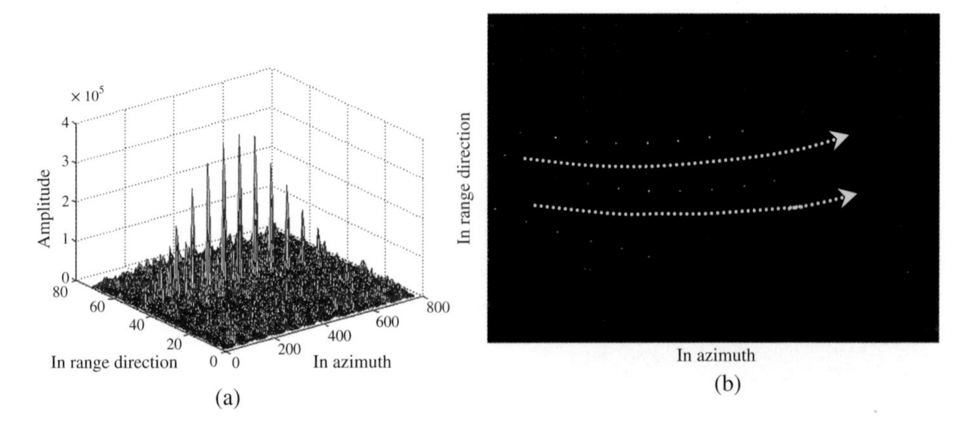


**Figure 6.19** Results of GMTI. (a) Single-frame moving target track; (b) track of the moving target.

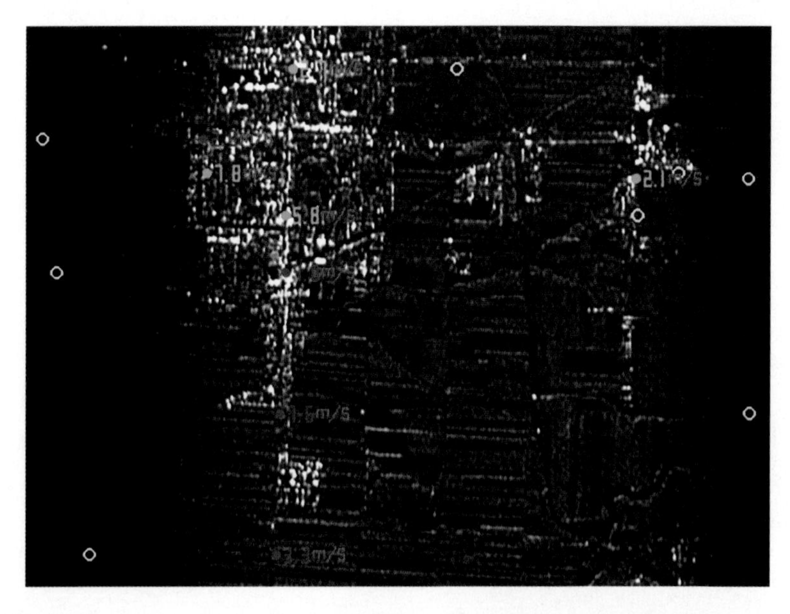

**Figure 6.20** Moving targets obtained by simultaneous SAR/GMTI overlaid on the SAR image.

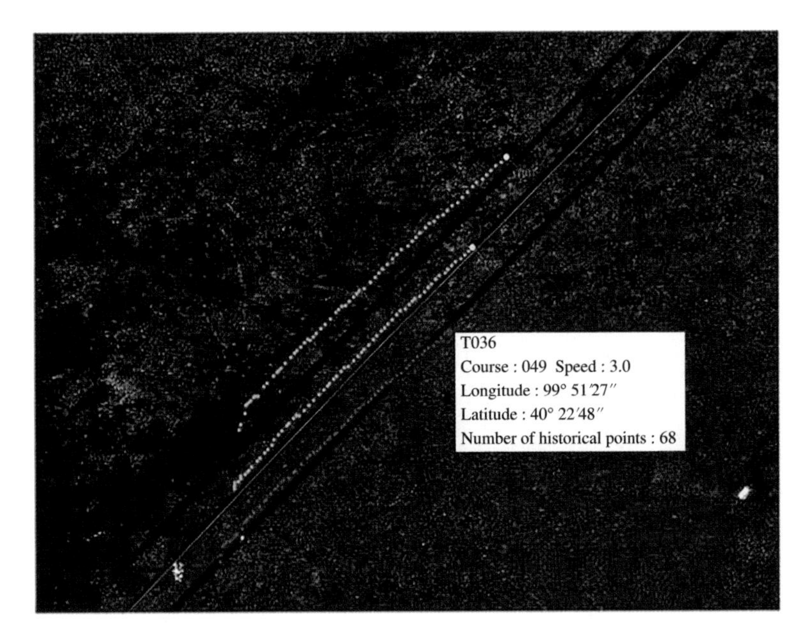


**Figure 6.21** Moving target track in wide-area GMTI.

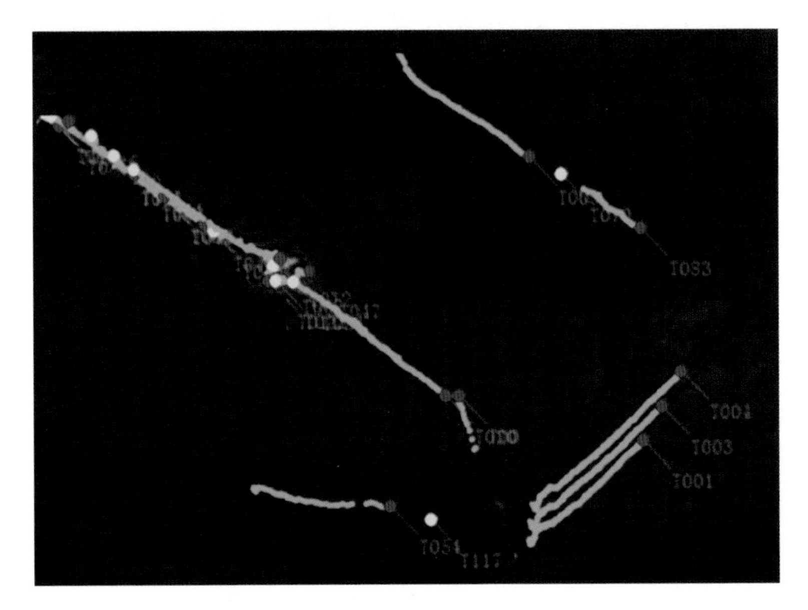


**Figure 6.22** Moving target track for GMTI.

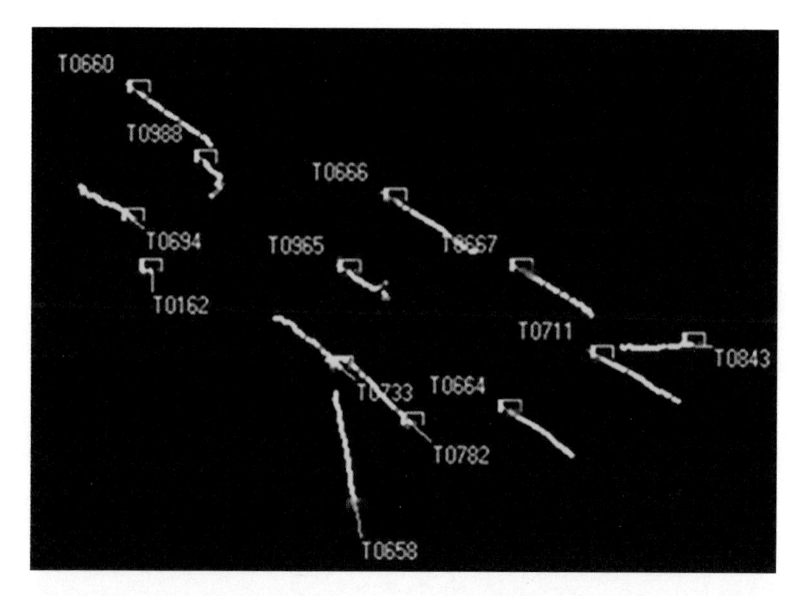


**Figure 6.24** The tracking finding of marine targets by an airborne SAR/MTI system.

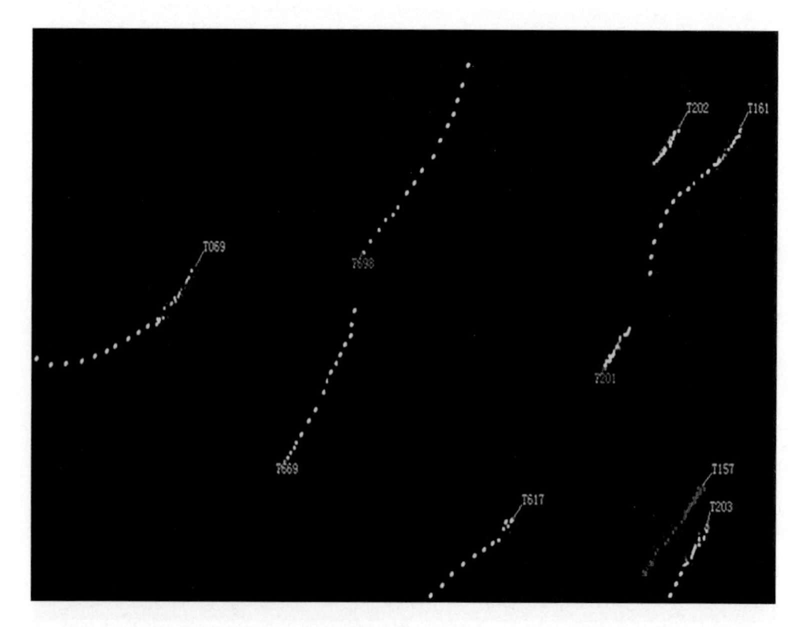


**Figure 6.25** Marine tracking finding in MMTI model.

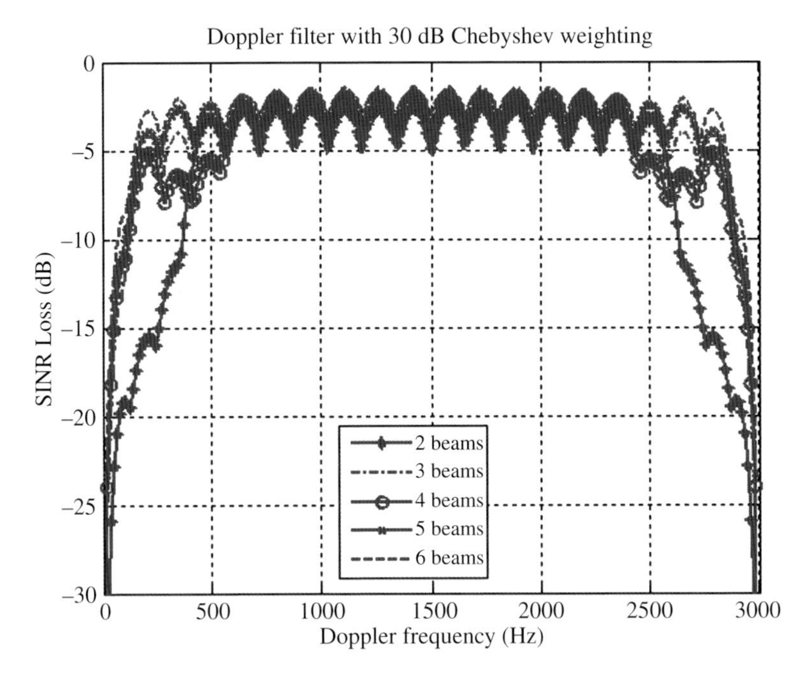


**Figure 6.29** Comparison of beam-domain STAP algorithms before Doppler filtering.

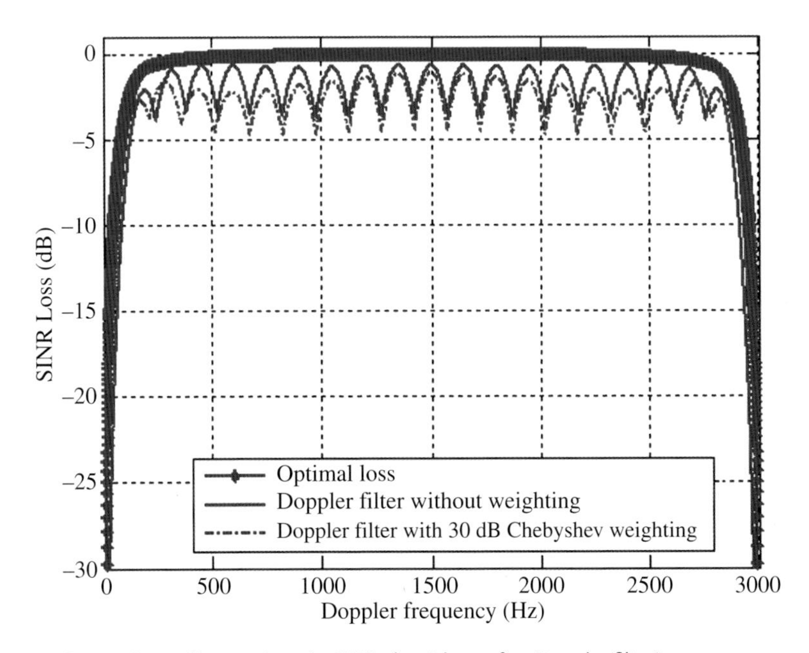


**Figure 6.30** Comparison of beam-domain STAP algorithms after Doppler filtering.

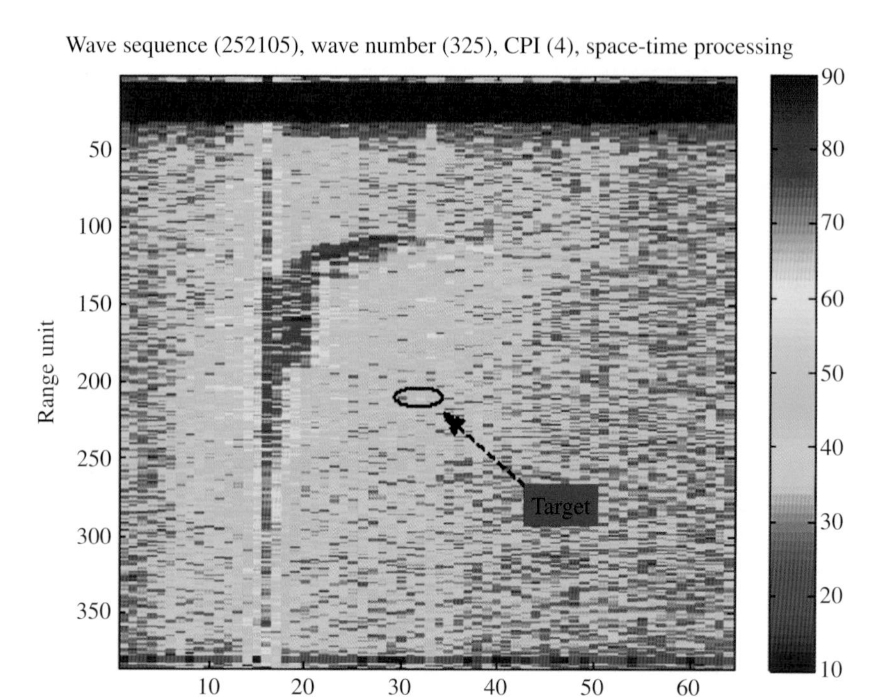


**Figure 6.31** Range-Doppler spectrum after PD processing.

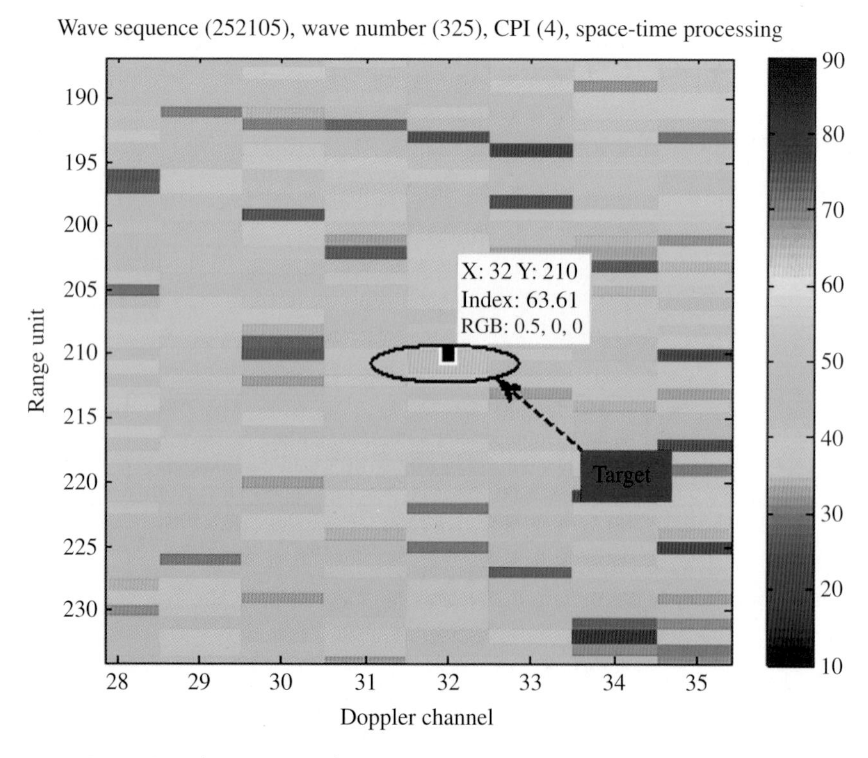


**Figure 6.32** Range-Doppler spectrum after STAP processing.

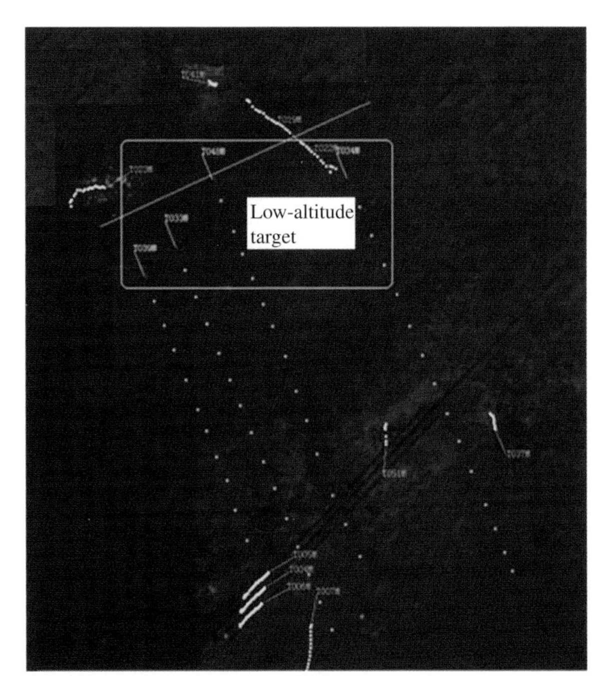


**Figure 6.33** Tracking map for low-altitude small target.

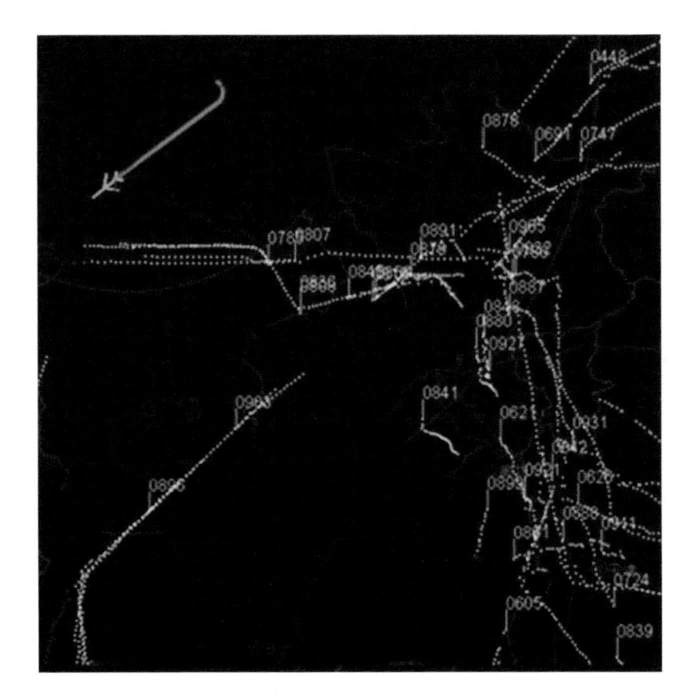


**Figure 6.34** Tracking map for AMTI.

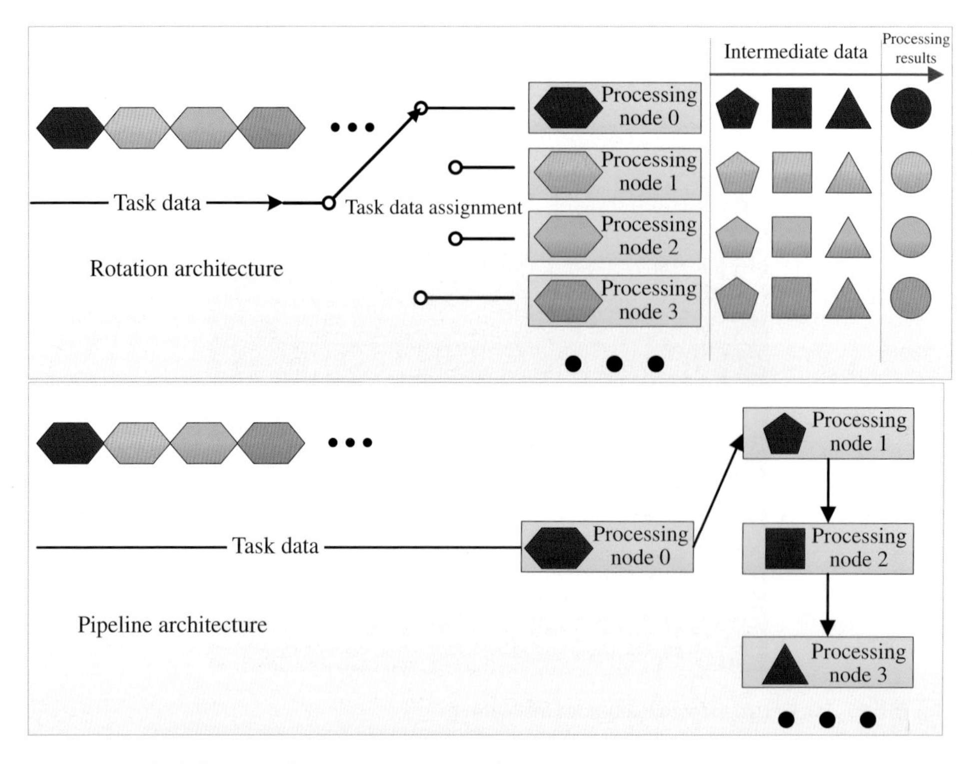


**Figure 6.37** Block diagram of system processing architecture.

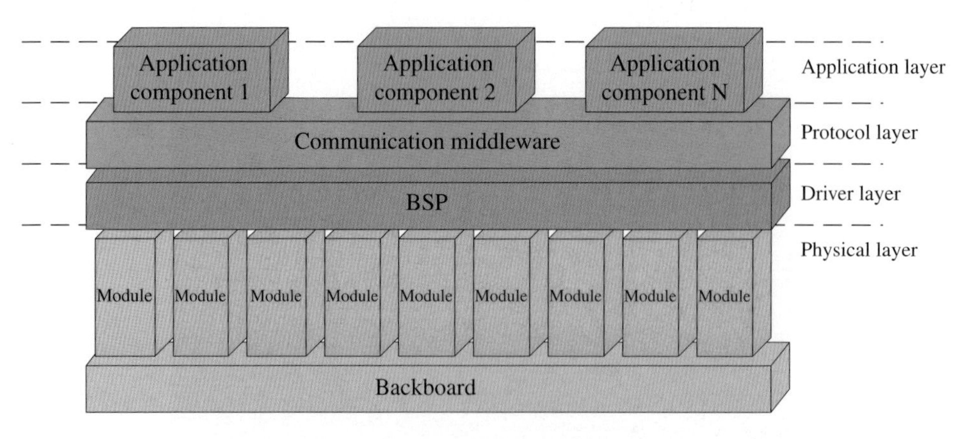


**Figure 6.38** Development architecture diagram of an integrated processing unit. BSP: board support package.

Figure 6.39 A typical SAR processing modue.

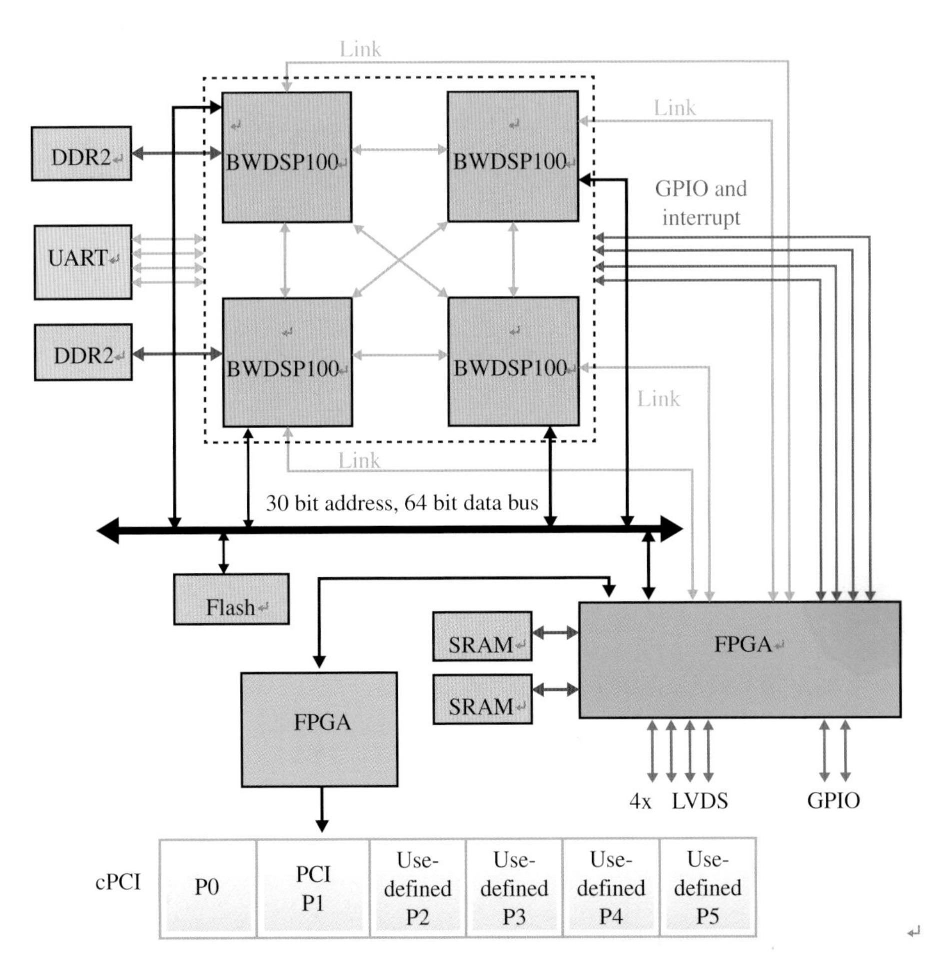


Figure 6.40 Internal structure of the processing module. GPIO: general purpose input/output; LVDS: low-voltage differential signaling; SRAM: static RAM; UART: universal asynchronous receiver/transmitter.

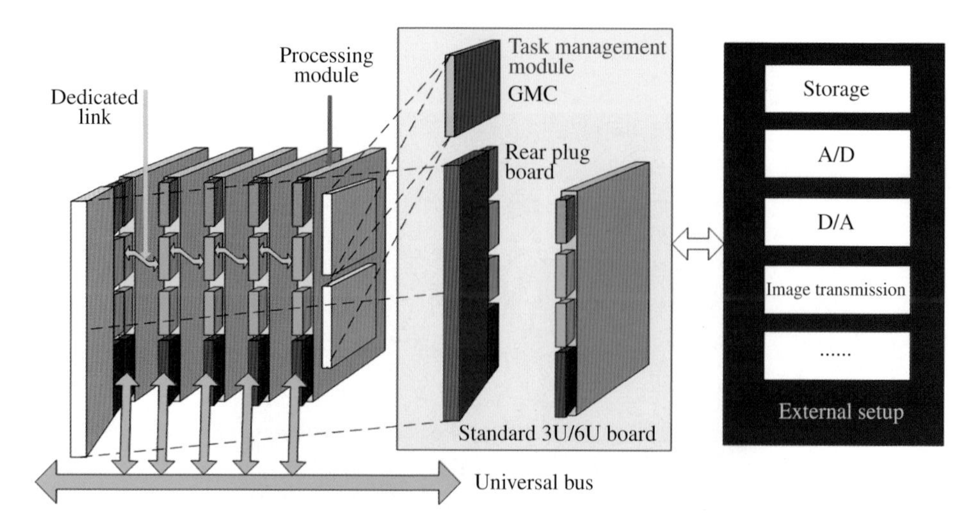


**Figure 6.42** Composition of the signal processor.

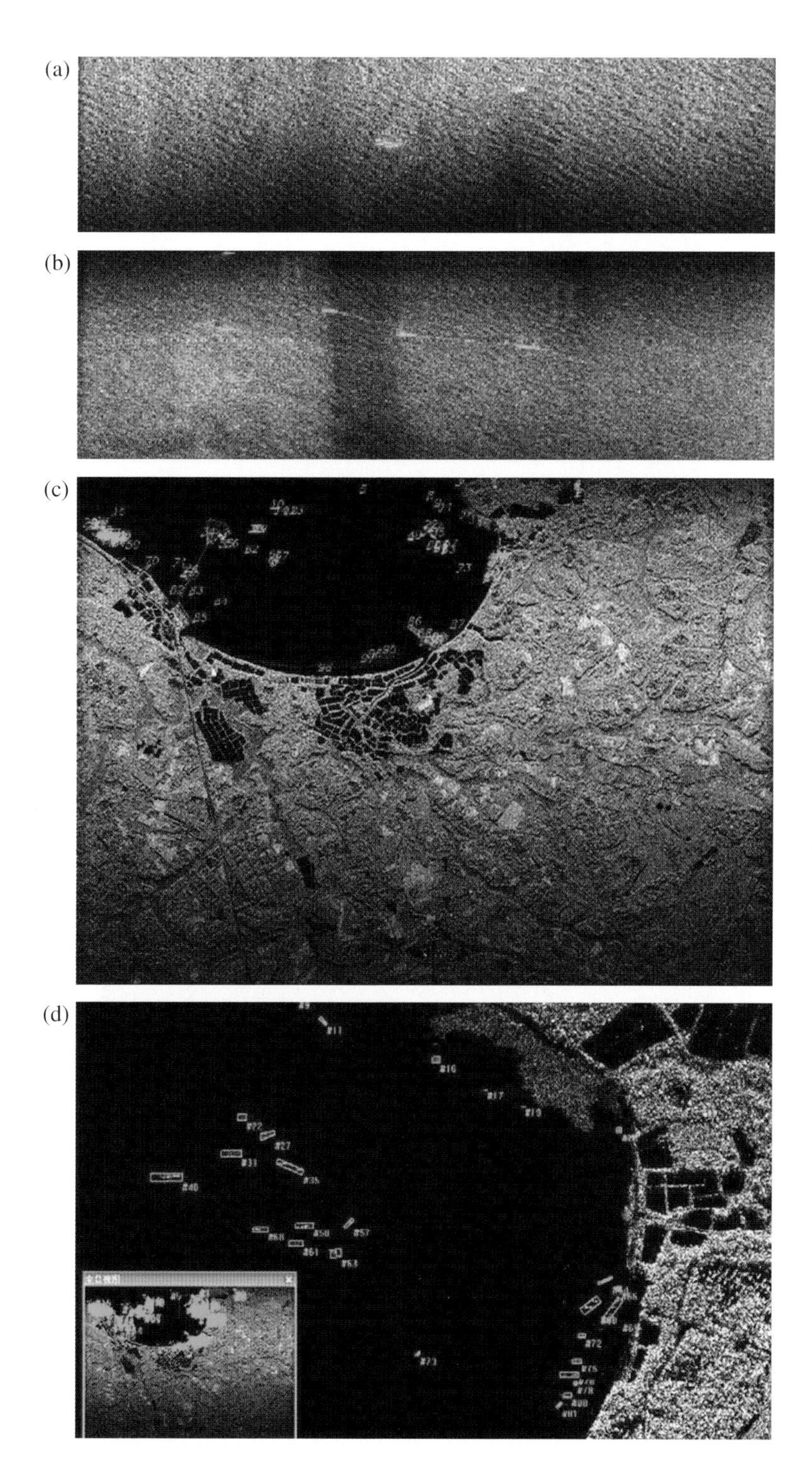

**Figure 7.8** Ship detection finding of airborne SAR image. (a) Image 1, resolution: 2 m, column × row: 4096 × 864; (b) image 2, resolution: 2 m, column × row: 5462 × 3936; (c) image 3, resolution: 2 m, column × row: 5419 × 1512; (d) local zoom-in of image 3.

**Figure 7.9** MSTAR detection results.

**Figure 7.10** Vehicle detection results of airborne SAR image.

**Figure 7.11** Spaceborne SAR bridge image (resolution about 10 m).

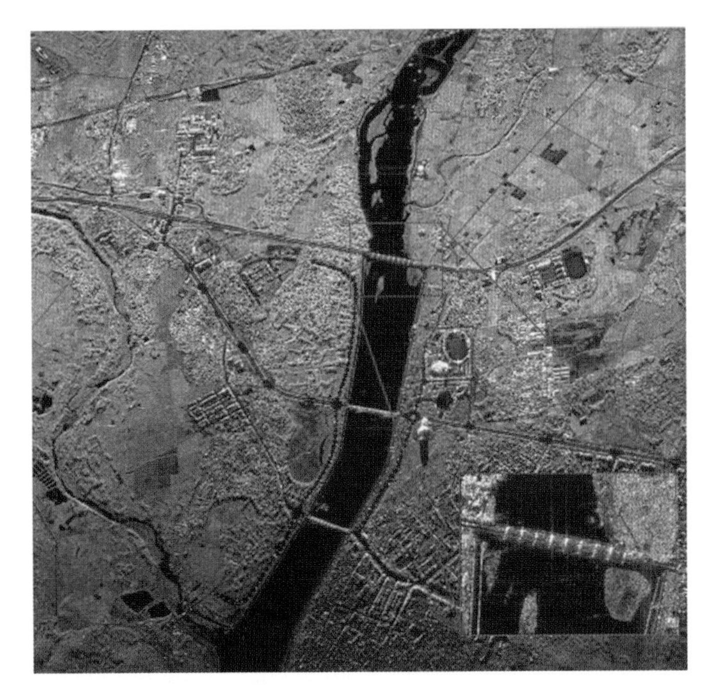

**Figure 7.12** 1 m resolution airborne SAR image of bridges.

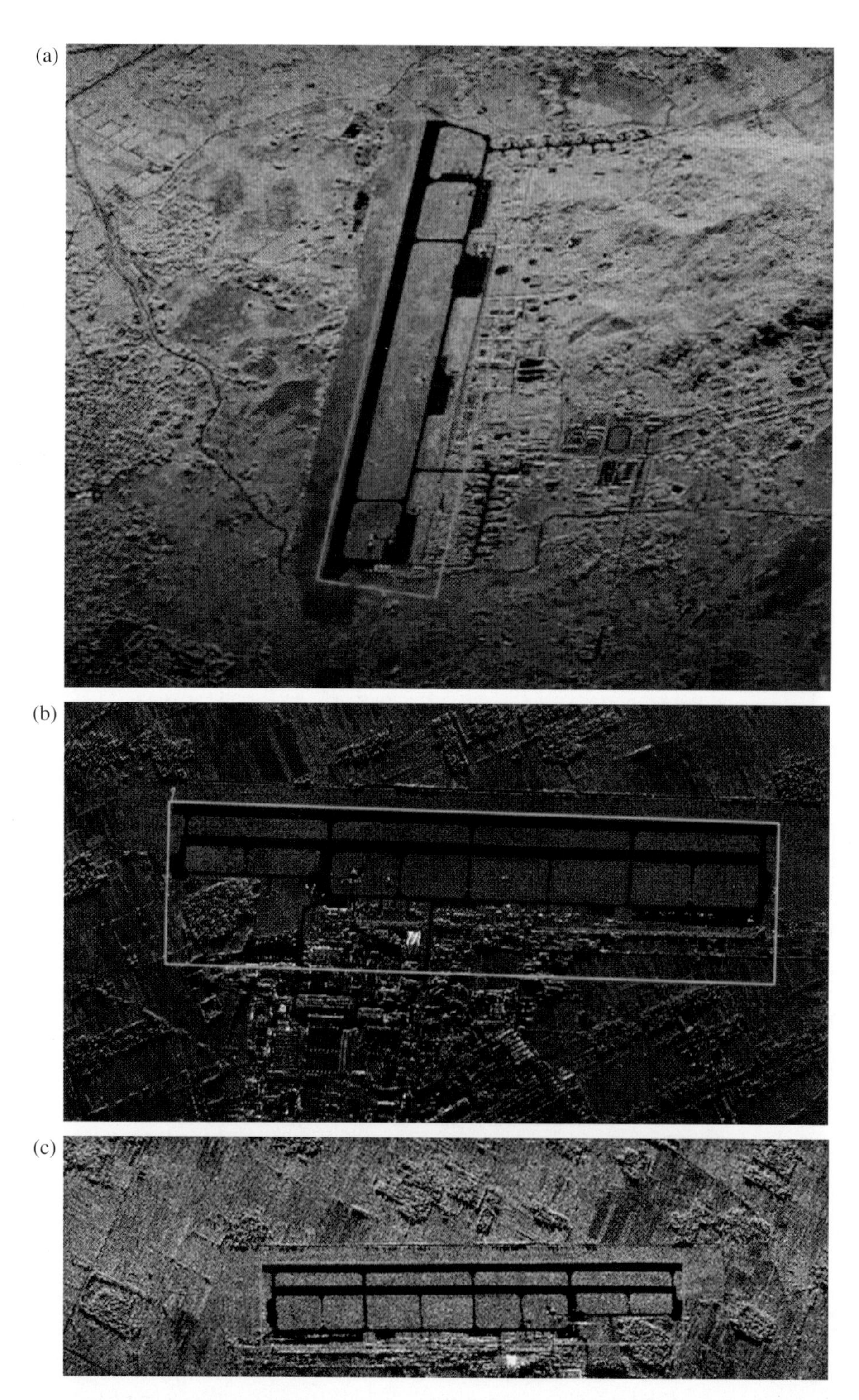

**Figure 7.14** Airport detection finding from airborne SAR images. (a) 2 m resolution; (b) 1 m resolution; (c) 0.6 m resolution; (d) 1 m resolution.

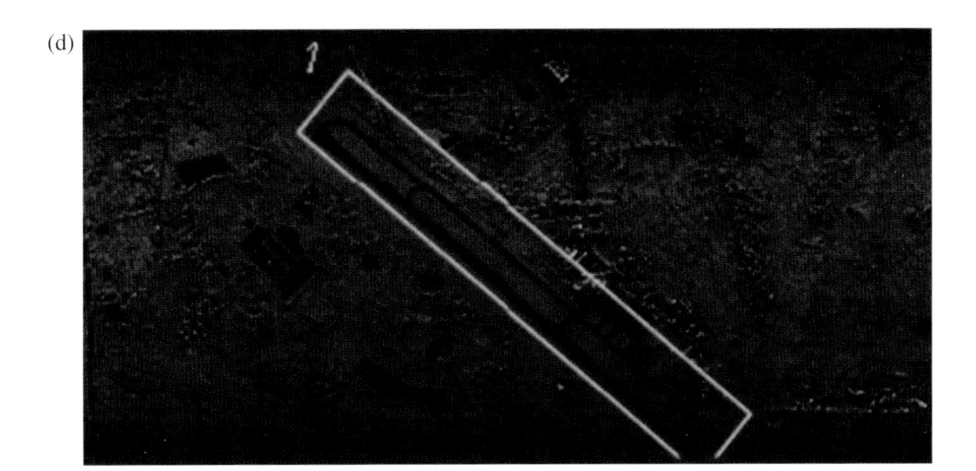


Figure 7.14 (*Continued*)

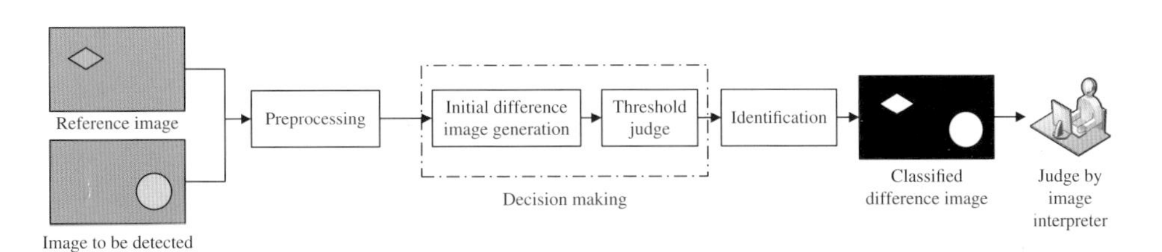


Figure 7.15 Change detection procedure.

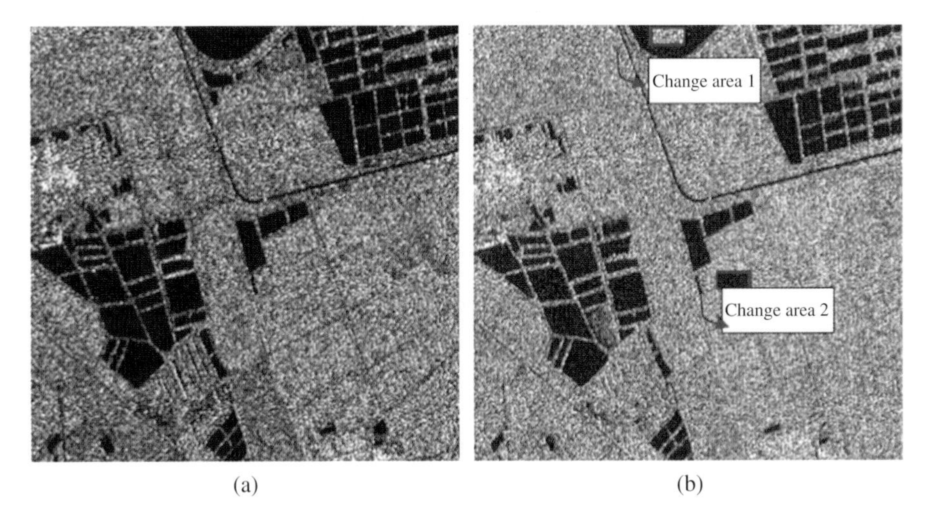


Figure 7.16 Bitemporal SAR data of one suburb (512 × 512). (a) June 25, 2005; (b) September 29, 2005.

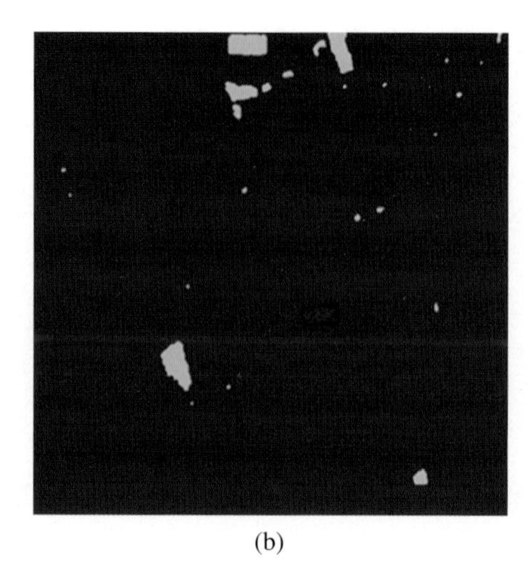

(b)

**Figure 7.17** Difference image and change detection results. (b) detection results of MRF segment change.

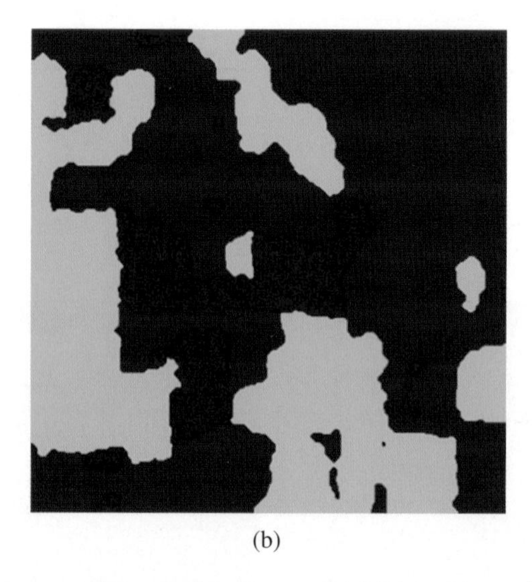

(b)

**Figure 7.19** Difference image and change detection results. (b) detection results of MRF segmentation change.

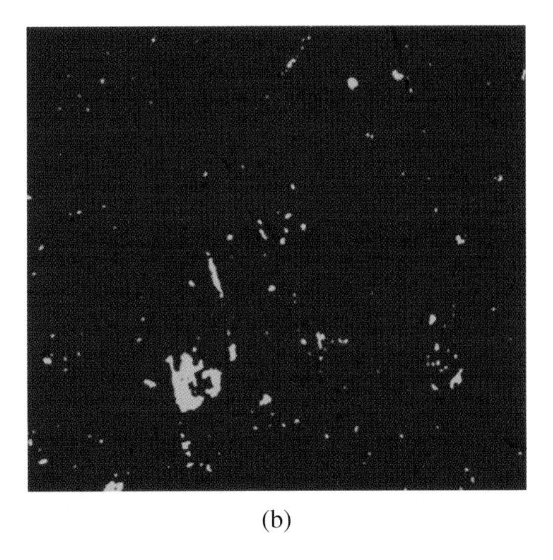

(b)

**Figure 7.22** Difference image and change detection results. (b) detection results of MRF segment change.

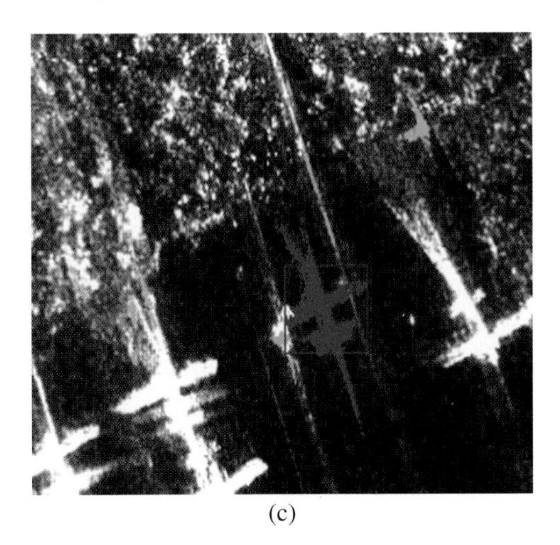

(c)

**Figure 7.25** Part 1: One plane disappears on the tarmac (see box). (c) change detection results.

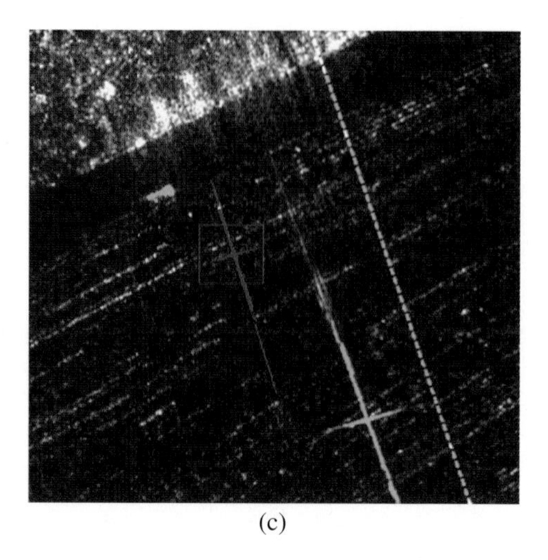
(c)

**Figure 7.26** Part 2: The upper corner of the airport lawn moved, and a new vehicle appears on the runway. (c) change detection results.

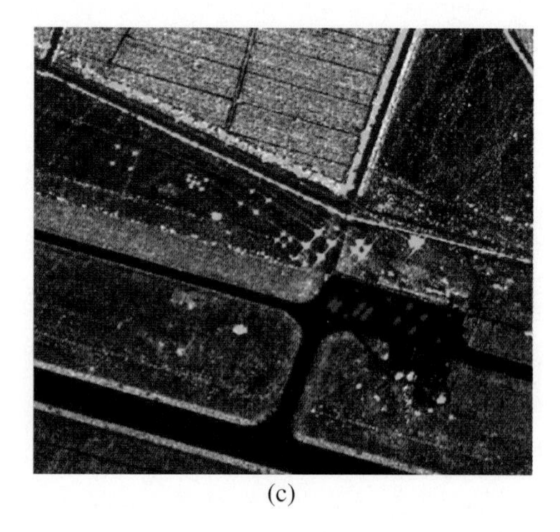
(c)

**Figure 7.27** The angle reflector change detection of an airport. (c) change detection results.

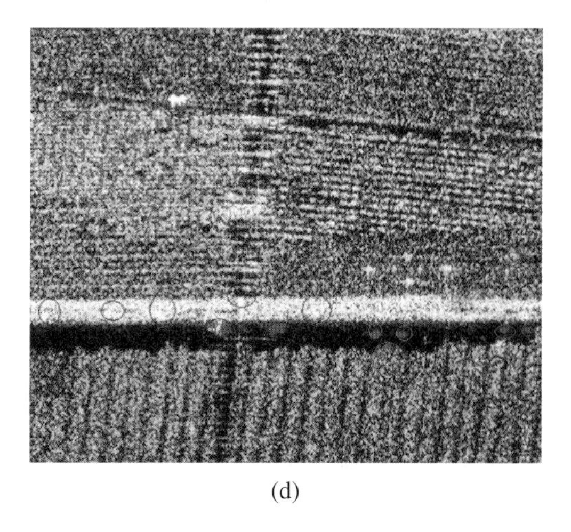

(d)

**Figure 7.32** Experimental images used for hidden target detection. (d) Ku, X, L triband fusion multiband data; (e) P band.

(d)

**Figure 7.36** Multiband characteristic analysis and fusion effect. (d) fused image.

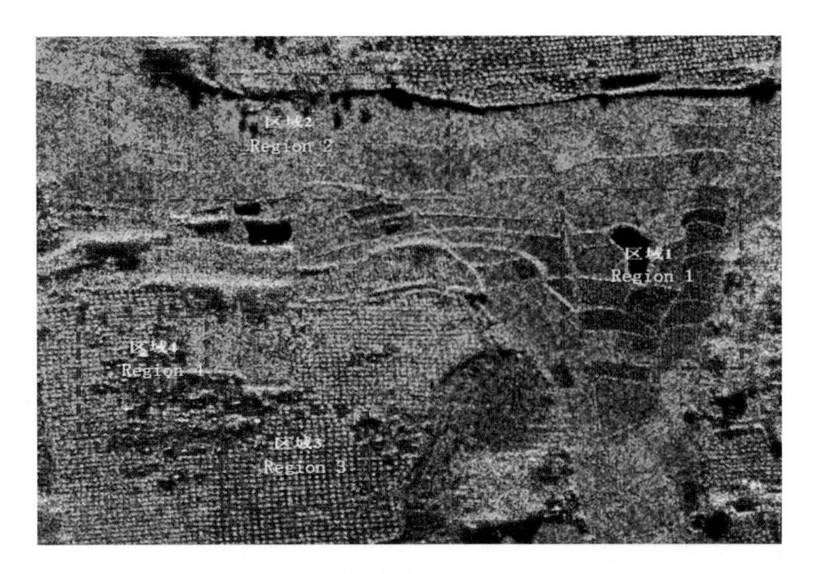


**Figure 7.40** Multiband pseudocolor synthesis of vegetation area.

**Figure 7.42** Postfusion pseudocolor SAR image.

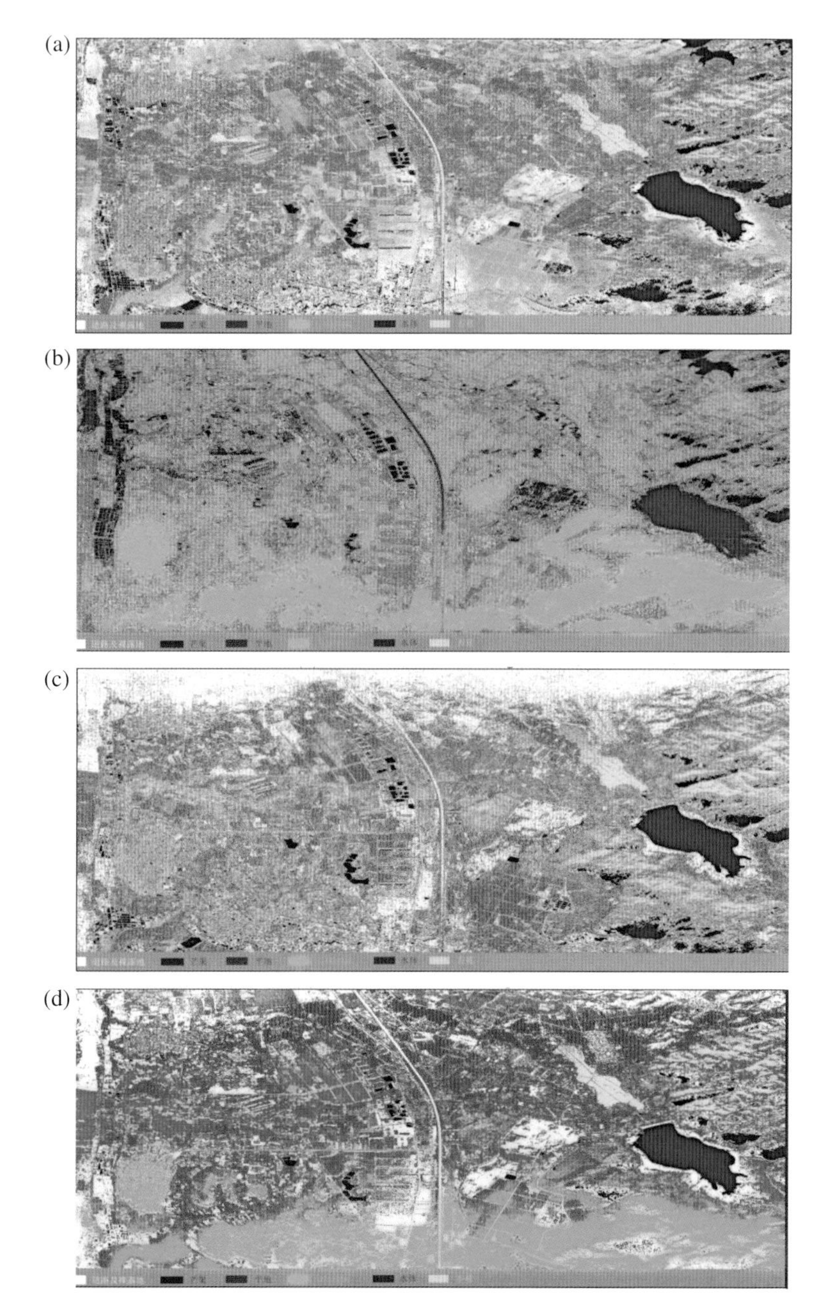


**Figure 7.43** Classification results of single-band SAR images and fused SAR image. (a) Classification image at Ku band; (b) classification image at L band; (c) classification image at X band; (d) classification image after multiband fusion.

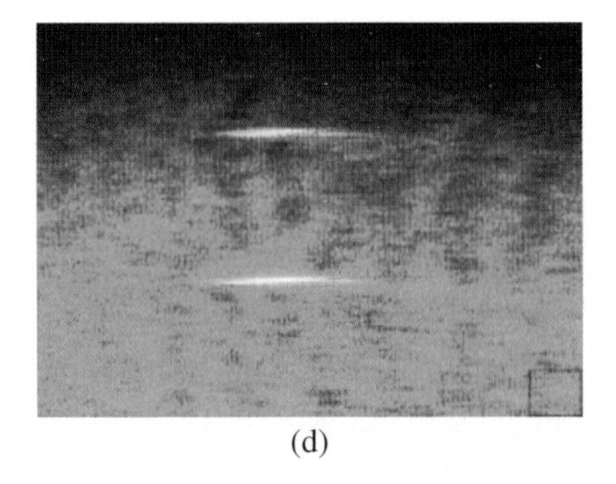

(d)

**Figure 7.44** Ship target comparison 1 in multiband SAR image. (d) fusion.

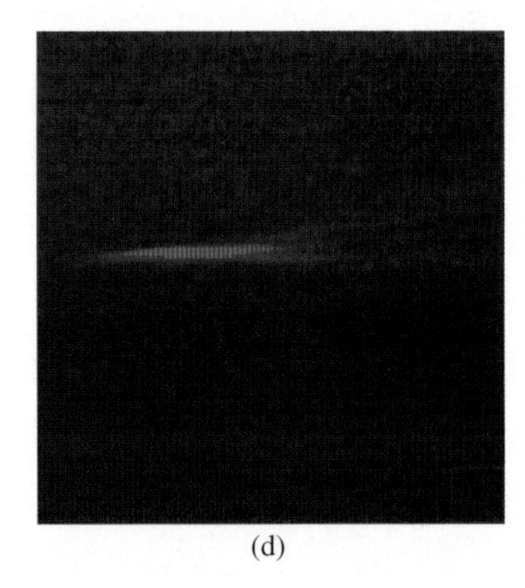

(d)

**Figure 7.45** Ship target comparison 2 in multiband SAR images. (d) fusion.

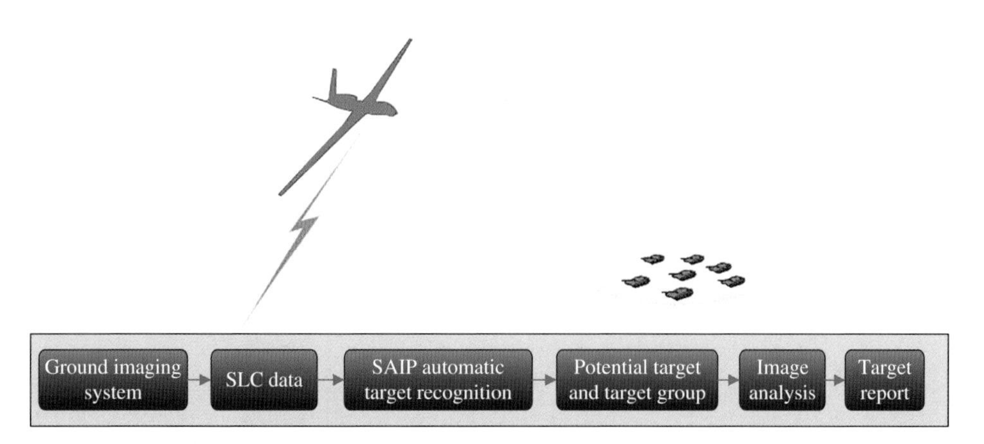


**Figure 7.51** Processing flow of SAIP SAR image information. SLC: single-look complex.

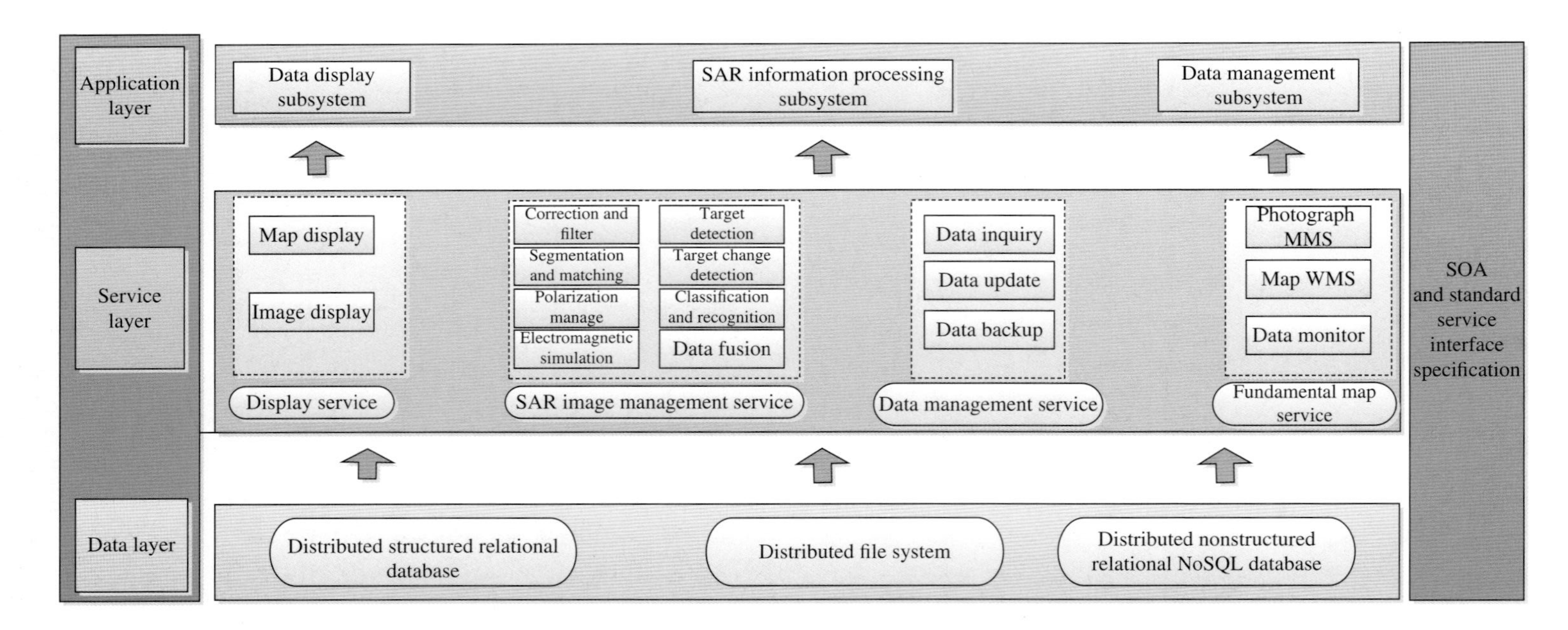


**Figure 7.52** SAR image information application system logic diagram. SOA: service-oriented architecture.

# 6

# Signal Processing

## 6.1 Overview

High-range resolution synthetic aperture radar (SAR) images can be achieved by transmitting wideband instantaneous signals. High-azimuth resolution can be achieved by transmitting signals and receiving echoes of the target at multiple positions by using a moving platform and synthesizing a large aperture via signal processing. The azimuth signal in the SAR is generally a linear chirp signal whose Doppler parameters may be estimated by the motion measurement system mounted on the platform and data of the echoes. In the meantime, range-Doppler (RD) coupling caused by the range migration needs to be solved. Properly choosing the Doppler parameter estimation method, imaging algorithm, and motion compensation technique is fundamental to obtain high-quality SAR images.

## 6.2 SAR Signal Processing Method

For chirp signals, the essence of digital processing is the process of pulse compression. According to different implementation methods, pulse compression can be divided into three types: time domain correlation, frequency domain matched filtering, and spectrum analysis.

### 6.2.1 Time Domain Correlation

Let the transmission signal $s(t)$ have the following form:

$$s(t) = \text{rect}\left(\frac{t}{T}\right)\exp\{j\pi\gamma t^2\} \tag{6.1}$$

where $t$ is time variable, $T$ is signal duration, and $\gamma$ is chirp rate.

Then the target signal received after the delay $t_0$ can be expressed as

$$s_r(t) = \text{rect}\left(\frac{t-t_0}{T}\right)\exp\{j\pi\gamma(t-t_0)^2\} \tag{6.2}$$

The time domain matched filter at $t_0 = 0$ is

$$h(t) = \text{rect}\left(\frac{t}{T}\right)\exp\{-j\pi\gamma(-t)^2\} = \text{rect}\left(\frac{t}{T}\right)\exp\{-j\pi\gamma t^2\} \tag{6.3}$$

*Design Technology of Synthetic Aperture Radar,* First Edition. Jiaguo Lu.

The output of the matched filter is

$$s_{out}(t) = s_r(t) \otimes h(t) = \int_{-\infty}^{+\infty} s_r(u)h(t-u)du$$
$$= T \sin c\{\gamma T(t - t_0)\} \quad (6.4)$$

### 6.2.2 Frequency Domain Matched Filtering

Pulse compression may also be achieved by designing high-precision matched filtering directly in the frequency domain.

Find the spectrum of $s_r(t)$ by using the stationary phase principle, approximated as

$$S_r(f) = \text{rect}\left(\frac{f}{|\gamma|T}\right) \exp\left\{-j\pi\frac{f^2}{\gamma}\right\} \exp\{-j2\pi f t_0\} \quad (6.5)$$

where || is the absolute operator, $\gamma$ is the chirp rate, $t$ is the time variable, $t_0$ is the delay, and $f$ denotes operating frequency.

The frequency domain matched filter is

$$H(f) = \text{rect}\left(\frac{f}{|\gamma|T}\right) \exp\left\{j\pi\frac{f^2}{\gamma}\right\} \quad (6.6)$$

The matched filtered signal spectrum is:

$$S_{out}(f) = S_r(f)H(f) = \text{rect}\left(\frac{f}{|\gamma|T}\right) \exp\{-j2\pi f t_0\} \quad (6.7)$$

Inverse fast Fourier transform (IFFT) is performed on the $S_{out}(f)$ to obtain a compressed signal:

$$S_{out}(t) = |\gamma| T \sin c[\gamma T(t - t_0)] \quad (6.8)$$

Except for the coefficient $|\gamma|$, the result is the same as the pulse compression in time domain.

### 6.2.3 Spectrum Analysis Method

The spectrum analysis method is to conjugate and multiply the echo chirp signal and the chirp reference signal with the same slope to obtain an equal frequency signal. Then, the equal-frequency signal is subjected to a fast Fourier transform (FFT) to perform pulse compression on the chirp signal. The spectrum of the equal frequency signal corresponds to different positions of the target.

## 6.3 Operating Mode and Signal Property

According to the antenna scanning mode, the imaging mode of SAR can generally be classified into azimuth antenna scanning, range antenna scanning, and two-dimensional (2D) antenna scanning in both range and azimuth directions. Azimuth antenna scanning can be divided into strip-map mode, spotlight mode, and sliding spotlight mode, as per different focus points of the azimuth antenna beam. ScanSAR is commonly seen in range antenna scanning mode whereas terrain observation by progressive scans (TOPS) is commonly seen in 2D antenna scanning mode. Every operating mode from the signal processing point of view is briefly introduced in the following sections.

### 6.3.1 Azimuth Antenna Scanning

In strip-map SAR, the azimuth central beams are parallel in all azimuth positions during data acquisition, that is, the azimuth central beams focus on infinity. In spotlight SAR, the azimuth central beams focus on the center of the imaging scene during data acquisition. The sliding spotlight SAR is between them, that is to say, during data acquisition, the azimuth central beams focus on a certain position that is farther than the center of the imaging scene. The data acquisition relationship of these three modes is shown in Figure 6.1, where $W_a$ is the azimuth imaging range, $L_s$ is the coverage width of the azimuth antenna beam, and $L_a$ is synthetic aperture length. Figure 6.2 shows

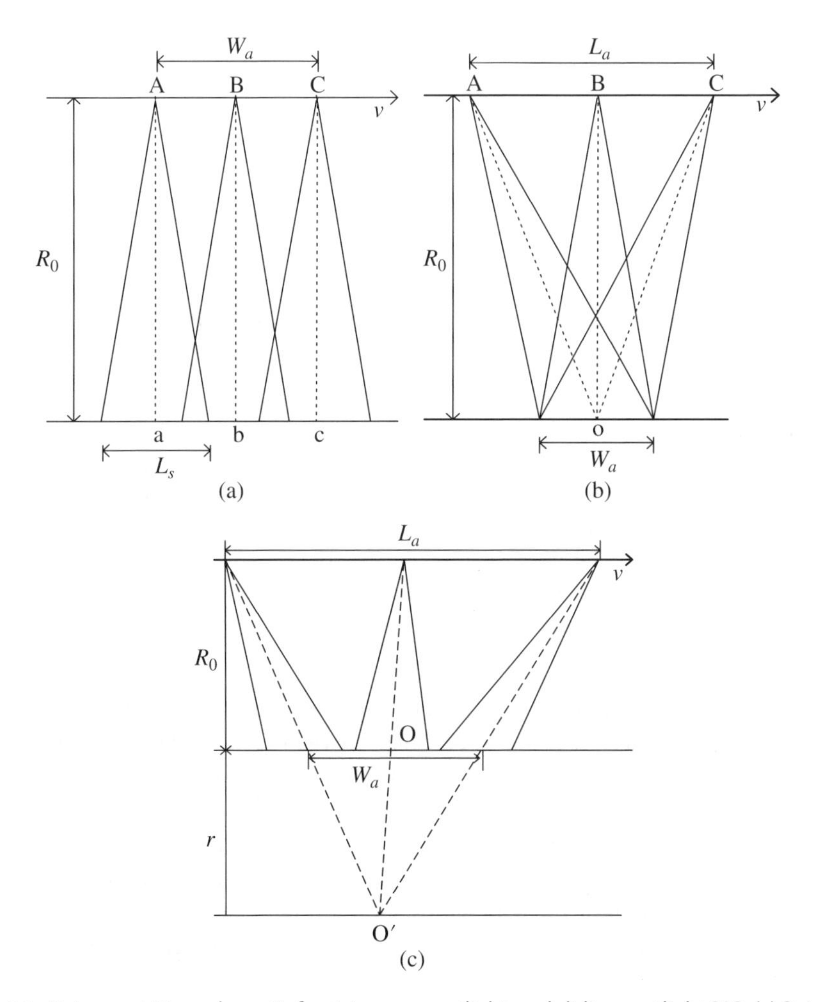


**Figure 6.1** Data acquisition schematic for strip-map, spotlight, and sliding spotlight SAR. (a) Strip-map SAR; (b) spotlight SAR; (c) sliding spotlight SAR.

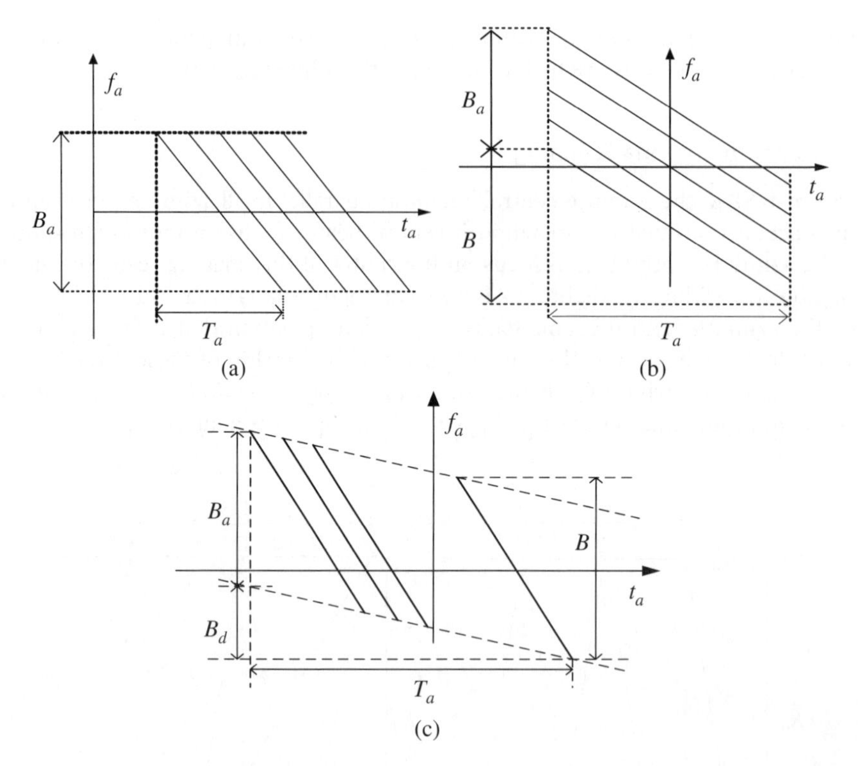


**Figure 6.2** The Doppler history of strip-map, spotlight, and sliding spotlight SAR modes. (a) Strip-map SAR; (b) spotlight SAR; (c) sliding spotlight SAR.

the Doppler history of multiple point targets at the same range but different azimuth in strip-map, spotlight, and sliding spotlight modes. In Figure 6.2, $B_a$ is the instantaneous Doppler bandwidth, $B_d$ is the variation range of Doppler center frequency, $B$ is the Doppler bandwidth in a point target, and $T_a$ is the synthetic aperture time. Generally, for Doppler bandwidth in strip-map SAR, $B_{\text{strip}}$, that of spotlight SAR, $B_{\text{spot}}$, and that of sliding spotlight SAR, $B_{\text{slide}}$, $B_{\text{strip}} < B_{\text{slide}} < B_{\text{spot}}$.

In sliding spotlight SAR, the sliding spotlight factor is defined as

$$A = \frac{r}{r + R_0} = \frac{v_g}{v} \tag{6.9}$$

where $v$ is the platform velocity, $v_g$ is the moving speed of antenna beam on the ground, and $r$, $R_0$, and $r + R_0$ are the shortest distance between scene and focal point, between SAR and scene, and between SAR and focal point, respectively. The sliding spotlight factor determines mode type:

(1) When $A = 0$, it is the spotlight mode. Here $r = 0$, the antenna central beam points to the center of the imaging scene, and the moving speed of the antenna beam on the ground is 0. In this case, the bandwidth of the azimuth signal is the maximum. And the best azimuth resolution is achieved. However, the coverage of azimuth imaging is strictly limited.
(2) When $0 < A < 1$, it is the sliding spotlight mode. The antenna central beam points to a position farther than the center of the imaging scene. The moving speed of the

antenna beam on the ground $v_g$ is smaller than platform velocity $v$. The moving direction of the antenna beam is the same as that of the platform. The azimuth coverage is larger than the spotlight mode and smaller than the strip-map mode. The azimuth resolution is worse than the spotlight mode but better than the strip-map mode.

(3) When $A = 1$, it is the strip-map mode. The antenna central beam focuses on infinity. The moving speed of the antenna beam on the ground is equal to platform velocity. The azimuth resolution is half of the antenna length in azimuth.

Thus, the strip-map mode, the spotlight mode, and the sliding spotlight mode are all unified by the sliding spotlight factor $A$.

### 6.3.2 Range Antenna Scanning

ScanSAR obtains wide swath through range antenna scanning at the cost of azimuth resolution. Figure 6.3 shows operation of a three-beam ScanSAR.

As the targets in the same substrip but in different azimuth are illuminated by different parts of the antenna pattern, ScanSAR shows an obvious scalloping effect in azimuth. In addition, the severe inconsistency of signal-to-noise ratio (SNR) and ambiguity in azimuth restricts the application of ScanSAR.

### 6.3.3 2D Antenna Scanning

TOPS is based on conventional ScanSAR, where the azimuth beam actively scans from back to front, which means the range beam switches among different substrips. In the meantime, it scans in azimuth, as shown in Figure 6.4a. Figure 6.4b shows the azimuth antenna scanning of TOPS. Between point A and point C, the central beam in azimuth antenna always points to O′. Wide swath is achieved by scanning the beam in azimuth on

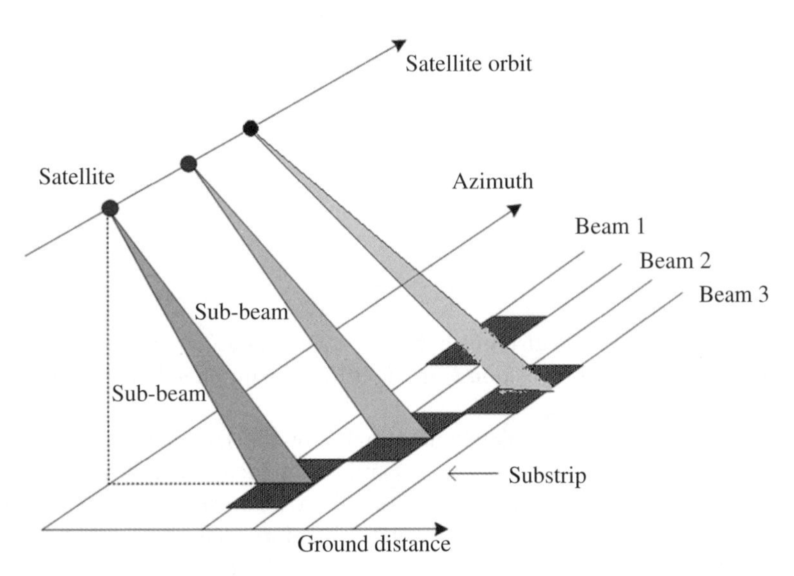


**Figure 6.3** Operating principle of ScanSAR.

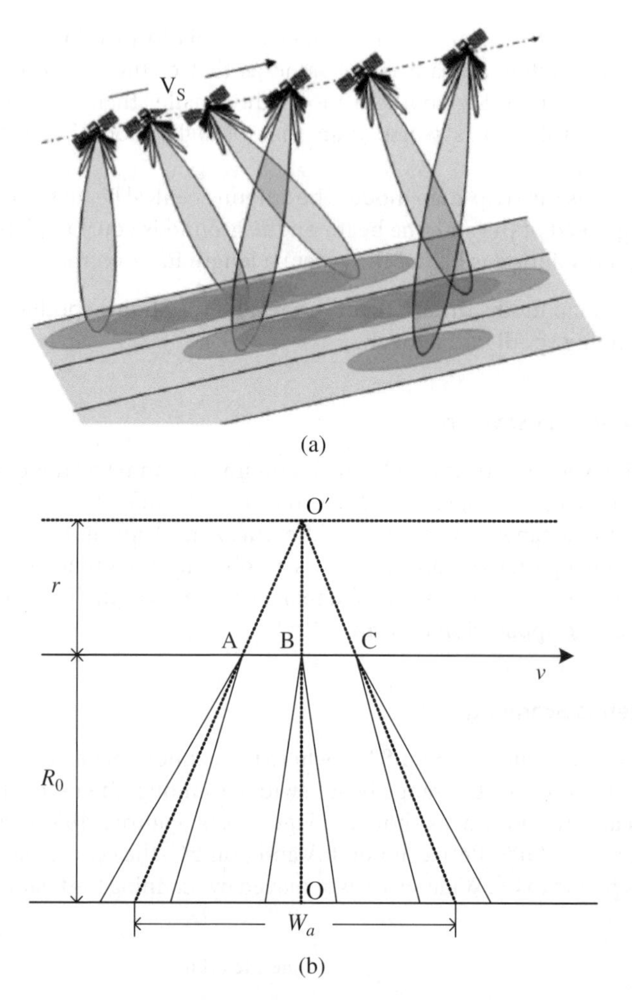


**Figure 6.4** Geometrical relationship of data acquisition in TOPS. (a) Schematic of TOPS 2D antenna scanning; (b) schematic of TOPS azimuth antenna scanning.

the ground. If the previously mentioned sliding spotlight factor is used to define TOPS, we have

$$A = \frac{r + R_0}{r} = \frac{v_g}{v} \tag{6.10}$$

where $r + R_0$ is the shortest distance between scene and focus point, $r$ is the shortest distance between SAR and focus point, and $v_g$ is the moving speed of the antenna beam on the ground. Here, $A$ is larger than 1.

In TOPS, every target is completely illuminated by the same antenna pattern. Thus the scalloping effect presented in ScanSAR, and severe inconsistency of ambiguity and SNR in azimuth can be avoided. However, the active beam scanning in azimuth compresses synthetic aperture time of the target. Therefore, spaceborne TOPS also sacrifices azimuth resolution to achieve wide swath. TOPS requires the antenna to have high-speed scanning both in range and azimuth directions. Generally, 2D active phased array is used in TOPS.

In ScanSAR and TOPS, individual substrips are generated by range direction beam scanning. These substrips are spliced together to produce a wide swath in the range direction. It is relatively easy to handle the range-scanning mode. However, for azimuth scanning, the imaging modes differ greatly due to different focus positions in the antenna beam centers. Nevertheless, different operating modes can be unified by the sliding spotlight factor defined in Eq. (6.11):

$$A = \frac{v_g}{v} \tag{6.11}$$

It is spotlight mode when $A = 0$, sliding spotlight mode when $0 < A < 1$, strip-map mode when $A = 1$, and TOPS azimuth scanning mode when $A > 1$. For the above operation modes, the resolution decreases with the increase of $A$. Thus spotlight SAR has the highest resolution and TOPS has the lowest resolution.

## 6.4 SAR Imaging

### 6.4.1 SAR Echo

Radar transmits pulse signals and then receives echoes reflected by the target. It is assumed that the radar operation satisfies "stop-go-stop" mode, namely, both the antenna and the target are relatively static during transmitting and receiving. For a SAR system, this assumption is valid, because the moving speed of an antenna platform is much slower than microwave transmission speed. It is assumed that the medium in which the radar wave transports is homogeneous, isotropic, and nondispersive, which means the space domain velocity is constant. Generally, SAR transmits a chirp signal. Assume complex expression of the signal is

$$p(\hat{t}) = w_r\left(\frac{\hat{t}}{T_P}\right)\exp(j2\pi f_c\hat{t} + j\pi\gamma\hat{t}^2) \tag{6.12}$$

where $w_r(\cdot)$ is envelope of the transmission pulse, $\hat{t}$ is time variable, $f_c$ is the center frequency of carrier, $\gamma$ is chirp rate, and $T_P$ is pulse duration of the chirp signal. Based on the above assumptions, echo of sliding spotlight SAR is given by

$$ss_r(\hat{t}, t_a) = \sigma \cdot w_r\left(\frac{\hat{t} - \tau_0}{T_P}\right) w_a\left(\frac{X - Avt_a}{L_s}\right)\exp[j2\pi f_c(\hat{t} - \tau_0) + j\pi\gamma(\hat{t} - \tau_0)^2] \tag{6.13}$$

where $\sigma$ is amplitude of the ground backscattering, $w_a(\cdot)$ is gain of the antenna in azimuth, $\hat{t}$ and $t_a$ are the time in range and azimuth, respectively, $\tau_0$ is the delay of echo, $\tau_0 = 2\sqrt{(X - vt_a)^2 + R_0^2}/C$, $R_0$ is the shortest slant range of the target, $X$ is position of the target in azimuth, $C$ is light speed, $A$ is the sliding spotlight factor defined above, and $L_s$ is width covered by the antenna beam in azimuth. In fact, for different azimuth antenna scanning modes, the echo form is similar to Eq. (6.13). The only difference is the antenna pattern in azimuth, i.e. sliding spotlight factor $A$ is different. Therefore, for all azimuth antenna scanning modes, the echo is in a unified form.

In ScanSAR, echoes in the same substrips are the same as that of strip-map SAR. The difference lies in that data exists in part of the aperture or burst.

In practice, the above echoes are quadrature modulated to baseband:

$$ss(\hat{t}, t_a) = \sigma \cdot w_r\left(\frac{\hat{t} - \tau_0}{T_P}\right) w_a\left(\frac{X - Avt_a}{L_s}\right) \exp[-j2\pi f_c\tau_0 + j\pi\gamma(\hat{t} - \tau_0)^2] \tag{6.14}$$

### 6.4.2 Imaging Algorithm

SAR imaging algorithms can be divided into time domain and frequency domain. The conventional time domain imaging algorithm includes the back projection (BP) algorithm [1] and its fast imaging approach, such as fast factorized back projection (FFBP) [2]. The BP algorithm originated in computer tomography and is an accurate time domain imaging approach. It is relatively simple in principle. It assumes that an impulse spherical wave is transmitted and a high-resolution image is realized by superposing the echoes in time domain coherently. The advantage of the BP algorithm is that it can process imaging while receiving data. This is substantially different from other imaging algorithms and is very useful in such applications as reconnaissance and monitoring. In addition, compared with frequency domain algorithms, the BP algorithm is easier to realize motion compensation and is more suitable for SAR imaging in arbitrarily moving platforms. However, due to point-by-point computation in the BP algorithm, the calculation amount is huge for high-resolution imaging in a large scene, and the efficiency is far lower than that of frequency domain algorithms.

FFBP is a fast algorithm proposed to solve redundancy computation in the BP algorithm. Its fundamental principle is as follows: azimuth resolution is proportional to aperture length. The azimuth bandwidth is small for subapertures divided in the early stage. The corresponding azimuth sampling rate may be low according to Nyquist sampling theorem. Consequently, an image with rough azimuth resolution can be obtained with a relatively small number of sample points. That is, there is no need to project the data of each aperture position (one aperture position corresponding to one azimuth sample point) to each pixel in the imaging grid. Based on this design idea, FFBP establishes a butterfly-like algorithm structure, greatly decreasing the computation burden. In the meantime, the FFBP algorithm retains advantages of conventional BP imaging, i.e. segmental imaging and easy to compensate. However, even so, the FFBP algorithm is still time-consuming for large scene computation. This makes it not suitable for situations with high time efficiency requirements, such as real-time processing.

The frequency domain imaging algorithm uniformly compresses in RD domain or in 2D frequency domain, based on the fact that Doppler frequency characteristics of targets in the scene are the same or similar. Because it compresses all targets at one time, its computation efficiency is far higher than that of point-by-point time domain computation. Common imaging algorithms in the frequency domain include RD, chirp scaling, wave number domain ($\omega - k$), and spectral analysis (SPECAN). All of these imaging algorithms in the frequency domain are developed from strip-map SAR. Some special imaging algorithms have been proposed for different operation modes. For example, the polar format algorithm is used in spotlight SAR for early imaging [3].

All frequency domain algorithms, except SPECAN, need to transform the echo to the frequency domain in the azimuth. Therefore, a prerequisite of these algorithms is that the signal is not aliased in the azimuth frequency domain. For spotlight SAR, sliding spotlight SAR, and TOPS, due to rotation of the beam, the spectrum of the azimuth signal is extended [4, 5]. If relatively low pulse repetition frequency (PRF) is used, the spectrum of full aperture azimuth signal will exceed the range (-PRF/2, PRF/2), meaning that the full aperture azimuth signal will be aliased in the frequency domain. If a relatively high PRF is used, the data amount in the system will increase, which brings about a burden to the system. In practice, PRF is normally only larger than the instantaneous azimuth bandwidth and is much smaller than full Doppler azimuth bandwidth. Therefore, signal processing is required to realize unambiguous imaging of the scene. The idea of SPECAN is used for broadside looking or low-squint imaging; the azimuth Doppler spectrum alias can be effectively removed by time domain convolution.

## 6.5 Doppler Parameter Estimation and Motion Compensation

### 6.5.1 Doppler Parameter Estimation

The precision of Doppler parameter estimation directly affects the imaging quality, despite the type of imaging algorithm. For high-precision SAR image processing, Doppler parameters defined as per parameters of inertial navigation systems (INS) and antenna direction cannot meet the precision requirements. Instead Doppler parameters need to be estimated based on the raw data of the echo.

#### 6.5.1.1 Estimation of Doppler Centroid

The SAR system is characterized with a large time-bandwidth product. There is fixed correspondence between Doppler frequency and azimuth time. The energy of the echo at a certain Doppler frequency must come from the target of the radar beam in a particular direction. The echo of every point target is chirp signal in the azimuth [6], whose center frequency $f_{dc}$ is modulated by the antenna pattern. The SAR image consists of a large number of scattered points. If they have the same scattering cross section, the azimuth power spectrum density of the echo has the same shape as that of the antenna pattern. Based on this principle, Doppler center can be estimated. Currently, there are many methods for estimating SAR Doppler center. The estimation accuracy of these methods depends to some extent on consistency between the azimuth spectrum shape of the echo/image and the antenna pattern. Therefore, it is closely related to statistical characteristics of targets.

(1) *Azimuth spectrum peak method*. In homogeneous scenes, the azimuth power spectrum of the echo, $|S_a(f_a, r)|^2$, is consistent with the spectrum of antenna pattern $|W_a(f_a)|^2$,

$$|S_a(f_a, r)|^2 = \sigma_0^2 |W_a(f_a)|^2 \tag{6.15}$$

with coefficient difference $\sigma_0^2$ only, where $\sigma_0$ is the mean value of the backscattering coefficient of the homogeneous scene. Therefore, the Doppler centroid is the center of the azimuth spectrum. As the peak of the antenna pattern is at the center, the peak of the azimuth spectrum corresponds to the Doppler centroid. However, the target scene in general is nonhomogeneous. The azimuth spectrum is a weighted antenna pattern without conforming to Eq. (6.15). There will be a large deviation if the Doppler centroid is estimated by the peak method. When there is a large fluctuation in the scattering coefficient of targets inside the antenna beam, the azimuth spectrum will be distorted, the center will deviate, and the estimation precision will be decreased.

(2) *Clutter-locked method*. There is a large randomness using azimuth spectra to measure the Doppler centroid, which increases estimation error, especially when the power spectrum in the antenna pattern is relatively flat. Consequently, the clutter-locked method is proposed.

As the echo of the point target in the azimuth is approximately chirp signal, the power spectrum is Fresnel, while amplitude of ground clutter follows Gaussian distribution, with its power spectrum Gaussian. These two spectrums are both symmetrical to Doppler centroid $f_{dc}$. Therefore, the Doppler centroid can be obtained by seeking the energy center of the azimuth spectrum, as shown in Figure 6.5, where $E_1$ and $E_2$ are the energy of the left-side spectrum and right-side spectrum, respectively. If $E_1 = E_2$, then $f_{dc} = f_0$. Therefore, averaging multiple azimuth power spectrums can reduce the instability caused by nonhomogeneous scenes. It can also make the azimuth spectrum more similar to the antenna pattern and can improve estimation precision. This method cannot solve the distortion of the azimuth spectrum caused by nonhomogeneous scenes. In addition, the range cell migration effect should have the azimuth spectrum distribution on several range cells. Therefore, the accuracy of this method is still not high. A very large azimuth spectrum window is needed to improve accuracy. Nevertheless, this method is commonly used to roughly estimate the Doppler centroid of spaceborne SAR imaging, due to its simple method and the small number of computations required.

(3) *Energy difference approximation method*. This method is an improved clutter-locked method. It has overcome the azimuth spectrum distortion of the echo caused by randomness of the reflection energy in nonhomogeneous scenes. In the meantime, it has finished correction of the range cell migration. The accuracy of the Doppler centroid estimation can be greatly improved. As shown in Figure 6.6, when the satellite flies for one aperture duration Ta, i.e. from point C to point D, only one point of the

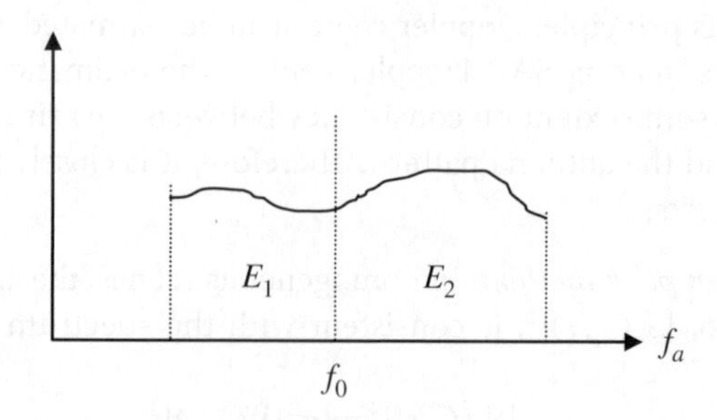


**Figure 6.5** Schematic diagram of clutter-locked method.

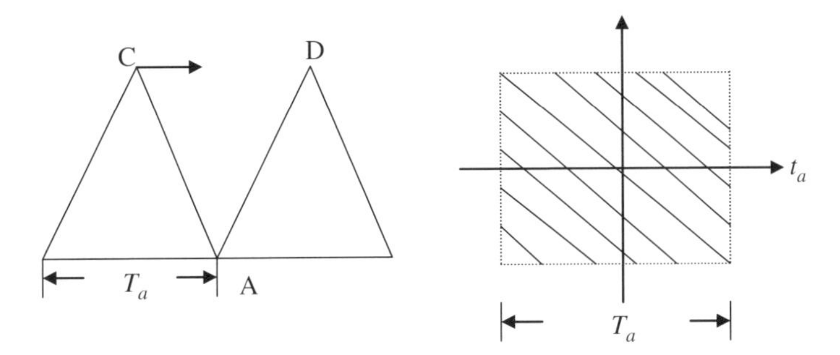


**Figure 6.6** Schematic diagram of the relationship between beam and target and the Doppler history.

echo data, point A, has one complete Doppler history. When the satellite flies for two aperture durations, only one aperture length has one complete Doppler history. When azimuth compressing, the echo data points with complete Doppler history may be extracted and points with partial Doppler history are removed.

The energy difference approximation method can be divided into the azimuth spectrum energy difference approximation method and the image energy difference approximation method. The former is an algorithm based on single-look azimuth processing. It obtains an image with complete Doppler history by constructing an azimuth reference function with initial Doppler parameters. Then the Doppler centroid is estimated. This newly estimated Doppler centroid is used to repeat the above computation. The accuracy of the Doppler centroid estimation is improved through continuous iteration. The latter is an algorithm based on multilook image processing. The Doppler centroid is estimated by balancing the energy difference of multiple subimages.

#### 6.5.1.2 Estimation of Doppler Ambiguity

Because the SAR signal is intrinsically transmitted pulse by pulse, its echo is discrete in the azimuth, and sampling rate is the PRF in the SAR system. Therefore, the azimuth spectrum of the echo is periodic. If $f_{dc}$ is goes beyond PRF, then only the mapped value in the primary period can be obtained, which might be several PRF different from the actual Doppler centroid. This problem is commonly called *azimuth ambiguity of the Doppler centroid*. Although azimuth ambiguity of the Doppler centroid has no influence on azimuth reference function (because azimuth reference function uses PRF as the sampling rate), this ambiguity may cause difference in range cell migration correction, which will decrease the positioning accuracy of the image.

One approach to solve Doppler ambiguity is to transmit signal with multiple pulse repeat frequency. This principle is similar to pulse radar, which uses multiple repeat frequency to solve the range ambiguity. It is obvious that this approach dramatically complicates the system devices. In practice, signal processing can be used to estimate Doppler ambiguity [7], such as in range subimage-related methods. If Doppler ambiguity exists, the track of the target is probably located in multiple range cells, and the range cells of the point target at different azimuth frequency are different as well. Therefore, Doppler ambiguity can be calculated by range deviation of the images. Assuming that the image is processed into four looks, the range deviation between the first look and the fourth look caused by Doppler ambiguity is given by

$$\Delta R_{1,4} = \Delta R(f_4) - \Delta R(f_1) = \frac{\lambda}{2} \cdot n \cdot PRF \cdot \frac{(f_4 - f_1)}{f_{dr}} \tag{6.16}$$

where $f_1$ and $f_4$ are the Doppler centroid in the first and the fourth look, respectively, $f_{dr}$ is the Doppler rate, and $n$ is the Doppler ambiguity number. This range deviation can be obtained by the cross-correlation function of the subimages in the range direction. Then Doppler ambiguity and the real value of the Doppler centroid can be calculated. Of course, if the default value of the Doppler centroid calculated is relatively accurate based on satellite ephemeris or INS data, it is not necessary to calculate the Doppler ambiguity when the previous iteration method is used for Doppler centroid estimation, as the ambiguity number $n$ is already included in the default value. Generally, the higher the frequency of the signal transmitted by the SAR system (the shorter wavelength), the more important the Doppler centroid ambiguity is.

#### 6.5.1.3 Estimation of Doppler Rate

The deviation of Doppler rate may lead to image defocus. To ensure accurate image focus, the Doppler rate must be calculated accurately.

(1) *Subimages correlation.* As shown in Figure 6.7, the solid line is the correct Doppler rate $f_{dr}$, while the dotted line is the Doppler rate with error $f'_{dr}$. If $f'_{dr}$ is used for azimuth compression, the images corresponding to times $t_1$ and $t_4$ will shift in opposite directions, with the time displacement given by

$$\Delta t_{1,4} = (t_4 - t'_4) - (t_1 - t'_1) = (f'_{dr} - f_{dr})(t_4 - t_1)/f'_{dr} \tag{6.17}$$

If $t_1$ and $t_4$ are the time centers of the first and the fourth subimages, respectively, and the synthetic aperture time is $T_a$, then $t_1 = -3T_a/8$ and $t_4 = 3T_a/8$. The azimuth displacement between these two subimages is given by

$$\Delta x_{1,4} = \Delta t_{1,4} \cdot v = \frac{3T_a \cdot v}{4} \frac{(f'_{dr} - f_{dr})}{f'_{dr}} \tag{6.18}$$

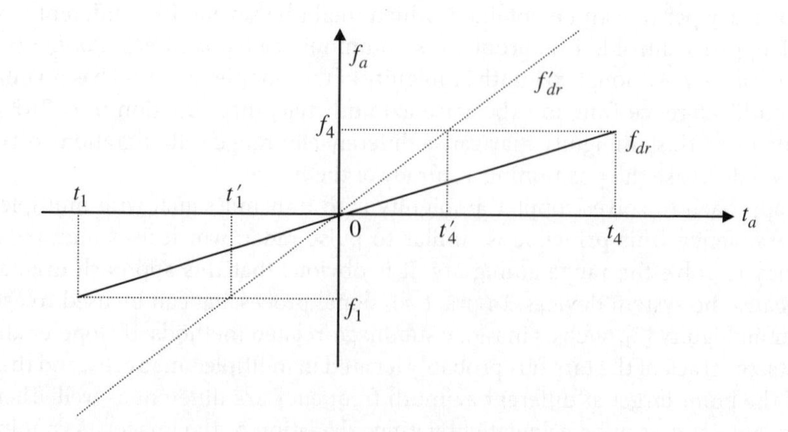


**Figure 6.7** Schematic diagram of $f_{dr}$ estimation by subimages correlation.

where $v$ is the platform velocity. The displacement between the first and the fourth subiages can be calculated by subimage correlation. Consequently, the deviation of the Doppler rate $\Delta f_{dr} = f'_{dr} - f_{dr}$ can be specified. Then the new Doppler rate is given by

$$
\begin{aligned}
f_{dr}^{(i+1)} &= f_{dr}^{(i)} - \Delta f_{dr}^{(i)} \\
\Delta f_{dr}^{(i)} &= \frac{4}{3}\Delta x_{1,4} f_{dr}^{(i)} / (T_a \cdot v)
\end{aligned}
\tag{6.19}
$$

Iterating the above procedure, the final $f_{dr}$ can be obtained by making $f_{dr}^{(i+1)}$ approximate to $f_{dr}^{(i)}$.

(2) *Maximum contrast method.* When the illuminated scene is quasi-homogeneous, with lack of strong targets, there is no obvious correlation peak between the subimages, and the randomness is large. Consequently, there is a relatively large deviation in estimation of Doppler rate. In this case, the maximum contrast method can be used to estimate $f_{dr}$. The image contrast is defined as

$$C = (m + v)/m \tag{6.20}$$

where $m$ is mean amplitude of the image, and $v$ is mean square deviation of the image. When the Doppler rate is correct, the image offers the highest contrast. Therefore, the rate of Doppler frequency can be accurately determined by images contrast obtained by different Doppler rates on the same area.

The maximum contrast method has higher accuracy over the subimage correlation method [8]. However, the maximum contrast method compares the $C$ value of the images obtained with different $f_{dr}^{(i)}$. Therefore, the initial value of $f_{dr}$ is of great importance. To estimate $f_{dr}$ accurately, the number of the $f_{dr}^{(i)}$ value must be sufficient. Considering the calculation amount, the maximum contrast method is more complicated compared to the previous one. In addition, variation of the $C$ value is very slight in a complete homogeneous scene. In this case, the estimation deviation of the maximum contrast method is increased dramatically. Therefore, this is not suitable for the homogeneous scene.

(3) *Autofocus based on phases.* Many autofocus approaches based on phases have been developed for SAR imaging, in which the phase gradient autofocus (PGA) is most frequently used [9]. In most cases, autofocus is not directly used in Doppler rate. Therefore, this can be classified to motion estimation and compensation, which will be discussed in the next section.

### 6.5.2 Motion Compensation

Due to a variety of factors, the platform always deviates from the ideal trajectory during motion, leading to motion error. There are several types of motion error. One is velocity error along the track, which makes the Doppler rate change and the spatial sampling nonuniform. Another is the variation of velocity along the line-of-sight direction, which changes the distance between antenna phase center (APC) and target, causing distance deviation and phase error. The third is attitude error in triaxial directions (pitch, roll, and yaw), which causes variation of the illuminated area and phase error.

According to different acquisition approaches of motion error, motion compensation can be divided into three types: based on sensors, based on raw echo data, and based on image data.

#### 6.5.2.1 Motion Compensation Based on Sensors

Global Positioning System (GPS), INS, or an inertial measurement unit (IMU) are used in sensor-based motion compensation to obtain motion parameters of the airborne platform. Combined with certain geometry models, motion error in the APC is calculated, and the influence on motion error can be removed from the radar data. INS and IMU have high data rates, short-term accuracy, and easily drifting accuracy in the long term, while GPS has low accuracy in the short term, low data rate, and good stability in the long term. It is proved in practice that the best navigation system in the motion compensation system is GPS/INS or an IMU integrated navigation system. This combination can accurately estimate the absolute position and velocity. It has a wide bandwidth and a high data rate as well as a long-term stability. By using a Kalman filter, information from various navigation systems, motion sensors, and some other known data sources are combined to estimate the error of every subsystem. Then the system is corrected by the optimal estimated value of the error to improve accuracy of the whole system, so that flight status in the platform can be more accurately estimated and the SAR imaging quality can be improved.

The geometrical relationship of SAR with motion error is shown in Figure 6.8, in which the $X$ axis is flight direction. The coordinates of the actual platform track are $(X_s, Y_s, Z_s)$, and those of the ideal track are $(Vt_m, 0, 0)$. $t_m$ is slow time. The difference between the two sets of coordinates are three error components in the radar APC. For the ideal track, APC lines uniformly run along the $X$ axis with an interval of $VT_r$, because the radar operates in a repeat period of $T_r$, and $t_m$ varies discretely in integral multiple of $T_r$. Assuming coordinates in the arbitrary point target are $(X, Y, Z)$, the distance between actual APC

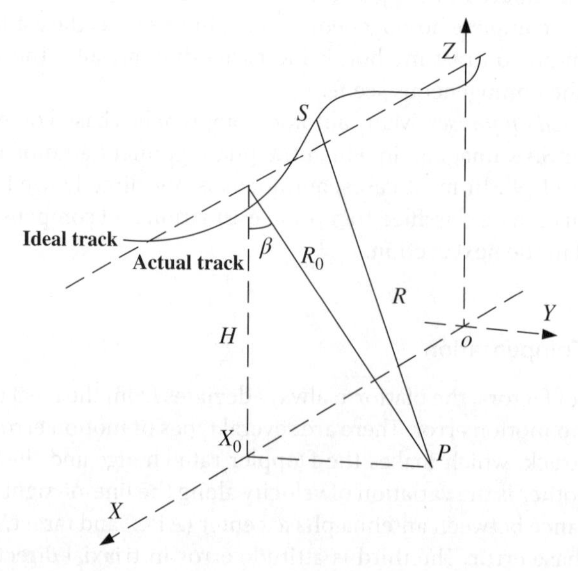


Figure 6.8 Diagram of geometrical relationship in SAR with motion error.

of the radar (point $S$) and point $P$ is given by

$$R = \sqrt{(X_s - X)^2 + (Y_s - Y)^2 + (Z_s - Z)^2} \approx R_0 + \frac{(X_s - X)^2}{2R_0} - Y_s \sin(\beta) - Z_s \cos(\beta) \tag{6.21}$$

where $R_0 = \sqrt{Y^2 + Z^2}$ is the shortest slant range from radar to target $P$, without velocity perturbation, and $\beta$ is the viewing angle. Compared with slant range in the ideal track, the deviation of slant range caused by motion error is given by

$$\Delta r_R = Y_s \sin(\beta) - Z_s \cos(\beta) \tag{6.22}$$

This deviation requires correction by two types of compensation. One is range resampling, which corrects variation of echo delay caused by $\Delta r_R$. The other is $\frac{4\pi\Delta r_R}{\lambda}$ phase correction. Generally, $\Delta r_R$ is related to not only motion error of the platform but also the viewing angle of the target in the normal plane. Because $\Delta r_R$ changes with the viewing angle, and the scene is normally covered far less than the distance between the center of the scene and track, it can be divided into first-order compensation component $\Delta r_{R1}$, which is the compensation component in the centerline of the scene, and second-order compensation component $\Delta r_{R2}$, which is the compensation component of the targets beyond the center distance.

There are two main problems with this method [10–12]. One is that accuracy of the INS parameters cannot meet compensation requirements. In particular, it is required that the measured position error needs to be smaller than $\lambda/4$ in the radar line-of-sight direction. It is not possible for Ku- and X-band SAR systems to compensate accurately. The other problem is that the error calculation model is influenced by terrain fluctuation. The radar incident angle of the target is affected by terrain fluctuation in the illuminated area. When terrain fluctuation is unknown, the accuracy cannot be guaranteed if motion error is calculated based on spatial geometry of the platform. The advantage of this method is that it is sensitive to high-frequency components of motion error of the platform, and it is suitable to compensate high-frequency phase error.

#### 6.5.2.2 Motion Compensation Based on Raw Data

Raw-data-based motion compensation mainly uses Doppler rate estimation to calculate disturbance acceleration of the platform, and then disturbance error is obtained by quadratically integrating the disturbance acceleration. The echo phase is $\varphi(t_m) = -\frac{4\pi R}{\lambda}$. By calculating the second-order derivative of it with respect to time, the instantaneous Doppler rate of the echo is obtained:

$$f_{dr} = -\frac{2}{\lambda}\frac{\mathrm{d}^2(R)}{\mathrm{d}t^2} = -\frac{2V^2(t_m)}{\lambda R_0} - \frac{2[X_s - X]}{\lambda R_0}\alpha(t_m) - \frac{2}{\lambda}\alpha_R(t_m) \tag{6.23}$$

where $a(t_m) = \frac{d}{dt}V(t_m)$ is acceleration along the track, and $\alpha_R(t_m) = \alpha_Y(t_m)\sin(\beta) - \alpha_Z(t_m)\cos(\beta)$ is acceleration along the line of sight in the normal plane. The deviation of the slant range can be obtained by firstly estimating Doppler rate based on multiple data sets in the range direction, and then by fitting velocity along the track and acceleration in the normal plane.

This method requires certain samples in the azimuth, so it is suitable for estimating motion error at relatively low frequency. The disadvantage of this method is that Doppler rate variation caused by velocity variation along the track of the platform couples with that caused by disturbance perpendicular to the track. Variations caused by these two factors can be separated by high-/low-frequency filtering. However, there is no fixed mode affected by airflow during the actual flight of the platform. Though it can be improved to some extent, this separation is not accurate, especially when the synthetic aperture time is longer, because the platform forms a curved track after long time movement, and in this separation method, disturbance error perpendicular to the track is equivalent to the track velocity error.

#### 6.5.2.3 Motion Compensation Based on Image Data

In image-data-based motion compensation, the phase error extracted from a complex image data block is compensated after imaging. Typical algorithms includes PGA. PGA is an iterative algorithm that can compensate both high-order and second-order phase error. Its basic steps are as follows: circular shifting is performed on the amplitude of the complex image after initial focusing, so that the peak of the amplitude is located in the center of the azimuth image at selected range, to eliminate Doppler frequency offset, and to improve the SNR of the estimated phase error. Windowing after shifting is performed on the image, and the phase error is estimated using data with high SNR inside the window. The window size can be chosen to be 1.5 $W_\alpha$, where $W_\alpha$ is twice the width between peak and point at 10 dB below peak. After circular shifting and windowing, FFT is performed on the data to obtain azimuth spectrum $G_n(u)$, where $n$ is the range gate. The phase derivative is given by

$$\dot{\varphi}_e(u) = \frac{\sum_n Im[\dot{G}_n(u)G_n^*(u)]}{\sum_n |G_n(u)|^2} \tag{6.24}$$

where Im(·) indicates extracting imaginary part, and $\dot{G}_n(u)$ and $G_n^*(u)$ are the derivative and conjugate of $G_n(u)$, respectively. $\varphi_e(u)$ is obtained by integrating $\dot{\varphi}_e(u)$. The complex phase after range compression is multiplied by $\exp[-j\varphi_e(u)]$. Then the phase correction is completed by FFT in the azimuth. Repeat above steps until $\varphi_e(u)$ is smaller than the preset threshold.

It can be seen from the execution steps that there is a large amount of computation required in PGA. However, there is no need to process in entire image; it is only required to process in the entire azimuth for a certain number of range gates. Therefore, the computation burden is very small compared with the amount of computation for generating a full image. As this approach is performed in image domain, and the phase error extraction is based on focusing the point target signal belonging to a type of blind correction, it is suitable for processing arbitrary orders of low-frequency motion error. Its disadvantage is that the compensation effect is limited by the illuminated scene. The compensation effect will be decreased when the scene lacks a point-like target.

Motion error obtained from the former two methods needs to be corrected in the echo domain, except the third method. If the synthetic aperture time is long, it will be difficult to extract and compensate motion error. None of the above methods can accurately obtain motion error alone. Generally, two or three of them are combined to attain relative ideal result.

## 6.6 Typical Examples

### 6.6.1 High-Resolution Imaging

More target information can be acquired by improving resolution, so that the target shape and the fine structures can be displayed more clearly, and consequently the target recognition ability can be improved significantly. High resolution imaging requires large instantaneous bandwidth (about 4 GHz). Currently, it is difficult to generate such large instantaneous bandwidth, so stepped frequency waveform is generally adopted; namely, a wideband signal is generated by synthesizing multiple sub-band signals in the range direction. Although the stepped-frequency waveform method has solved the problem of large instantaneous bandwidth, the complexity of signal processing increases, which requires reconstruction of the target spectrum in the range direction and consistency calibration between sub-bands. Taking airborne SAR radar as an example, the practical airborne platform in the azimuth is prone to air turbulence and other environmental conditions during flight, causing nonuniformity along the flight track and disturbance perpendicular to the flight track. In high resolution imaging, high motion compensation accuracy is required, and it is necessary to study highly accurate motion estimation and compensation methods.

#### 6.6.1.1 Ultra-wideband Synthesis in Range

The stepped-frequency SAR achieves high range resolution by intrapulse compression and interpulse coherent synthesis, and the signal transmitted by radar is a sequence of linear frequency modulated pulses with stepped carrier frequency. Assuming whole bandwidth in the stepped-frequency chirp signal is $B$, the number of subpulses is $N$, the sub-bandwidth is $B_m = B/N$, duration of the subpulse is $T_p$, chirp rate is $\gamma = B_m/T_p$, pulse repeat period is $T_r$, frequency step is $\Delta f$, and center frequency of the carrier for the first subpulse is $f_0$, then for the stepped frequency waveform, the carrier center frequency of the $n^{\text{th}}$ pulse is $f_n = f_0 + (n-1)\Delta f$, where $n = 1, 2, \ldots, N$; the frequency relationship of the subpulses is shown in Figure 6.9. Assuming the initial phase of the $n^{\text{th}}$ pulse is $\theta_n$, the $n^{\text{th}}$ pulse signal transmitted by the radar is given by

$$\begin{aligned} X_n(t) &= A_n u(t - nT_r) \cdot \exp[j(2\pi f_n t + \theta_n)] \\ &= A_n \text{rect}\left(\frac{t - nT_r}{T_p}\right) \cdot \exp[j\pi\gamma(t - nT_r)^2] \cdot \exp[j(2\pi f_n t + \theta_n)] \end{aligned} \tag{6.25}$$

where $u(t) = \text{rect}\left(\frac{t}{T_p}\right) \cdot \exp(j\pi\gamma t^2)$ is linear frequency modulated pulse at baseband, and $A_n$ is amplitude of the $n^{\text{th}}$ pulse. Then the $n^{\text{th}}$ pulse received by the radar is given by

$$\begin{aligned} y_n(t) = B_n \text{rect}\left[\frac{t - nT_r - \tau(t)}{T_p}\right] \cdot \exp\{j\pi\gamma[t - nT_r - \tau(t)]^2\} \\ \cdot \exp\{j[2\pi f_n(t - \tau(t)) + \theta_n]\} \end{aligned} \tag{6.26}$$

where $B_n$ is amplitude of the $n^{th}$ received pulse, and $\tau(t)$ is time delay of the echo.

Frequency domain synthesis is generally used for wideband synthesis in the range direction. In this method, the modulated chirp signal of every stepped frequency is matched and filtered in the frequency domain. Then coherent synthesis is performed

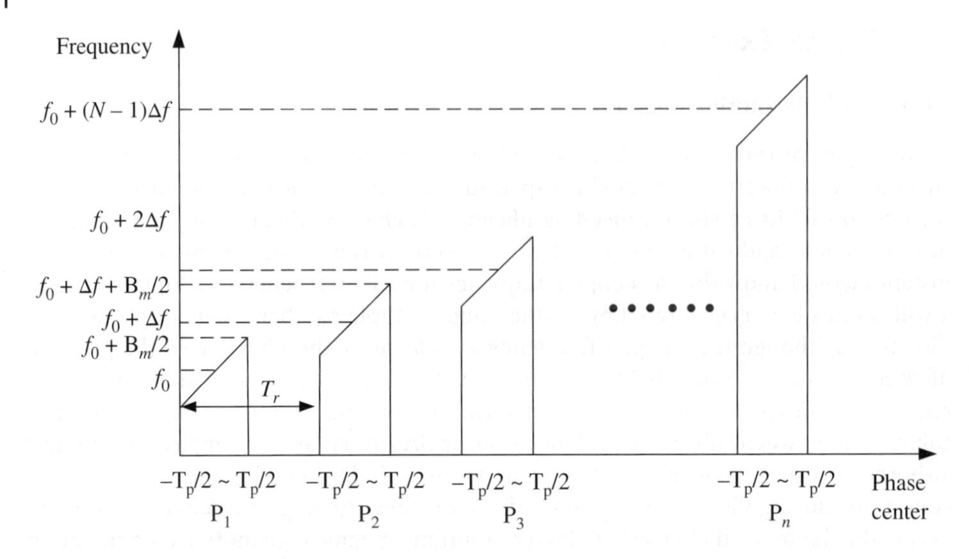


**Figure 6.9** Frequency distribution of a linear stepped-frequency chirp waveform.

in the frequency domain after spectrum shifting. Finally, inverse Fourier transform is used to improve the range resolution.

Assume the echo model is

$$s_n(t) = \int_{r_1}^{r_2} g_0(r) \exp[j\omega_n(t - 2r/c) + j\pi\gamma(t - 2r/c)^2] \mathrm{d}r \tag{6.27}$$

where $s_n(t)$ is the received echo when transmitting the $n^{th}$ pulse, $n = 1, 2, \ldots, N$, $\omega_n = 2\pi f_n$, $g_0(r)$ is target reflectivity function between the range of $[r_1, r_2]$, and $r$ is the distance between target and radar. Let $\tau = 2r/c$ and $g(\tau) = g_0(c\tau/2)$; Eq. (6.27) can be expressed as

$$s_n(t) = \int_{r_1}^{r_2} g_0(\tau) \exp[j\omega_n(t - \tau) + j\pi\gamma(t - \tau)^2] \mathrm{d}\tau \tag{6.28}$$

Furthermore, let $b_n(t) = \exp(j\omega_n t + j\pi\gamma t^2)$, and obviously, Eq. (6.28) can be expressed as the following convolution form:

$$s_n(t) = g_n(t) * b_n(t) \tag{6.29}$$

where $g_n(t)$ is the observation signal of $g(t)$ in the $n^{\text{th}}$ subfrequency band. After down-conversion in the receiver, the echo signal in Eq. (6.29) is given by

$$s_n(t) \cdot e^{-j\omega_n t} = [g_n(t) * b_n(t)] \cdot e^{-j\omega_n t} \tag{6.30}$$

After Fourier transform, Eq. (6.30) is given by

$$S_n(\omega - \omega_n) = G_n(\omega - \omega_n) \cdot B_n(\omega - \omega_n) \tag{6.31}$$

where $(\omega - \omega_n) \in [-\pi\gamma T_P, \pi\gamma T_P]$ is the frequency range at baseband, $S_n(\omega - \omega_n)$ is the baseband spectrum of the received signal when transmitting the $n^{\text{th}}$ pulse, $B_n(\omega - \omega_n)$ is baseband spectrum of the transmitted $n^{\text{th}}$ pulse, and $G_n(\omega - \omega_n)$ is baseband spectrum of the $n^{\text{th}}$ observed signal of the range target reflectivity function. After the second frequency conversion, Eq. (6.30) is given by

$$s_n(t) \cdot e^{-j\omega_n t} \cdot e^{j(n-1)\Delta\omega t} = [g_n(t) * b_n(t)] \cdot e^{-j\omega_n t} \cdot e^{j(n-1)\Delta\omega t} \tag{6.32}$$

where $\Delta\omega = 2\pi\Delta f$. After Fourier transform, Eq. (6.32) is given by

$$\begin{aligned} S_n(\omega - \omega_n + (n-1)\Delta\omega) = G_n(\omega - \omega_n + (n-1)\Delta\omega) \\ \cdot B_n(\omega - \omega_n + (n-1)\Delta\omega) \end{aligned} \tag{6.33}$$

$$\begin{aligned} G_n(\omega - \omega_n + (n-1)\Delta\omega) &= \frac{S_n(\omega - \omega_n + (n-1)\Delta\omega)}{B_n(\omega - \omega_n + (n-1)\Delta\omega)} \\ &= S_n(\omega - \omega_n + (n-1)\Delta\omega) \\ &\quad \cdot B_n^*(\omega - \omega_n + (n-1)\Delta\omega) \end{aligned} \tag{6.34}$$

where $B^*(\omega)$ is the complex conjugate of $B(\omega)$. $S_n(\omega - \omega_n + (n-1)\Delta\omega) \cdot B_n^*(\omega - \omega_n + (n-1)\Delta\omega)$ represents the matched filtering in the frequency domain. With coherent synthesis $G_n$ obtained from the $n = 1 \sim N$ pulses, we have

$$\begin{aligned} G(\omega - \omega_1 + (N-1)\Delta\omega) &= \sum_{n=1}^{N} G_n(\omega - \omega_n + (n-1)\Delta\omega) \\ &= \sum_{n=1}^{N} S_n(\omega - \omega_n + (n-1)\Delta\omega) \cdot B_n^*(\omega - \omega_n + (n-1)\Delta\omega) \end{aligned} \tag{6.35}$$

Then, the final one-dimensional (1D) range profile is given by

$$g_0\left(\frac{ct}{2}\right) = g(t) = FFT^{-1}[G(\omega - \omega_1 + (N-1)\Delta\omega)] \tag{6.36}$$

The above process seems complicated, but the actual algorithm is very simple. Because the first downconversion can be realized completely by the receiver hardware, and the second one can be realized by only shifting the spectrum, after matched filtering echoes of the $n = 1, 2, \ldots, N$ pulses, finally the spectrum of range function can be obtained by coherent summation of the $N$ converted spectrums. Figure 6.10 shows a simple schematic diagram of spectrum synthesis. A 1D range profile can be obtained by performing inverse Fourier transform on the synthesized spectrum.

#### 6.6.1.2 High-Resolution Compression in Azimuth

For high-resolution imaging, it is very difficult to apply the conventional imaging algorithm, due to the requirement of high-resolution long synthetic aperture time to cause a large range cell migration. The large range cell migration requires a space-variant reference function in the azimuth direction. In addition, the Fresnel approximation is not established. These not only cause Doppler frequency variation to be nonlinear in the azimuth but also lead to severe coupling between the range and azimuth directions, increasing complexity and computation burden of the processing.

The wave number domain algorithm can automatically correct range cell migration by precise interpolation, and it solves the coupling problem by azimuth processing in the 2D frequency domain. In case of 0.05 m resolution, at a slant range of 15 km, the range curvature reaches 75 m, much larger than the range resolution. Then both curvature and

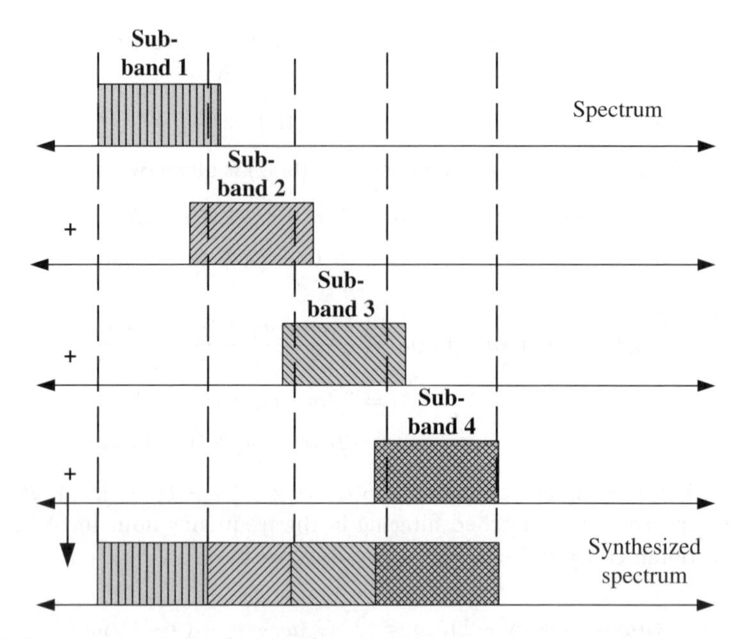


**Figure 6.10** Spectrum synthesis at baseband.

coupling between the range and azimuth directions cannot be neglected. The 2D wave number domain algorithm can be decoupled, which makes it capable to process SAR data with large range cell migration. Therefore, the wave number domain algorithm with subaperture-based motion estimation and autofocus are selected, as the range resolution is severely affected by Doppler rate mismatch, and speed variation is the main cause of Doppler rate variation. According to the focus depth criteria, if the phase difference caused by mismatch at the Doppler frequency edge is smaller than $\pi/4$, the mismatch can be neglected. Therefore, to avoid defocusing in azimuth, speed variation within the subaperture should meet

$$\Delta v \leq \sqrt{2\lambda R_0}/2T \tag{6.37}$$

where $T$ is the synthetic aperture time. The approximate velocity variation curve can be obtained through INS, so that the range of subaperture time can be estimated.

There are two kinds of processing for range bandwidth synthesis in high resolution SAR. One is that the range bandwidth between sub-bands is firstly synthesized, and the 2D imaging is considered. The influence of the Doppler effect can be neglected within several pulses, and direct spectrum summation can be performed. Second, low-range resolution and high-azimuth resolution images are generated for the corresponding sub-bands. Then these images are registered in the azimuth. Finally, a high resolution range image is obtained by spectrum synthesis of the range bandwidth. In the first method, the influence of Doppler between subpulses is neglected, which may broaden the main lobe and increase the side lobe, and the impulse response function is left–right asymmetry, although the amount of computation is small. If the second method is chosen, it is necessary to interpolate the images when registering the subimages, as the azimuth difference between subimages is less than a pixel, and the amount of computation is too large at this moment. The amount of computation can be reduced by performing Fourier transform on subimages requiring to be aligned during time shift or frequency shift. If the range synthesis is performed directly without image registration, the azimuth resolution will decrease, and the image quality will deteriorate.

To generate a precise image, the second imaging method is commonly used, namely, subpulse compression and image registration are firstly performed, and with the $N$ registered subimages, a high resolution range image is obtained by shifting and coherently combining the spectrum.

#### 6.6.1.3 Motion Error Estimation and Compensation

It is not possible to obtain an image with 0.05 m azimuth resolution without precise motion extraction and compensation. The motion estimation can be performed as follows: highly accurate sensor data is acquired with multisensor integration technology, such as GPS/INS or IMU. The line-of-sight displacement of the APC is calculated. The residual phase error accurately extracted from raw data can be eliminated, and a high-resolution image can be acquired.

The platform motion error can be divided into displacement error and attitude error. The displacement error causes SAR to generate phase error in the azimuth, and the attitude error causes the radar echo modulation to change and the Doppler centroid to shift. Due to the long distance of the operating radar and relatively long synthetic aperture time, the errors of the platform are more complicated. These errors may cause the Doppler spectrum distortion, which decreases the focusing accuracy and influences resolution and imaging quality. Therefore, compensation based on motion sensors as well as the raw echo data is implemented for high-resolution imaging. In front end, the parameters provided by motion sensors are used for coarse compensation (including uniform sampling in the azimuth by PRF variation). In back end, motion estimation and compensation based on SAR echo is used to realize precise focus. Because the problem of speed instability in front end is overcome by PRF adjustment and uniform azimuth sampling is ensured, it can be assumed that the platform flies constantly in a straight line along an ideal track after motion error compensation.

#### 6.6.1.4 High-Resolution Imaging Process and Results

High-range resolution may be obtained by ultrawideband synthesis technology. And high-azimuth resolution can be obtained by motion compensation technology. Figure 6.11 shows a block diagram of ultra-high-resolution SAR imaging. A spotlight image with 0.1 m resolution is obtained by Ku-band airborne SAR, as shown in Figures 6.12 and 6.13.

### 6.6.2 Ground Moving Target Indication

The signal processing of ground moving target indication (GMTI) mainly includes ground clutter suppression, moving target extraction, single-frame moving target processing, moving target plot centroid processing, track generation, and information extraction. The ground clutter suppression is one of the critical steps in GMTI signal processing. The moving target indication (MTI) capability can be significantly improved if ground clutter is effectively suppressed.

According to the number of receiving channels, GMTI systems can be divided into single channel and multichannel. High PRF is usually used in traditional single-channel

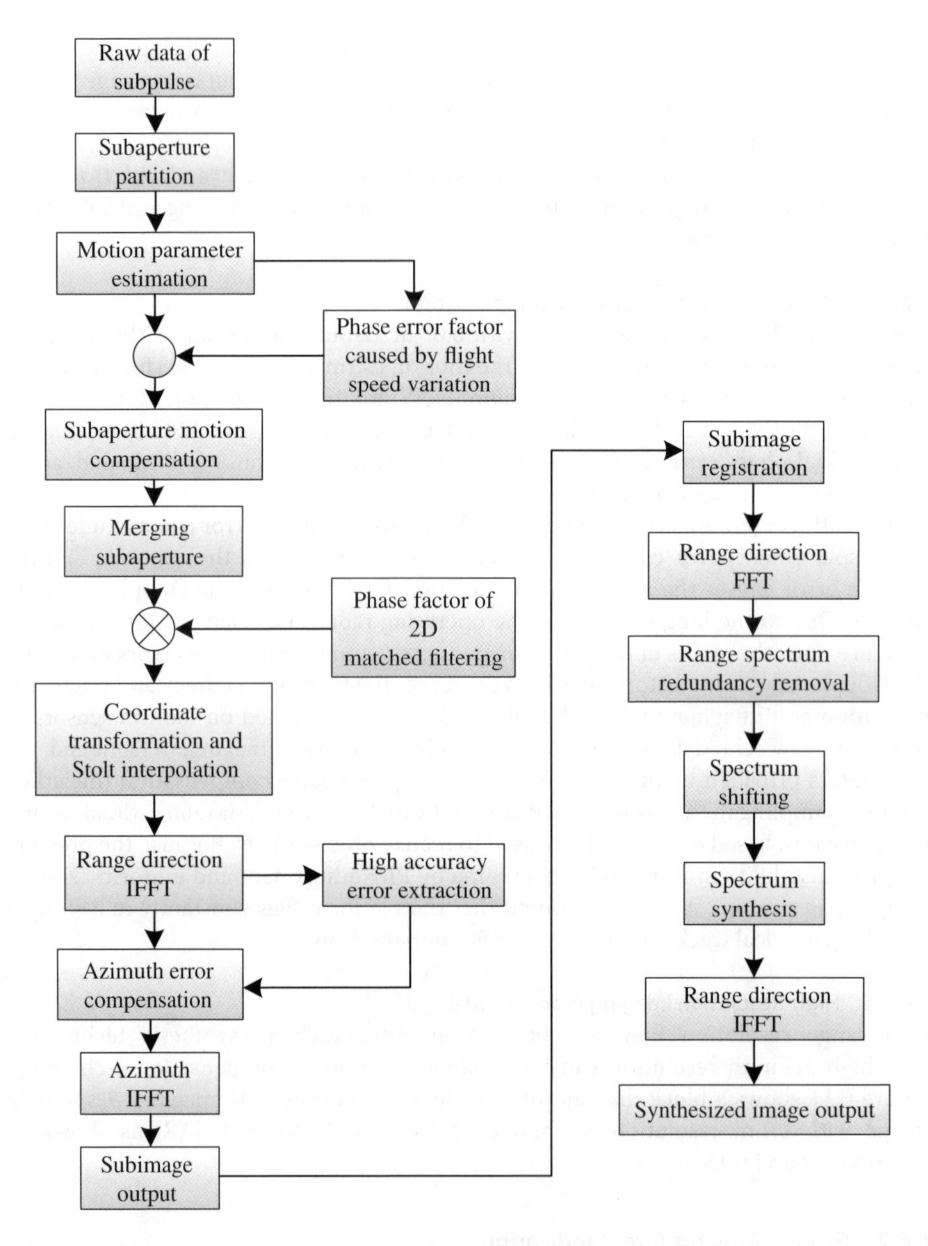


**Figure 6.11** Block diagram of ultra-high-resolution SAR imaging.

GMTI systems to obtain a large Doppler clean area for moving target detection (MTD). Due to limitation of the azimuth antenna size and the platform speed, in most cases, it is difficult to apply high PRF, so clean area is limited and multichannel GMTI must be used. Figure 6.14 shows the transmitting and receiving diagram of three-channel GMTI. The entire antenna aperture for transmitting and the antenna aperture for receiving are equally divided into three subapertures. The antenna beam is extended during transmitting to ensure that the illuminated area is the same for both transmitting and receiving. Currently, conventional multichannel GMTI processing technologies include

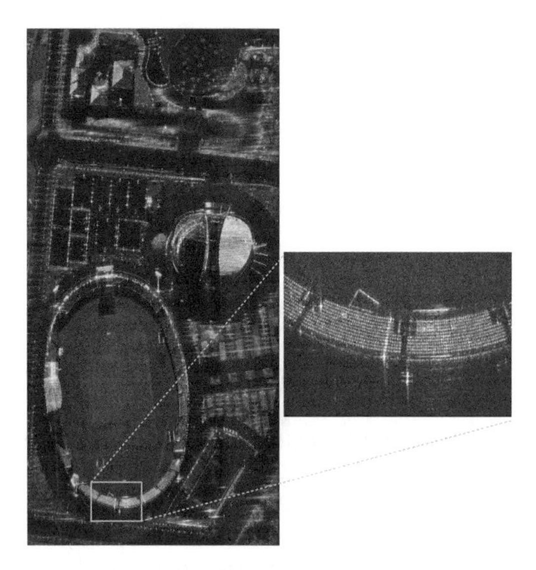

Figure 6.12 SAR image of a stadium with high resolution of 0.1 m × 0.1 m.

Figure 6.13 Typical high-resolution SAR image (0.1 m × 0.1 m).

displaced phase center antenna (DPCA) signal cancellation, along-track interferometry (ATI), clutter-suppressing interferometry (CSI), space-time adaptive processing (STAP) [13–15], etc.

#### 6.6.2.1 DPCA and ATI

DPCA uses multichannel receiving antennas to suppress the clutter. The fundamental requirement of this technology is that PRF in the radar system and the platform speed should meet a certain relationship. And the main lobe clutter is suppressed by using correlation of the received echoes in every channel. DPCA can be divided into two approaches. One is time domain DPCA, which cancels the clutter before generating an image (i.e. time domain). The other is frequency domain DPCA, which cancels

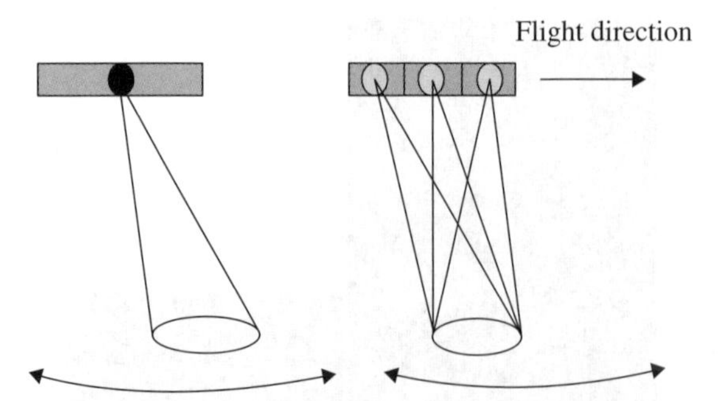


**Figure 6.14** Schematic diagram of transmitting and receiving for three-channel GMTI.

the clutter in the image domain. Ideally, there is no difference in clutter cancellation between these two approaches. However, when system error (such as mismatch between channels) exists, the frequency domain DPCA is better in side lobe clutter cancellation, though the amount of computation and the system complexity will increase.

The basic principle of ATI is that two antennas are assembled on one platform and are aligned along the track to observe the same ground scene. If the target is static, the images acquired from two antennas will be exactly the same, except for deviation along the track, whereas a phase shift in the corresponding pixels will be generated for a moving object. After imaging two echoes acquired by two antennas, an interferometric phase map will be created. The phase difference $\Delta\Phi$ is related to Doppler frequency shift $\omega_d$ and $\Delta t$, meaning it is proportional to radial velocity of the moving target. The moving target can be detected by comparing $\Delta\Phi$ with the threshold $\Phi_{threshold}$.

#### 6.6.2.2 CSI

CSI, similar to frequency domain DPCA, is a kind of STAP, with slightly higher computation burden than that of DPCA. But it does not need to meet the constraints of DPCA, and it has a better clutter cancellation capability. Its fundamental principle is that subimages in the channels are firstly generated. Then the subimages are canceled two-by-two to improve the SNR of the moving target. The phase interference of the two subimages after cancellation is used to extract the Doppler position and radial velocity information of the moving target. The implementation of this approach can be divided into four main steps: channel equalization, channel subimage generation, channel cancellation, and relative positioning and velocity measurement of the moving target [16].

Channel equalization compensates inconsistency between channels and improves the correlation of channel signals to facilitate channel cancellation. This is an indispensable step of multichannel GMTI processing. First, a test signal in each channel will be collected, and Fourier transform is performed on these test signals. Then both amplitude-frequency and phase-frequency characteristics of the test signal in each channel are obtained. Finally, one channel is chosen as a reference, and the amplitude and phase errors of the other channels are calculated and compensated.

For channel subimage generation, the signals from each channel are channel corrected and are used to generate an image with the conventional strip-map SAR algorithm. Note that Doppler shift in the channel due to deviation of the phase center in the receiving antenna must be compensated. After range compression and migration compensation, the channel in the middle is used as a reference to compensate Doppler shift in the other channels. The value to be compensated is $\frac{v_a d}{\lambda R}$, where $v_a$ is the platform speed, $d$ is the distance between APCs in two channels, $\lambda$ is wavelength, and $R$ is slant range. After Doppler shift compensation and azimuth compression, Doppler subimages in each channel can be obtained.

During channel cancellation, the phase difference between channels needs to be compensated because the receiving APC in each channel is inconsistent. Ideally, only the phase error given by Eq. (6.38) requires compensation during two-by-two channel cancellation.

$$\Delta\varphi(n) = -2\pi\frac{d}{v_a}f_a(n) - \pi\frac{d^2}{\lambda R}\left(1 - \frac{f_{DC}^2 \cdot \lambda^2}{4v_a^2}\right) \tag{6.38}$$

where $f_a(n)$ is the Doppler frequency corresponding to the $n^{\text{th}}$ pixel of the Doppler subimage, and $f_{DC}$ is the central frequency of the Doppler subimage. Practically, because of variation of the platform speed and existence of the motion error, direct cancellation produces a bad effect. In practice, coefficients of both the linear term and the constant term in Eq. (6.38) need to be extracted by self-estimating complex data of the subimage in each channel.

Regarding relative positioning and velocity measurement for the moving target, after channel cancellation, the clutter has been effectively suppressed, which is convenient for MTI. There are many approaches for detecting the moving target, which will not be discussed here. Once the moving target is detected, by subtracting phases of the moving target in two residual subimages, the interferometric phase $\Delta\Phi$ of the moving target can be obtained. According to the moving target positioning principle, we have

$$Y = \frac{\lambda R \cdot \Delta\Phi - \pi d^2}{2\pi d} \tag{6.39}$$

where $R$ is slant range between the moving target and radar. An interferometric phase of the moving target $\Delta\Phi$ has a periodic ambiguity of $2\pi$. Therefore, the moving target azimuth position $Y$ calculated according to Eq. (6.39)) is also ambiguous, and the corresponding azimuth positioning ambiguous period is $\lambda R/d$(m). If it is converted into an azimuth angle, the ambiguous period is the antenna beam width $\lambda/d$. Knowing the actual azimuth position of the moving target, the Doppler frequency of the target caused by the platform movement can be calculated. Subtracting the Doppler frequency from the corresponding Doppler frequency of the pixel in the Doppler image where the moving target locates, the Doppler difference is caused by radial velocity of the moving target $v_r$. So radial velocity of the moving target is given by

$$v_r = \frac{f_m \lambda}{2} - \frac{v_a Y}{R} \tag{6.40}$$

where $v_a$ is the platform speed, and $f_m$ is the Doppler frequency corresponding to the pixel in the Doppler image where the moving target locates.

Figure 6.15 shows the basic principle diagram of CSI-based four-beam GMTI. With the adaptive CSI, findings of real data acquired in four-channel airborne wide-area MTI mode are displayed in Figure 6.16.

#### 6.6.2.3 Three Doppler Transform STAP

Although full STAP has the best clutter suppression performance in practice, the huge burden of both computations and equipments make it difficult to implement. In practical application, it is necessary to reduce the data dimension. Dimension reduction can

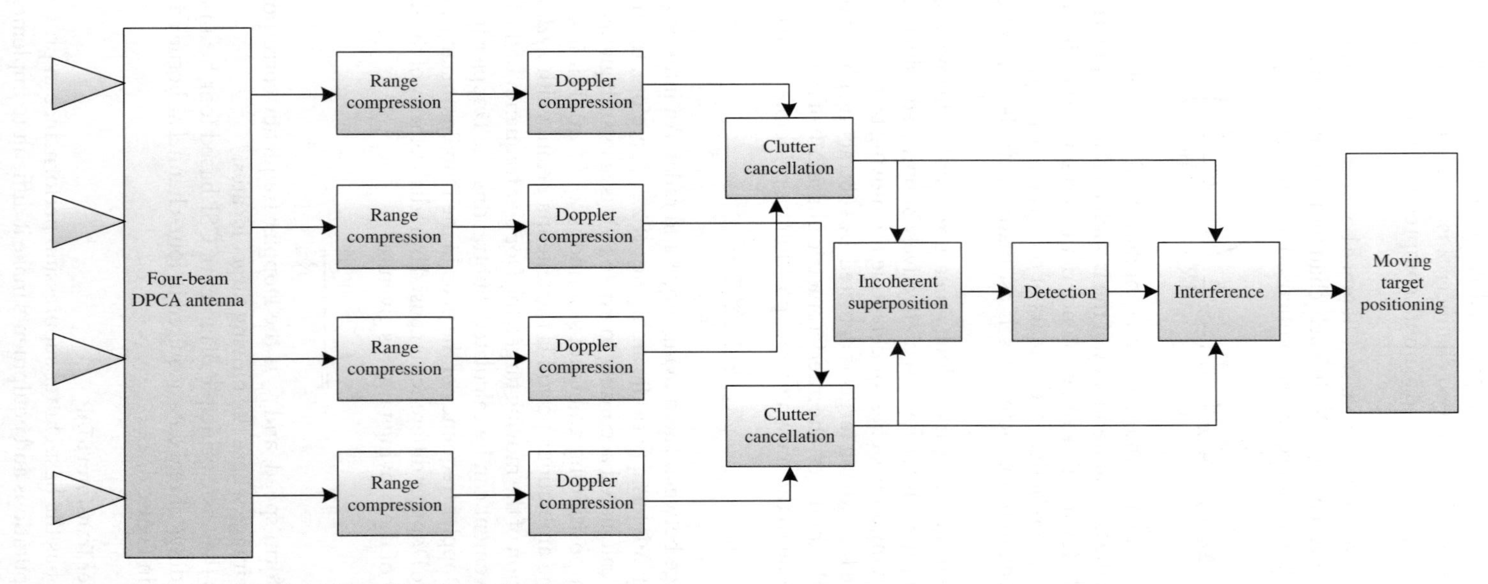


**Figure 6.15** Basic principle of CSI-based four-beam GMTI.

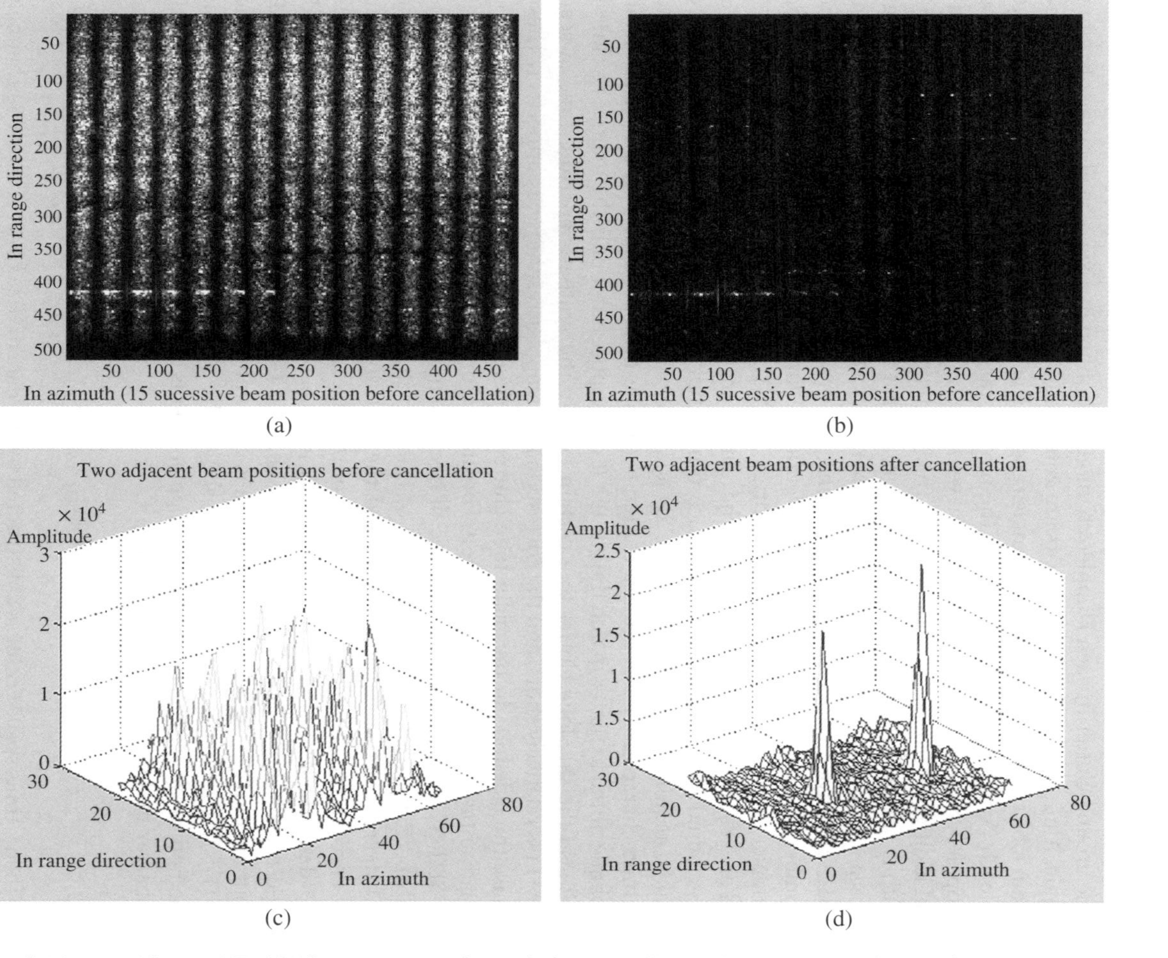


**Figure 6.16** Findings of airborne wide-area MTI. (a) Fifteen successive bursts before cancellation; (b) 15 successive bursts after cancellation; (c) two adjacent bursts before cancellation; (d) two adjacent bursts after cancellation. (See color plate section for the color representation of this figure).

be performed in spatial domain or its corresponding beam domain. It can also be carried out in time domain or its corresponding Doppler domain. Taking into account that the array number of the target is normally small in practice, dimension reduction in time domain (or Doppler domain) is generally used. And it is Doppler domain dimension reduction that is commonly preferred. The reason is as follows: there is a defined correspondence between Doppler frequency of the fixed ground target $f_a$ and squint angle. The ground clutter in each Doppler channel is confined in a very small range of squint angle. A number of Doppler outputs are equivalent to splitting illumination area of the main lobe clutter into many small angle ranges. The corresponding narrow angle range of each Doppler output is different. The moving target is output by the corresponding Doppler channel. However, due to radial velocity, the angle of moving target in this channel is different from that of the clutter. This is intended to make each Doppler channel output only a part of the clutter spectrum, so that the SNR of the output can be significantly improved. The signal in each channel is Doppler filtered through FFT. Then space adaptive processing is performed on a pair of Doppler outputs that have the same numbers in the subapertures. The conventional dimension reduction approach in the Doppler domain, three Doppler transform STAP (3DT-STAP), is described below.

For airborne phased array radar, the clutter in space-time 2D spectrum plane is distributed in strip, with a strong space-time coupling. Therefore, better performance can be acquired through space-time 2D adaptive processing. Specifically, according to certain characteristics of ground clutter, the space-time 2D approach can generate an oblique notch to better match the clutter. It can suppress not only the main lobe clutter but also the side lobe clutter to a certain extent. The synthetic aperture time of scan GMTI is relatively short, and the corresponding Doppler resolution is imperfect. These cause the side lobe in the main lobe clutter to cover other frequency areas, which is disadvantageous for target detection. For this reason, three adjacent Doppler channels in 3DT-STAP are used as the degree of freedom in the time domain, and the antenna subapertures are used as the degree of freedom in the spatial domain. A space-time 2D steering vector is constructed, and the corresponding weight vector is obtained by decomposing the covariance matrix of the clutter. After 2D filtering of space-time 2D data, the clutter may be suppressed, and the moving target signal is obtained. Figure 6.17 shows the diagram of four-channel GMTI processing with 3DT-STAP. Four channels are divided into two groups by combining three adjacent channels, and 3DT-STAP is performed. Figure 6.18 shows the architecture of 3DT-STAP. In this method, array data vector of the $k^{\text{th}}$ Doppler filter is output after preprocessing for each subarray in the time domain, given by

$$X_k = [x_{1k}, x_{2k}, \cdots, x_{Nk}]^T \tag{6.41}$$

Then a time-space vector is constructed by the data of the $(k-1)$ and $(k+1)$ channels

$$B_k = (X_k^T, X_{k-1}^T, X_{k+1}^T)^T \tag{6.42}$$

So the clutter covariance matrix is given by

$$R_k = E[B_k B_k^H] \tag{6.43}$$

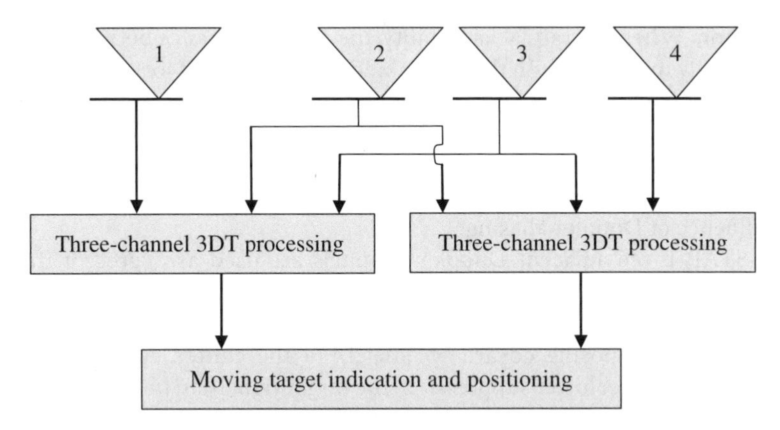


**Figure 6.17** Diagram of four-channel 3DT-STAP.

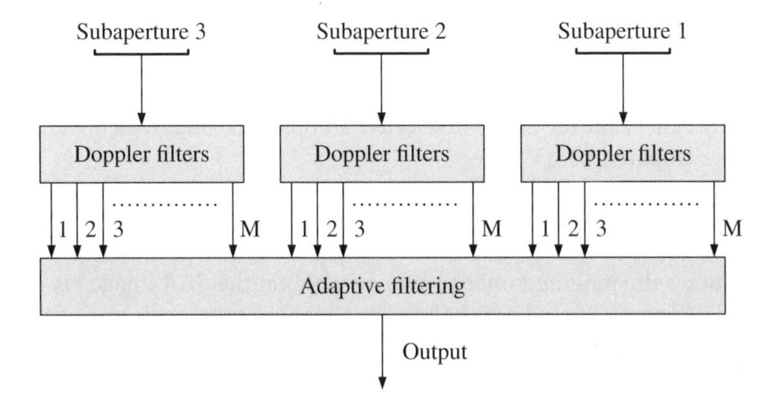


**Figure 6.18** Schematic architecture of 3DT-STAP.

According to the linearly constrained minimum variance criterion, to solve for the following optimization problem:

$$\begin{cases} \min\limits_{W_k} W_k^H R_k W_k \\ s.t. W_k S_0 = 1 \end{cases} \tag{6.44}$$

where $S_0$ is time-space 2D normalized steering vector, the optimum solution is found to be

$$W_{kopt} = \mu R_k^{-1} S_0 \tag{6.45}$$

where $\mu = 1/(S_0^{-1} R_k^{-1} S_0)$ is the normalization coefficient.

#### 6.6.2.4 Comparison

Due to the fact that DPCA and ATI are restricted by many conditions in practice, CSI and 3DT-STAP are usually considered.

Usually, CSI compensates the envelope offset and phase difference caused by channel spacing in the image domain and performs two-by-two cancellation among multichannels. There is no need to adaptively compute the weight vector, due to the small amount of computation. When aircraft lacks stability, the phase difference between two-by-two channels is firstly measured with the clutter and is compensated for cancellation, which eliminates motion error of the aircraft. The disadvantage is that it does not make full use of space-time freedom information. In addition, without satisfying the DPCA condition, the algorithm cannot suppress the clutter where Doppler aliased. Therefore, it is necessary to take into account the platform speed, the antenna pattern, optional PRF, and the influence of Doppler aliasing.

In 3DT-STAP, three adjacent Doppler channels are used as degree of freedom in time domain, the antenna aperture is used as the degree of freedom in spatial domain, a space-time 2D steering vector is constructed, and the corresponding weight vector is obtained by decomposing covariance matrix of the clutter. After 2D filtering of space-time 2D data, i.e. clutter suppression, the target signal can be obtained. In case of multiple channels, the notch is narrow but useful for low-speed ground MTD. Second, due to the use of more degrees of freedom, desired performance can be achieved in both main lobe and side lobe. The disadvantage is that weight vectors need to be adaptively calculated by data sample extraction when the amount of computation for sample extraction is large. When the moving targets are densely gathered, it is difficult to select samples. In addition, when three receiving channels are jointly processed, this algorithm may cancel clutter of the first-order Doppler aliasing, which will lower PRF requirements compared with CSI.

#### 6.6.2.5 Results of SAR/GMTI

Figure 6.19 shows the track of a moving target detected by dual-channel GMTI. Figure 6.20 shows the moving target results overlaid on the SAR image via simultaneous SAR/GMTI processing, where the round circles are false positions of the moving target, and the dots are true positions of the moving target with different radial speed. Figure 6.21 shows the ground trajectory generated, formed by real-time tracking of the moving target for a long time. A modified three-channel CSI is used in wide-area GMTI mode.

Under a ground surface environment with dense targets and complex clutters, it is difficult to detect and track ground moving targets. The probability of wrong tracking, mixed tracking, tracking by mistake, and loss of the radar target has increased. The

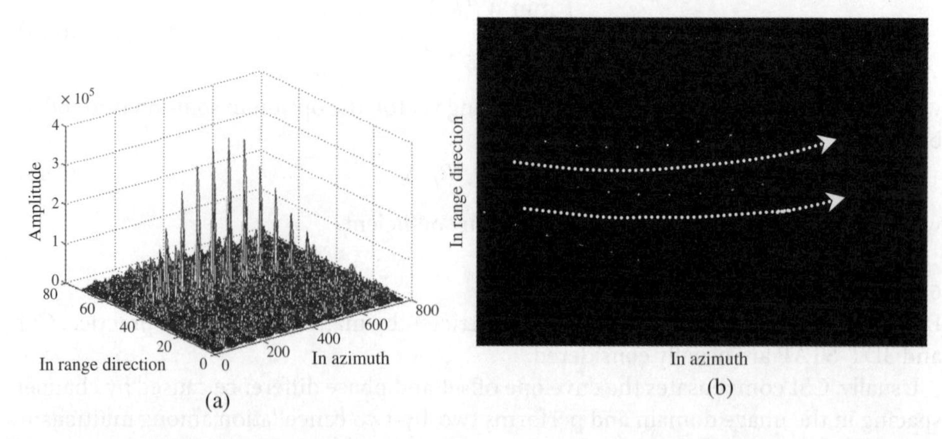


**Figure 6.19** Results of GMTI. (a) Single-frame moving target track; (b) track of the moving target. (See color plate section for the color representation of this figure).

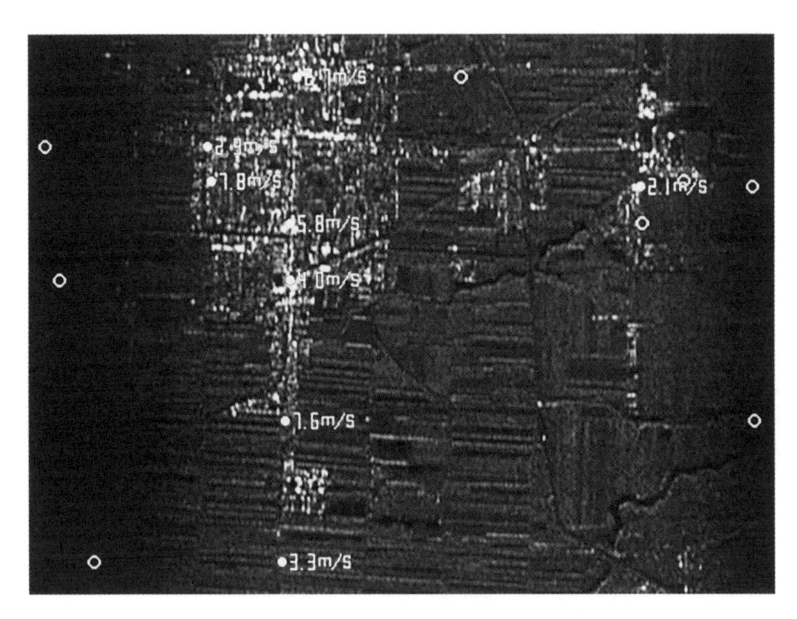


**Figure 6.20** Moving targets obtained by simultaneous SAR/GMTI overlaid on the SAR image. (See color plate section for the color representation of this figure).

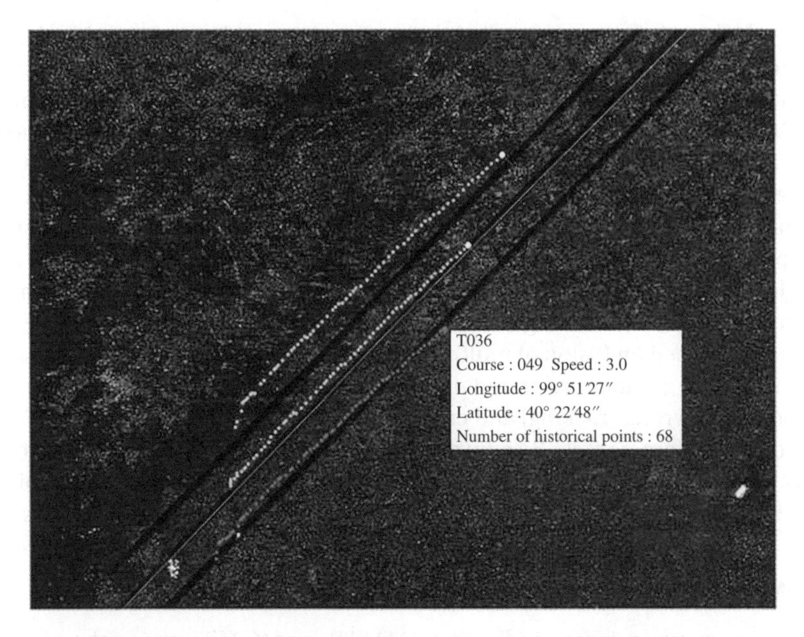


**Figure 6.21** Moving target track in wide-area GMTI. (See color plate section for the color representation of this figure).

amount of data associated computation increases geometrically with the intensity of the target. The continuity problem of ground target tracking is an important but difficult issue in the GMTI model for multiple target tracking. Joint probabilistic data association (JPDA) [17] is one of the most effective tracking algorithms in dense clutter and multiple targets. Its basic principle is as follows: the concept of data associated full spectrum is

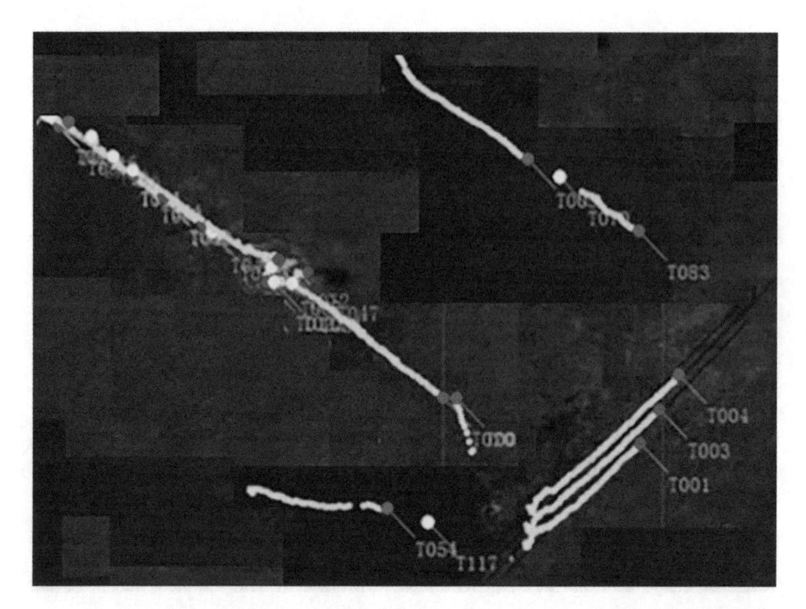


**Figure 6.22** Moving target track for GMTI. (See color plate section for the color representation of this figure).

used to calculate the posterior probabilities of each candidate echo inside the associated gate. The posterior probabilities are used as a weight to permute and combine the candidate echoes, from which effective joint events are selected to calculate joint probability. Other feature information acquired from radar measurements is introduced into the JPDA algorithm. Such information as the predicted target location, flight course, predicted Doppler value (predicted radial velocity of the target), and amplitude of the trace point is comprehensively used. This information is weighted to calculate the value of comprehensive correlation function, and association probability obtained with the JPDA algorithm is modified by association probability acquired after fusing multifactor parameters. Finally, the target state is updated by new association probability. The JPDA method based on fusion of multifeatures can effectively solve the data association problem of dense ground targets and maintain a stable tracking of flight course to avoid mixed tracking, wrong tracking, and cross tracking to a large extent. Figure 6.22 shows a GMTI moving target track. By adjusting the center of the scanning beam, a continuous tracking of ground moving target is achieved with excellent tracking stability and continuity. The main targets are distributed on the traffic network.

### 6.6.3 Marine Moving Target Indication

Both marine moving target indication (MMTI) and GMTI share the same basic principle. The major difference between them is background clutter. Within a certain time, characteristics of ground clutter remain the same, whereas characteristics of marine clutter are quite different in that they are time-varying. For long periods of time, the marine clutter feature under different marine conditions is also very different. Therefore, the methods commonly used in GMTI do not produce the desired effect in MMTI. To accurately extract the target information of the sea surface, it is

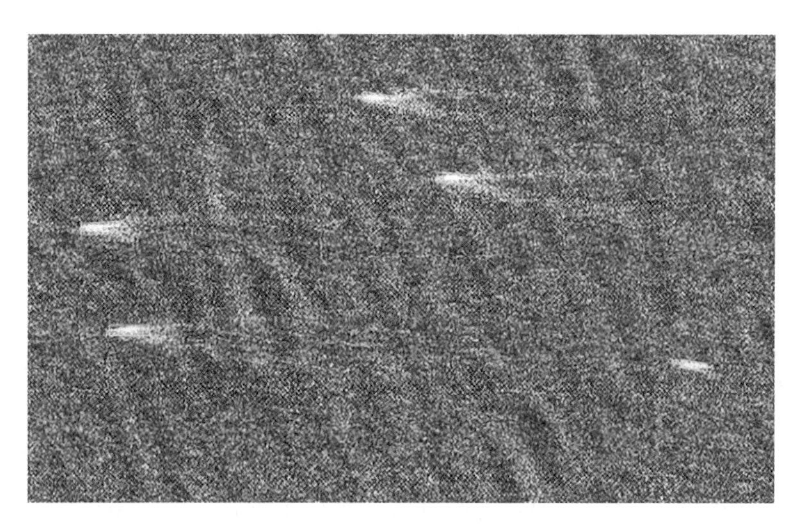

**Figure 6.23** The detection of moving vessels in a marine area.

necessary to suppress the sea clutter according to a sea clutter feature to weaken the influence of the sea clutter on target detection. There are two common approaches to suppress sea clutter. One is frequency-agile noncoherent processing. The other is fixed coherent frequency MTD filter bank. In addition to the above-mentioned marine target detection methods applied in the signal domain, a SAR system may be used to detect marine vessels by means of imaging the marine and target detection in the image domain. Due to the complicated motion characteristics of ship targets, long synthetic aperture imaging is difficult. Therefore, the marine surface image is usually obtained by airborne subaperture imaging or spaceborne imaging. It is necessary to separate the target from background clutter in the SAR image. The basic method will be introduced in Chapter 7. Figure 6.23 shows the detection of moving ships in some marine areas.

#### 6.6.3.1 Frequency-Agile Noncoherent Processing

The amplitude and phase of marine clutter satisfy a certain statistical distribution law. The sea clutter feature is closely related to the sea state, working frequency, polarization, radar illumination region, and other factors. According to different characteristics, it can be divided into temporal correlation, spatial correlation, and frequency correlation. Frequency-agile processing reduces the frequency dependence on marine clutter by changing the frequency of the carrier pulse to pulse. Define $R(i,j)$ as the normalized correlation coefficient between $i^{\text{th}}$ and $j^{\text{th}}$ clutter region:

$$R(i,j) = \frac{\sin(\pi\tau\Delta f)}{\pi\tau\Delta f} \tag{6.46}$$

where $\tau$ is pulse duration time, $\Delta f = f_i - f_j$, and $f_i$ and $f_j$ are the transmitted radar frequencies corresponding to the $i^{\text{th}}$ and $j^{\text{th}}$ clutter region, respectively. It can be seen from Eq. (6.46) that when variation of the carrier frequency is larger than $\frac{1}{\tau}$, basically there is no frequency correlation between sea clutters.

When frequency agility operation is used, fluctuation of the target echo becomes fast due to frequency decorrelation. After the accumulation of a certain number of echoes, the target amplitude tends to be the average value. However, after frequency decorrelation, the statistical characteristics of sea clutter are similar to those of the receiver noise, and the amplitude decreases significantly. Therefore, the signal-to-clutter ratio can be improved by noncoherent accumulation of the pulses.

#### 6.6.3.2 Filter Bank Method for Fixed-Frequency-Coherent MTD

When the radar system works in fixed-frequency mode, the MTD filter bank can be used to suppress the sea clutter effectively. The basic principle is as follows: the pass band of the Doppler filter bank covers a certain frequency range. The output of each filter is used to determine whether there is a moving target in this frequency band. If there is a moving target, relative velocity of the target may be calculated. MTD commonly uses an evenly distributed filter bank. Each filter only passes the in-band clutter, which reduces the influence of other frequency clutters effectively, so that the signal-to-clutter ratio is improved. However, the improvement factor of filters differs. The improvement factor is worse when it is close to the center of the clutter spectrum.

The MTD filter bank is usually realized by FFT in the frequency domain or by finite impulse response (FIR) filter in the time domain. In terms of computation burden, the FFT method has absolute advantage. For N-point FFT, the operation of each point corresponds to a bandpass filter. In each filter, the side lobe level can be reduced by the weighting window function. The FFT method is limited in the fixed notch of the filter. Especially at zero frequency and near the integer multiple of pulse repeat frequency, there is no notch, and the static sea clutter cannot be effectively suppressed. Compared with the FFT method, the FIR filter bank is designed flexibly. The frequency spectrum of the filter bank can be optimized by adjusting filter coefficients adaptively. There is no strict limit on the number of coherent pulses. Normally, when the number of coherent pulses is large, the FFT method is generally used. When the pulse number is small, and there is a special requirement on clutter suppression, the optimized FIR filter bank is generally preferred.

When the marine target is in low moving speed and the radar is far away, the limitation of azimuth resolution or angle error will lead to an unobvious target motion pattern in a short time. This makes it very difficult to initiate track and to track steadily and continuously. Taking into account the influence of sea clutter and the motion characteristics of marine targets, the data association method based on dynamic multifactor weighting is usually used in MMTI to track multiple targets [18]. The basic principle of the method is as follows: the local nearest neighbor method is used to associate the plot and track. At this time, multiple plots may be associated with multiple tracks, and all of them fall into the best target correlation gate. Then the motion characteristic of the target is used to calculate the dynamic correlation factor of each plot and relevant track to determine the optimal combination of plot and track and to remove the ambiguous association. The conventional motion characteristics of the target are location, flight course, speed, and so on.

(1) The deviation of plot and track is identified by the flight course, and the plot with minimum deviation is chosen as the best association plot. The mathematical model is

$$
\begin{aligned}
dx &= |(x_i - x_t)\sin k_t + (y_i - y_t)\cos k_t| \\
dy &= |(x_i - x_t)\cos k_t + (y_i - y_t)\sin k_t|
\end{aligned} \tag{6.47}
$$

where $(x_i, y_i)$ is the observed flight track, $(x_t, y_t)$ is the predicted flight track, $k_t$ is the course, and $dx$ and $dy$ are deviations in the $x$ and $y$ directions, respectively.

(2) Such information as track position (predicted position), course, and speed is comprehensively used and is weighted, processed to calculate the integrated correlation function value and to find the strongest points. The correlation function of the plot and the track is

$$\delta_i = \rho_1 R + \rho_2 V + \rho_3 K \tag{6.48}$$

where $\rho_1$, $\rho_2$, and $\rho_3$ are the weighting factors, whose value is related to characteristics of the track. Define a generic function as

$$f(x) = \begin{cases} 1 & x_1 \geq x \\ 1 - (x - x_1)/(x_2 - x_1) & x_1 < x \leq x_2 \\ 0 & x > x_2 \end{cases} \tag{6.49}$$

If $x = r$, $x_1 = r_1$, $x_2 = r_2$, Eq. (6.49) is position factor function $R(r)$. Similarly, if $x = v$, $x_1 = v_1$, $x_2 = v_2$ Eq. (6.49) is the speed factor function $V(v)$. If $x = k$, $x_1 = k_1$, $x_2 = k_2$, Eq. (6.49) is the course factor function $K(k)$. $r_1$, $r_2$, $v_1$, $v_2$, $k_1$, and $k_2$ are the corresponding correlation gates for location, speed, and course. Every function has two gates. When $\delta_i \geq \varepsilon$, the plot is correlated to the track. $\delta_i = \max_{i=1 \to n} \delta_i$ is selected as the plot that is correlated to the track.

Figure 6.24 is the continuous tracking finding of marine vessel targets by an airborne SAR/MTI system. It can be seen that the track is stable and continuous. Figure 6.25 shows the marine tracking finding in MMTI model.

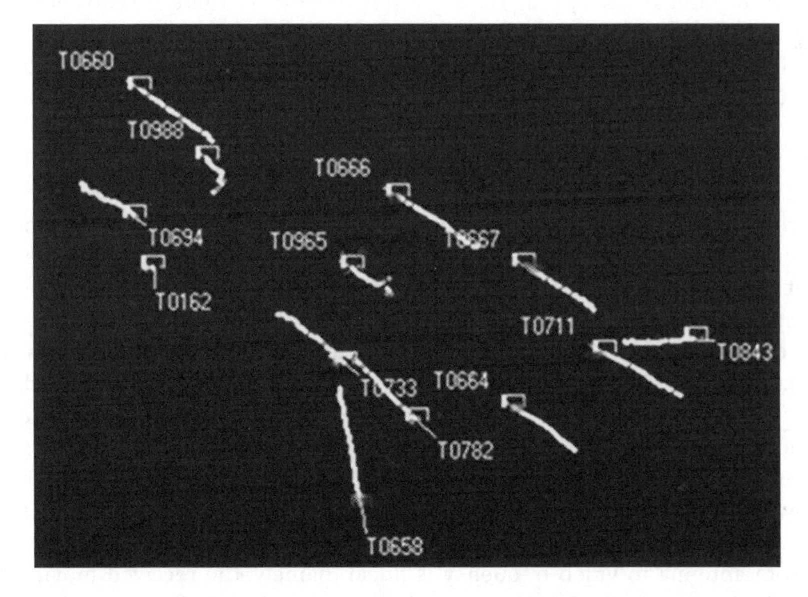


**Figure 6.24** The tracking finding of marine targets by an airborne SAR/MTI system. (See color plate section for the color representation of this figure).

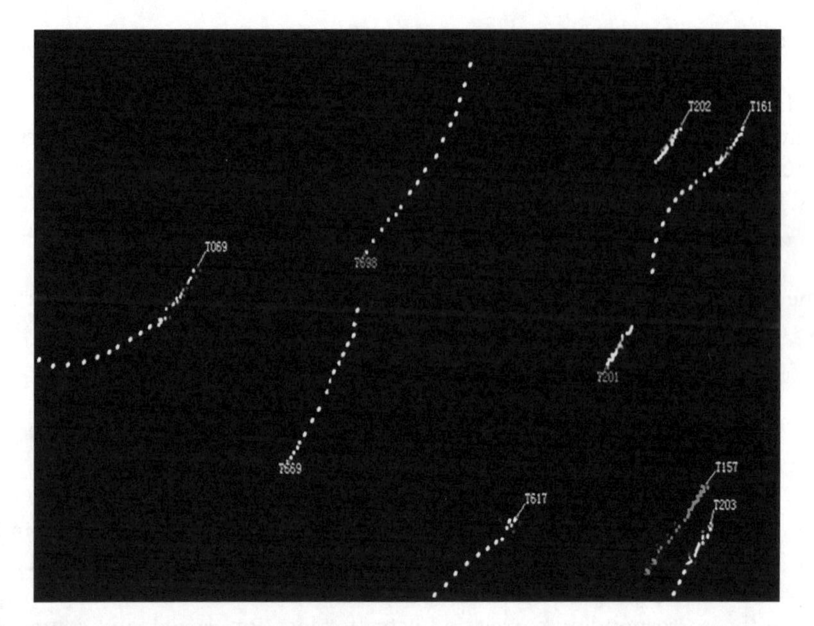


**Figure 6.25** Marine tracking finding in MMTI model. (See color plate section for the color representation of this figure).

### 6.6.4 Airborne Moving Target Indication

Airborne moving target indication (AMTI) of SAR/MTI radar is primarily used on airborne platforms. The main mission is to detect small targets at low altitude. The clutter characteristics in AMTI mode are fundamentally similar to those in GMTI or MMTI mode, but the target velocity range is larger in AMTI mode. There are two conventional methods for airborne MTD. One is pulsed-Doppler (PD) processing, and the other is STAP. The basic principle of PD is as follows: at first, the central frequency in the main lobe clutter spectrum is estimated as per the range-Doppler spectrum. This frequency center is used to compensate the echo, and the main lobe clutter location of the compensated signal is moved around zero frequency. Finally, the purpose of MTD is achieved by filtering the main lobe clutter through the designed filter bank (or FFT transform). Compared with 1D Doppler filter in PD, STAP is 2D joint processing based on array space and time domain Doppler. Mathematically, the array space and time domain Doppler-based 2D joint processing has a higher degree of freedom and has more abundant information of distinguishable target, clutter, and noise.

The echo signals from different pulses have different Doppler frequencies due to the radial velocity difference of target, jamming, and clutter relative to the radar. This Doppler frequency difference is used to realize the separation of target and clutter in the frequency domain, the moving target is detected from strong clutter background, and the target speed is accurately measured. In short, the core of Doppler filter processing is a narrow-band filter bank, which retains the required signal to be detected by filtering out interference and clutter in the frequency domain, assuming all conversion of the radar from antenna to video frequency is linear, namely, the received moving target information is saved in the receiver without distortion to video frequency. In this way, the Doppler filter bank can be implemented in video frequency (zero intermediate

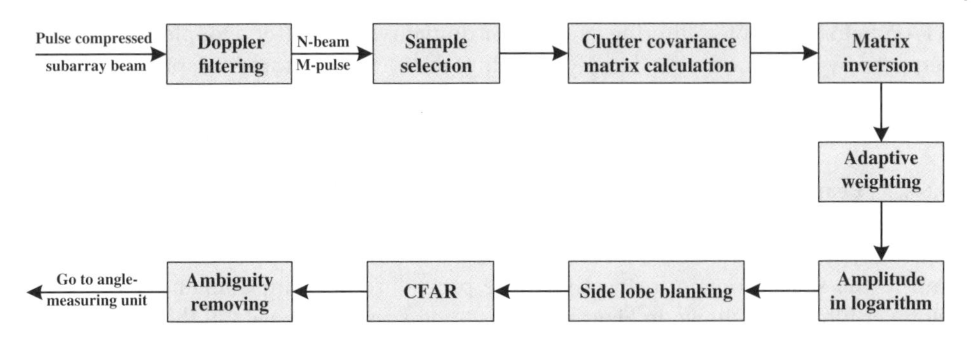


**Figure 6.26** Flowchart for moving target detection of STAP.

frequency). Compared with intermediate-frequency filter, video frequency filtering is easily realized, and the video frequency quantization of the digital processing is much easier. Therefore, video frequency Doppler filter is simpler and more reliable. Moreover, digital filtering has the advantages of high reliability, electromagnetic interference resistance, simplicity and flexibility in design, high precision, etc. The basic principle of digital Doppler filtering is to obtain spectrum of the received echo signal by using FFT. Thus, spectrum information of the echo signal is provided for target detection.

The essence of Doppler filtering is coherent accumulation of echo pulses. It can improve the SNR of the target but cannot suppress clutter. STAP can suppress clutter adaptively according to different spatial angle and improve the signal-to-clutter ratio of the output signal. In practical application, PD is used to detect targets in Doppler clean areas to reduce the difficulty of real-time processing, and STAP is used to detect targets in side lobe clutter. The following example is used for illustration. Figure 6.26 shows the MTD process of STAP.

(1) After Doppler filtering the compressed subarray beam data (or array data), a three-dimensional (3D) matrix, namely, beam, pulse (Doppler channel), and range, is produced. A clutter sample is selected at every filtered Doppler channel. The main purpose of sample selection is to make the sample distribution as uniform as possible and to remove those suspected targets, avoiding the effect of target cancellation caused by spatial suppression. Classic sample selection methods include generalized inner product, symmetrical window detection, and correlation dimension.
(2) Cumulative average is applied on the trained samples after selection to estimate the clutter covariance matrix. Multiplying the inversion of the clutter covariance matrix by the target-steering vector, the space-time adaptive weighting vector is obtained. This weighting vector is used to filter the beam–pulse–range 3D matrix, and the range-Doppler 2D spectrum is obtained after STAP.
(3) Taking the logarithm value of the range-Doppler spectrum amplitude, after side lobe blanking, constant false alarm rate (CFAR) detection, and range-Doppler disambiguation, target angle estimation by sum and beam forming is performed. Finally, such information as range, velocity, amplitude, and angle is smoothed and correlated.

In airborne target detection mode, PD processing is relatively simple, and it will not be discussed in this book. STAP processing can be generally divided into beam-domain

STAP before Doppler filtering and beam-domain STAP after Doppler filtering as required by different design. Simply speaking, the former performs adaptive weighting before Doppler filtering, and the latter performs adaptive weighting after Doppler filtering.

#### 6.6.4.1 Beam-Space STAP Before Doppler Filtering

Assuming the number of antenna elements is $N$, the number of coherent pulses is $M$, and $K_t$ and $K_s$ are the number of pulses and beams after rank reduction in the pulse domain and in the space domain, respectively, the procedure of beam-domain STAP before Doppler filtering is shown in Figure 6.27. Assume $\widetilde{x}_p$ is sample data of the $p^{\text{th}}$ $K_tK_s \times 1$ beam-space sub-coherent pulse interval (CPI), $x$ is the array sample data received by the antenna, and $\widetilde{G}$ is the beam-forming matrix, namely

$$\widetilde{x}_p = (J_p \otimes \widetilde{G})^H x \tag{6.50}$$

where $J_p$ is the pulse selection matrix of $M \otimes K_t$, and superscript $H$ is the matrix conjugate transpose. The sub-CPI adaptive weight of group $p$ is

$$W_p = \widetilde{R}_{up}^{-1}\widetilde{u}_t \tag{6.51}$$

where $\widetilde{R}_{up}$ is the estimated covariance matrix according to sub-CPI samples, and $\widetilde{u}_t$ is space-time steering vector in the expected target.

The adaptive weights of sub-CPI are rearranged into a $K_s \times K_t$ weight matrix, given by

$$W_p = [w_{p,0}, w_{p,1}, \dots, w_{p,k}] \tag{6.52}$$

After adaptive weighting, the output of sub-CPI is

$$y = [(I_M \otimes \widetilde{G}W)]^H x \tag{6.53}$$

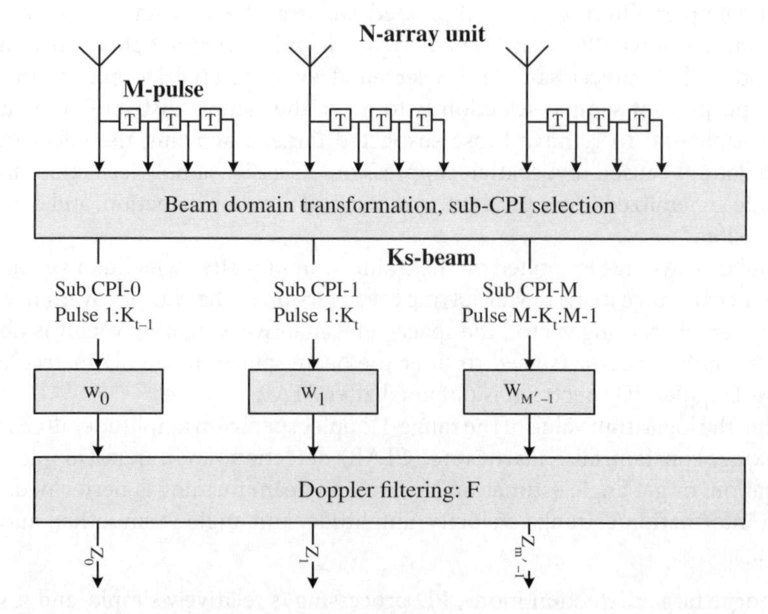


Figure 6.27 Beam-domain STAP before Doppler filtering.

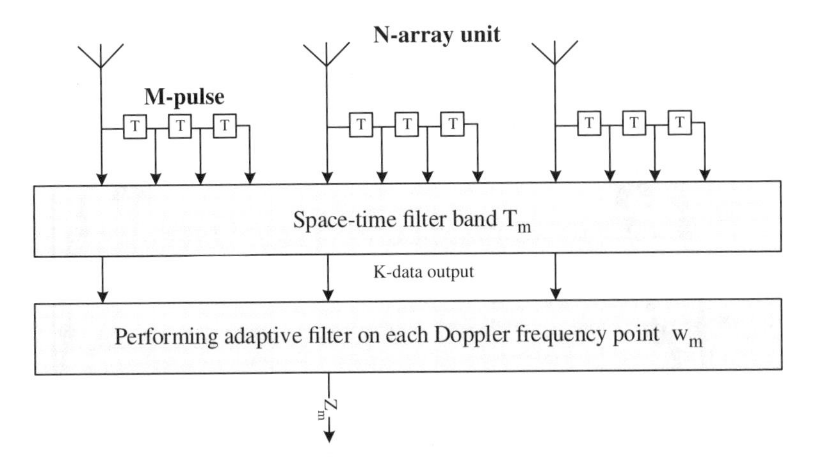


**Figure 6.28** Beam-domain STAP after Doppler filtering.

where $W = [\text{vec}(W_0 J_0^T), \text{vec}(W_1 J_1^T), \ldots, \text{vec}(W_{M'-1} J_{M'-1}^T)]$. Finally, the output of all sub-CPI passes through a Doppler filter with a length of $M'$ and the filter factor of $F = [f_0, f_0, \cdots, f_{M'}]$. The final output of the $m^{\text{th}}$ Doppler channel is

$$z_m = f_m^H y = w_m^H x = \left[ I_M \otimes \widetilde{G} W f_m \right]^H x \tag{6.54}$$

#### 6.6.4.2 Beam-Domain STAP After Doppler Filtering

The general procedure of beam-domain STAP after Doppler filtering is shown in Figure 6.28. Assuming $\widetilde{F}_m$ is the Doppler filtering matrix of $M \times K_t$ and $\widetilde{G}$ is the beam-forming matrix of $N \times K_s$, let $K = K_t K_s$ and $T_m = \widetilde{F}_m \otimes \widetilde{G}$, the data after space-time filter bank is

$$\widetilde{x}_m = T_m^H x = \left( \widetilde{F}_m \otimes \widetilde{G} \right)^H x \tag{6.55}$$

The space-time adaptive weight of the $m^{\text{th}}$ Doppler channel is

$$W_m = \widetilde{R}_{um}^{-1} \widetilde{u}_t \tag{6.56}$$

where $\widetilde{R}_{um}$ is the sample covariance matrix of $K_s K_t \times K_s K_t$, and $\widetilde{u}_t$ is the expected target of $K_s K_t \times 1$. The output of the $m^{\text{th}}$ Doppler channel after filtering is

$$z_m = W_m^H \widetilde{x}_m = [(\widetilde{F}_m \otimes \widetilde{G}) \widetilde{W}_m]^H \widetilde{x}_m \tag{6.57}$$

Based on the simulated data and real data of an airborne SAR/MTI system in the AMTI model, both PD and STAP are compared and analyzed in their performance. The parameters of the simulation are set as follows: the radar working frequency is 3 GHz, the transmitted peak power is 200 kW, duty cycle of the transmitted pulse is 6%, the number of antenna array elements is 18, and the gain in the transmitting and receiving antenna are 22 dB and 10 dB, respectively. The receiver bandwidth is 4 MHz and the radar system loss is 4 dB, the pulse repeat frequency is 3000 Hz, and the total of accumulated pulse is 20. Figure 6.29 shows a comparison of conditions where the beam-forming number changes from 2 to 6. It can be seen that as the beam number goes up, the signal-to-interference-plus-noise ratio (SINR) loss of small Doppler area

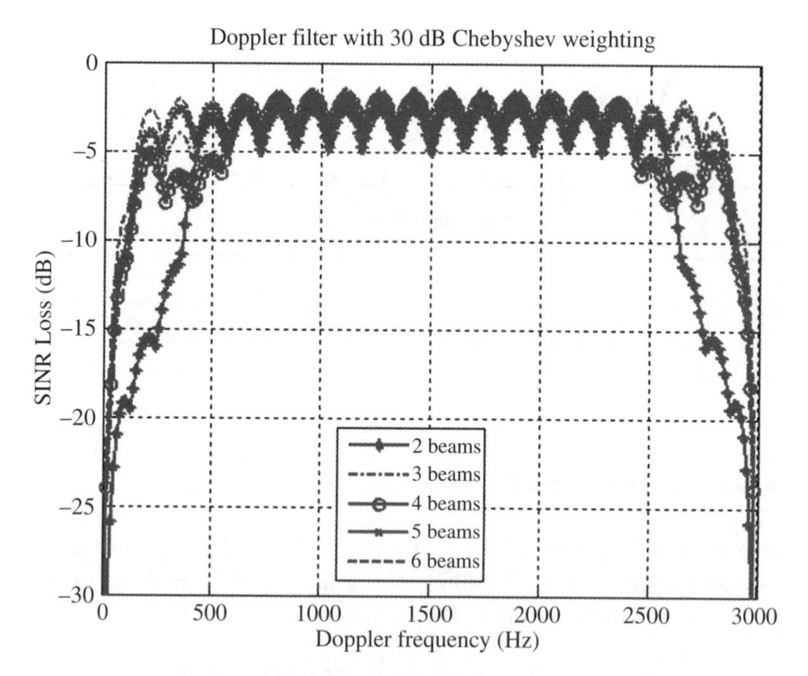


**Figure 6.29** Comparison of beam-domain STAP algorithms before Doppler filtering. (See color plate section for the color representation of this figure).

(close to main lobe clutter) decreases, and low-speed target detection performance is better. Figure 6.30 shows SINR loss of the Doppler filter bank with and without weighting. It can be seen from simulations that the loss near the main lobe clutter is obviously larger than that of the algorithm with 30 dB weighting, although the loss of the unweighted algorithm in the clean area is small. Therefore, weighting the Doppler filter bank is helpful to improve detection capability of a low-speed target.

The performance of PD and STAP is compared through airborne real data. The test data is located in the 252105 beam number, with burst number of 325 and CPI number of 4. Figure 6.31 is the range-Doppler spectrum after PD processing. It can be seen that residual clutter in the main lobe and side lobe clutter are relatively large, and the false alarm caused by target detection is high. After applying beam-domain STAP after Doppler filtering, the clutter around the target can be significantly suppressed, as can be seen from the local zoom in Figure 6.32, which improves the SNR, implying the residual clutter decreases dramatically, and the target is highlighted.

In AMTI mode, during tracking a low-altitude target, influence of the ground echo and high-speed ground target challenges the stability of low-altitude and small target tracking. Track before detect (TBD) is usually used. The basic principle is as follows: a large number of badly correlated plots and temporary tracks are discarded frame by frame. Several possible temporary tracks are correlated to every plot with an exhaustive method. The more accumulation frames of the associated temporary track there are, the longer the duration is. According to the number of false alarm tracks, the number of associated frames in TBD is automatically set, so that the real target is automatically detected to reduce the number of false tracks.

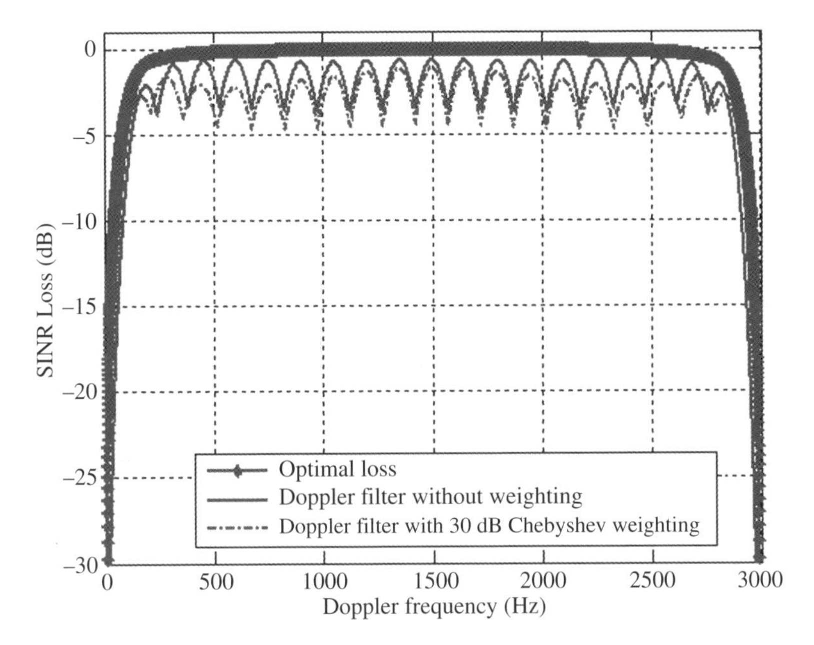


**Figure 6.30** Comparison of beam-domain STAP algorithms after Doppler filtering. (See color plate section for the color representation of this figure).

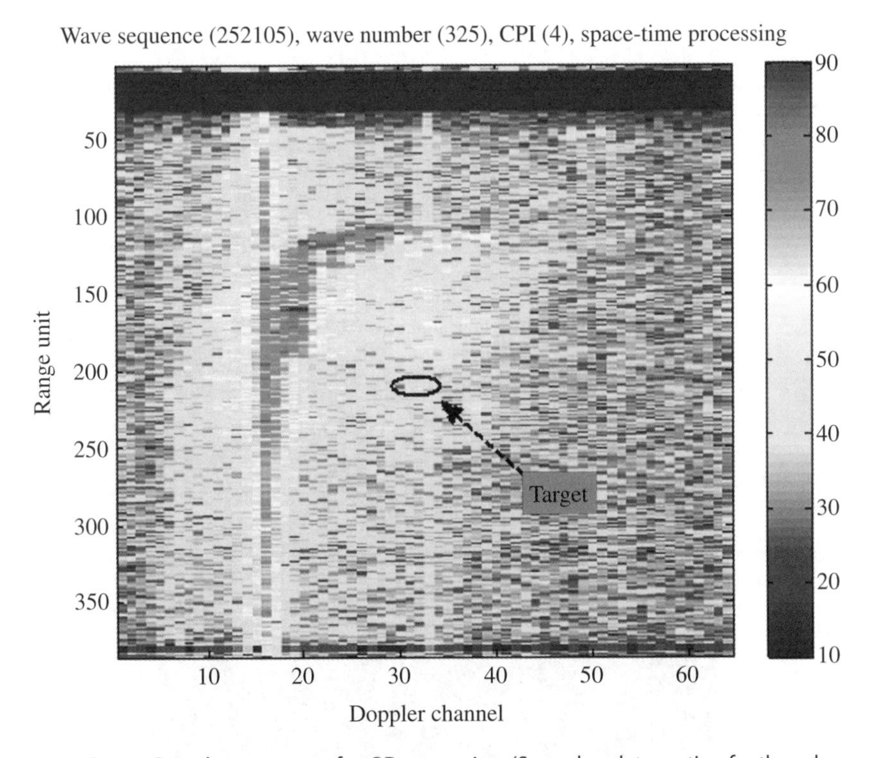


**Figure 6.31** Range-Doppler spectrum after PD processing. (See color plate section for the color representation of this figure).

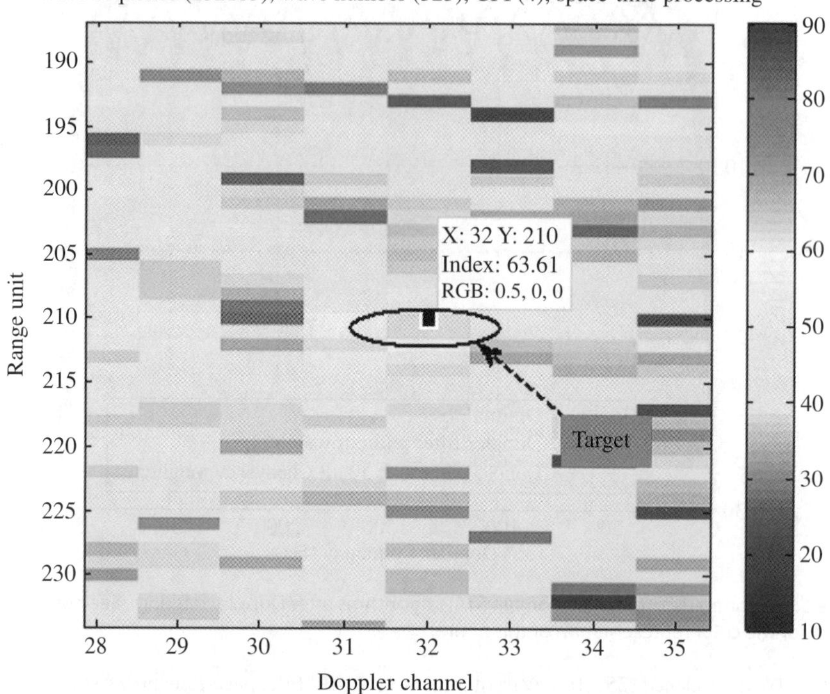


**Figure 6.32** Range-Doppler spectrum after STAP processing. (See color plate section for the color representation of this figure).

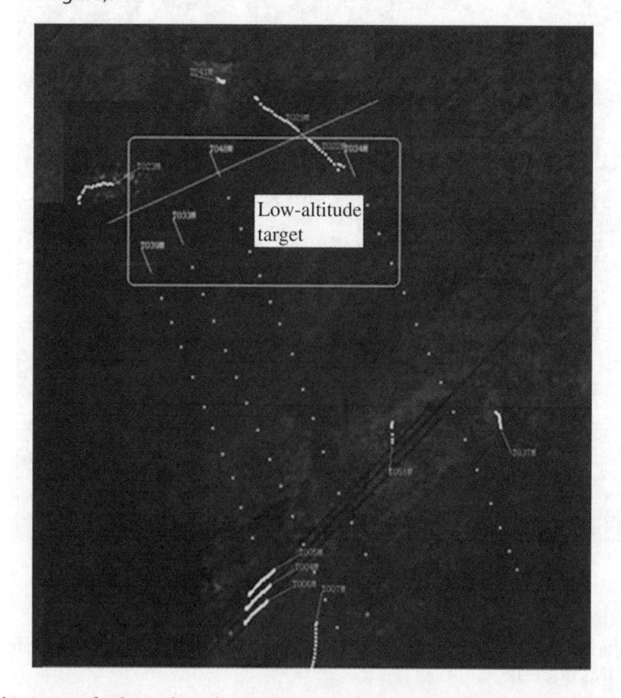


**Figure 6.33** Tracking map for low-altitude small target. (See color plate section for the color representation of this figure).

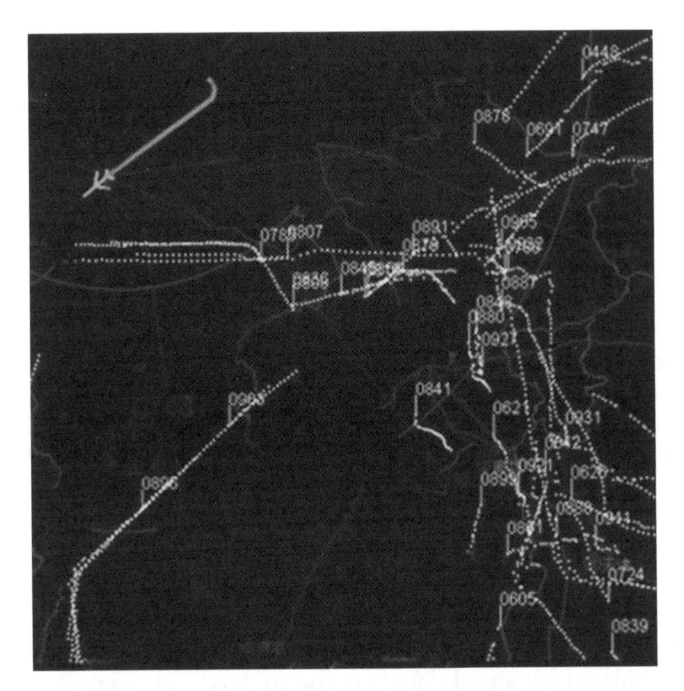


**Figure 6.34** Tracking map for AMTI. (See color plate section for the color representation of this figure).

During the tracking process, according to characteristics of the target (speed and flight course), terrain, and traffic network, low-altitude targets can be distinguished from ground targets. Figure 6.33 shows the actual ground surveillance map of airborne SAR/MTI radar, which not only displays the ground moving target track but also indicates tracks of four low-altitude small targets at a speed of about 100 m s$^{-1}$. Figure 6.34 is the tracking map in AMTI mode.

## 6.7 SAR Signal Processor

Compared with conventional radar, more information on the target can be acquired from SAR due to high resolution in both the range and azimuth directions, which also leads to massive data processing. How to complete the real-time processing of massive data has become a difficult problem for signal processors.

In the early stage, optical processing in SAR was adopted to obtain real-time processing results, and the signal processor was constructed by optical devices. Because of performance advantages of digital processing over optical processing, contributions have been made to reduce the implementation difficulty in digital processing. With the emergence of digital signal processors (DSPs), digital processing capability is enhanced dramatically. Currently, digital processing is used for real-time processing in modern SAR. The DSP constructed with high-speed digital devices can process gigabit-level data in a few seconds, and it has the ability to acquire SAR processing results in real time. Unless otherwise stated, "signal processor" refers to DSP in the following context.

The signal processor is the processing core of the SAR system. Signal processing, data processing, and system management of SAR under various operation modes are usually driven by software. Software usually adopts the idea of modular and componentized layered design, decomposes system complexity, reduces system coupling and dependence, improves cohesion, improves system reliability, and has the ability to reconstruct and upgrade software tasks.

### 6.7.1 System Architecture

To meet requirements for multiple operating modes of the SAR system, the signal processor follows the design concept of "reconstructing physical state and reconfigurable software mission" by modularization design and standardization design. Standardization design can make full use of shelf products and improve the system stability. In modularization design, the system architecture is not affected by increasing or decreasing the module number, making complex systems and simple systems compatible. System architecture based on chip interconnection is flexible and can meet different application requirements.

The signal processor composed of modules with standard sizes generally follows certain existing industrial standards (such as cPCI and OpenVPX), and data is exchanged through high-speed backplane. The internal processor construction generally includes three types of buses: high-speed data bus, control bus, and monitor bus. Taking the OpenVPX standard as an example, 4×RapidIO is usually used as a high-speed data bus, Gigabit Ethernet can be used as a control bus, and standard system management bus $I^2C$ can be used as a monitor bus. The system architecture is shown in Figure 6.35, where the interface module is used for external data interface, and the switching module is used for data interaction between modules within the unit. The interface module, switch module, and other modules are normally merged and implemented on a signal processing board to reduce the size and weight of the signal processor.

The hardware architecture of a signal processor is mostly based on the exchange structure. The data is exchanged through the internal communication mechanism inside the processing module. Between the processing modules, the data is exchanged through

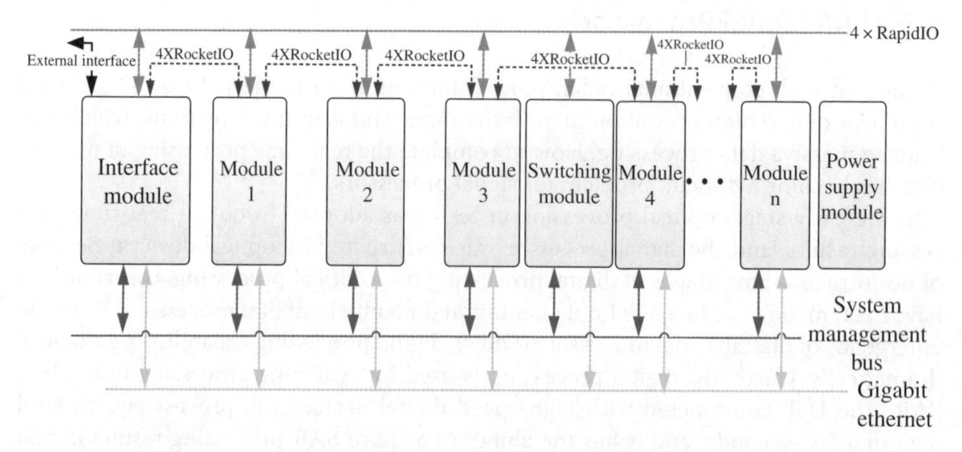


Figure 6.35 System architecture of a signal processor.

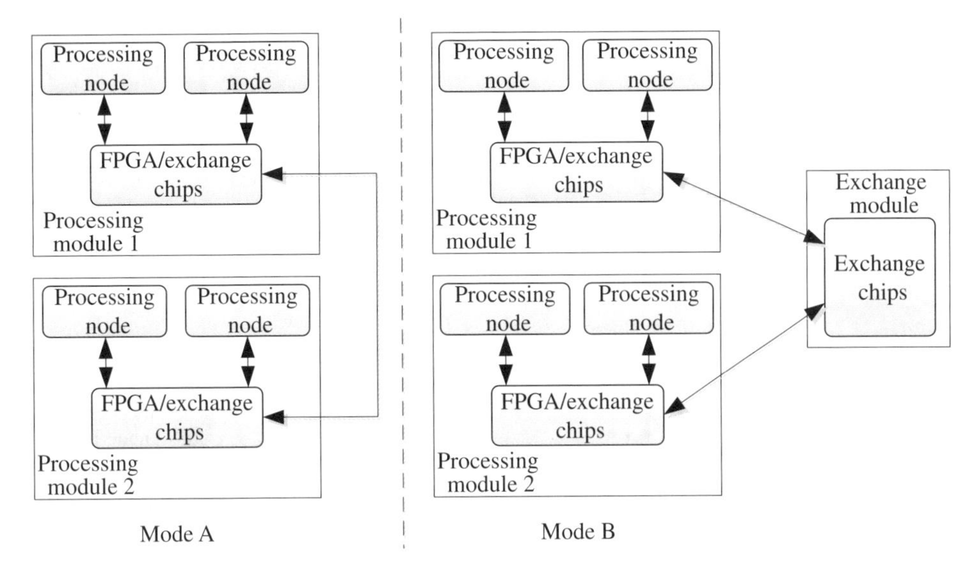


**Figure 6.36** Schematic of data exchange between processing modules.

field-programmable gate array (FPGA)/exchange chips and modules. In this way, the software function can be deployed dynamically to different processing modules, rather than concentrating on one certain module, which greatly improves the flexibility of system design. The data exchange between the processing modules is shown in Figure 6.36. A high-speed backplane bus in mode A is used to realize direct interconnection. The exchange chip in the exchange module in mode B is used for interconnection. Compared with mode B, mode A is simpler and occupies fewer resources, but its topology is fixed, making it difficult to generate a network array connection. Generally, a complementary of mode A and mode B is used to realize a flexible and reasonable distribution of the occupied resource.

### 6.7.2 Processing Architecture

In different SAR operating modes, processing capacity, memory size, and interface data rate in the signal processor have different requirements. Virtual node technology is generally introduced to improve the processing efficiency. The virtual node refers to a large processing node that contains several physical processing nodes where a high-speed communication link is used to exchange and share data between these nodes, with the following advantages of the flexible structure: the actual virtual mode is dynamically reconfigurable, a proper virtual mode can be automatically selected according to system operating mode, and optimal design can be achieved by finding the best balance point between system scale and absolute delay of processing. Its disadvantages are that the software programming is complex, the system hardware architecture is closely related, and the portability is poor.

Processing delay is a prominent parameter in real-time radar signal processing. It is limited by system size and power consumption. The virtual node technology can effectively solve these two problems. The virtual node technology is used in signal processors

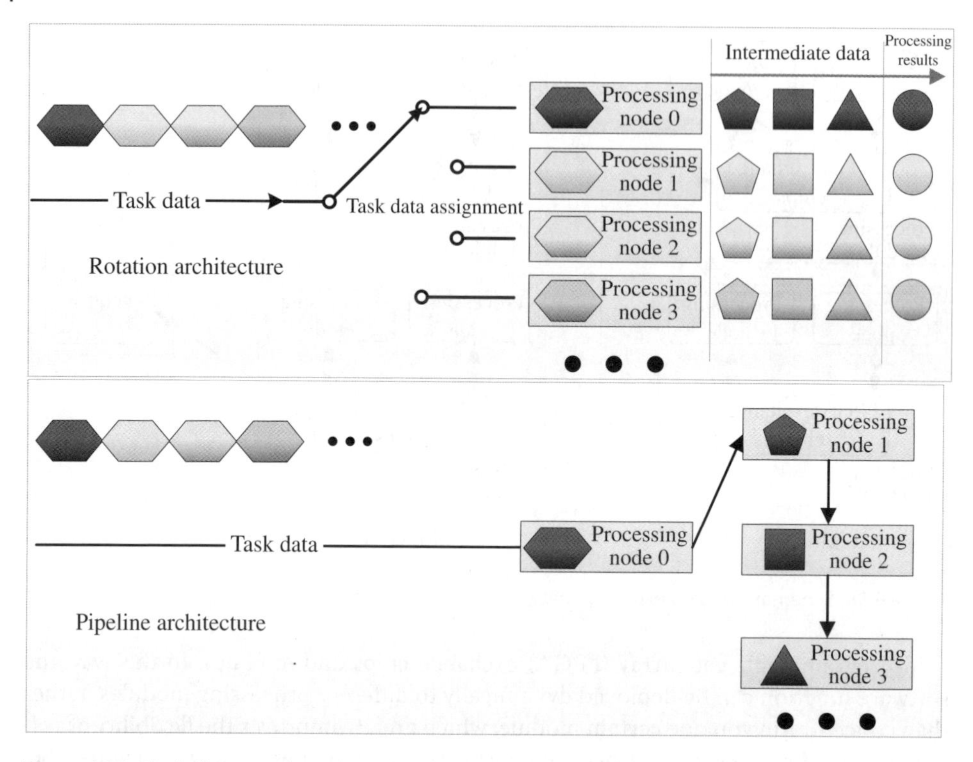


**Figure 6.37** Block diagram of system processing architecture. (See color plate section for the color representation of this figure).

to achieve optimal configuration in multiple modes, with configuration being dynamically defined according to computation amount, memory, and delay.

According to data flow of the system, the processing architecture can be divided into pipeline architecture and rotation architecture, as shown in Figure 6.37. In pipeline architecture, each node processes different submissions of the complete process flow one by one, and the data flow is divided as per the mission. In rotation architecture, each node completes the full process flow alone, and the data flow is divided as per the time. The characteristics of the two architectures are listed in Table 6.1. The table shows that rotation architecture has a shorter data transmission length than pipeline architecture, which improves the system reliability. Rotation architecture has higher node

**Table 6.1** Characteristics for comparison of rotation and pipeline architectures.

| Item | Rotation architecture | Pipeline architecture |
|---|---|---|
| Data transmission length | Short | Long |
| Processing node consistency | High | Low |
| Processing flexibility | High | Low |
| Board design difficulty | High | Low |
| 2D data read–write efficiency | Low | High |

consistency than pipeline architecture, which can not only reduce variety of the boards and maintenance difficulty but also can easily realize hot backup and improve reliability of the system. Rotation architecture is more flexible in processing than pipeline architecture, which makes it easy to achieve optimal system performance in multiple operation modes and to compress the system scale. For board design, the rotation architecture is more complicated and difficult than the pipeline architecture, because a high-capacity memory in rotation architecture needs to be equipped for every processing node. For 2D data read–write speed, the rotation architecture has lower efficiency over the pipeline architecture, because in rotation architecture, there is no memory board dedicated to solving the matrix transpose problem during image processing.

With the development of chip and circuit board manufacture, the two drawbacks of rotation architecture are gradually solved. Length-variable direct memory access (DMA) can be used to directly read and write 2D data. The read–write bandwidth of memory interface in a modern processor can reach Gbps in order of magnitude. The transmission bandwidth of a single high-speed serial bus is in the order of Gbps. Even though length-variable DMA causes a certain efficiency loss, the final read–write speed is comparable with interboard transmission speed. In addition, because DMA can work in parallel with the processor, with the improved memory integration, the memory capacity becomes larger and larger. A single chip with 4 GB memory capacity has emerged. With the development of such board manufacture technology as multilayer board, blind hole and surface mounting, and packaging, this problem will be addressed in more acceptable means.

### 6.7.3 Development Architecture

The development architecture determines performance of the signal processor. The processing platform can be divided into several layers, as shown in Figure 6.38. The first is the physical layer, a platform composed of modules and backplanes. The second is the driver layer, which provides low-level driving for every module, realizes hardware abstraction, and facilitates software development. The third is the protocol layer, which defines the intermodule and intramodule communication interface and constructs a

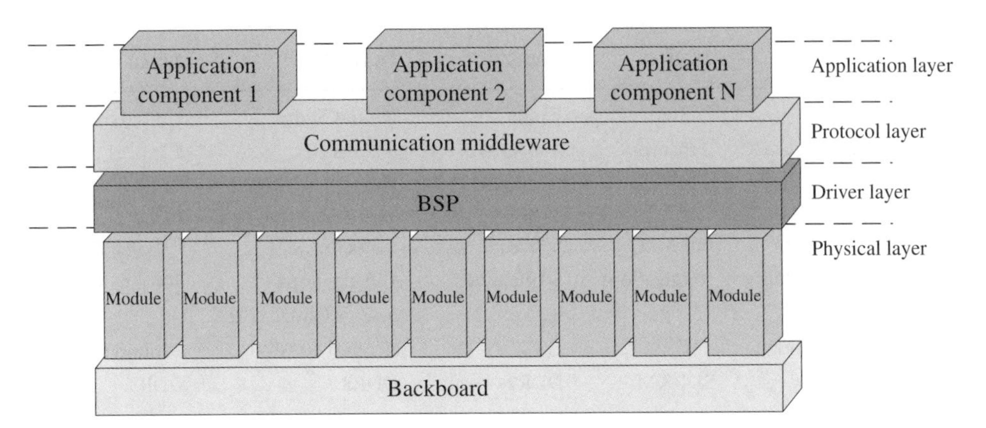


**Figure 6.38** Development architecture diagram of an integrated processing unit. BSP: board support package. (See color plate section for the color representation of this figure).

standard function library according to protocol. The fourth is the application layer, where different software algorithms are implemented. For modules that use FPGA for input–output management, the protocol layer has already been implemented into FPGA, and the driver layer is used to provide low-level communication support for the application layer.

### 6.7.4 Processing Module

A signal processor is usually composed of the following modules: signal processing, data processing, mission management, and power supply. According to application requirements, the appropriate modules in practical implementation can be selected to accomplish the system mission independently or jointly, which may make full use of the module performance, improve system efficiency, and reduce size and weight of the system.

#### 6.7.4.1 Signal Processing Module

The signal processing module usually uses DSP as the main processing chip. Compared with a multicore central processing unit and general-purpose graphics processing unit, the high performance and low power consumption of DSP makes it more suitable for SAR signal processing. The signal processing module is usually composed of multiple DSPs and high-capacity memory, which together constitute a high-speed parallel-interconnection multichip processing network, to meet the requirements for a large amount of data and computations in SAR signal processing.

(1) *Processing chip*. At present, typical processing chips are TigerSHARC chipsets from ADI, TMS320 chipsets from TI, and PowerPC chipsets from Freescale. In recent years, China also began to release a variety of DSP chips, and BWDSP100 is a representative one. BWDSP100 is a high-performance DSP that is completely domestically designed, from instruction set and system architecture to software development environment and high-level language compiler. This chip has excellent processing capability. The performance of different processing chips is shown in Table 6.2.

**Table 6.2** Performance comparison of mainstream processing chips.

| Items | TS201 | BWDSP100 | C6678 | T4240 |
|---|---|---|---|---|
| Country | United States | China | United States | United States |
| Internal core/macro | 2 macros | 4 macros | 8 cores | 12 cores |
| On-chip cache | 3 MB | 3 MB + 512 KB | (512 KB + 32 KB + 32 KB) × 8 + 4 MB | 128 KB × 12 + 2 MB × 3 |
| Operating frequency | 600 MHz | 500 MHz | 1.25 GHz | 2.0 GHz |
| Operational capability | 3.6 gigaflops | 30 gigaflops | 160 gigaflops | 240 gigaflops |
| FFT time at 1 K | 15.7 μs | 2.35 μs | 5.47 μs (single core) | — |
| Power consumption | 2 W | 6.5 W | Around 10 W | Around 50 W |
| External memory interface | SDRAM | DDR2 | DDR3 | DDR3 |
| Operating temperature | −40~80 °C | −55~85 °C | −40~105 °C | — |

(2) *Module architecture.* The signal processing module constitutes a high-speed parallel-interconnection multichip processing network. According to the types of buses, the high-speed parallel-interconnection multichip processing network can be divided into two types, tight coupling and loose coupling. In a tight coupling system, time division multiplex is used for every node to share the bus. In loose coupling, an independent distributed bus is adopted. A comparison of tight and loose coupling architectures is shown in Table 6.3.

#### 6.7.4.2 Data Processing Module

Most data processing modules are realized by PowerPC architecture-based high-performance embedded computers. PowerPC architecture is a type of reduced instruction set computer architecture. It has high performance and low power consumption and is mainly used in embedded systems. It can be used as a single-board computer for high-performance computing and image processing.

In the PowerPC family, G4 is currently the most widely used series. Compared to G3, there are two major improvements for G4. One is that it supports symmetric multiprocessor architecture. The other is that the first-class AltiVec technology is introduced in G4 to deal with vector operation. AltiVec technology is a 128-bit single instruction, multiple data vector processing engine, which may improve the performance 4.3 times or so.

The data processing module generally provides a variety of external interfaces, including full-duplex RS232 serial interface, full-duplex RS422 serial interface, onboard serial RapidIO exchange, and onboard PCIe exchange.

**Table 6.3** Comparison of tight and loose coupling architectures.

| Items | Tight coupling | Loose coupling |
|---|---|---|
| Architecture | One data bus is shared by several processing units, and the address segment occupied by the memory in different DSPs is normally the same | Every processing unit has an independent data memory, and the DSP is interconnected through high-speed serial interface |
| Connection | Single and linear | Linear, tree, star, and network matrix |
| Acceleration ratio | Only the quasi-linear acceleration ratio is available, when the computation amount is much larger than that of communication | Linear acceleration ratio |
| Feature | Simple in structure. A high acceleration ratio can be reached when the number of DSPs is small. When the number of processing units is relatively large, the shared bus will cause frequent bus collision and waiting | The processing capability of the system increases with increase of the processing scale, which is suitable for the construction of a large-scale parallel system |
| Applications | Suitable for large number of shared data applications, and it can avoid processing the shared data of every local memory in the distributed structure | Suitable for multichannel and pipeline processing, and every DSP processes data almost independently |

**Figure 6.39** A typical SAR processing module. (See color plate section for the color representation of this figure).

#### 6.7.4.3 Mission Management Module

The mission management module generally adopts the architecture of the exchange chip combined with FPGA, in which the control management function is completed by FPGA, and the data exchange function is realized by the exchange chip. The mission management module also provides a variety of multichannel external interfaces, including network interface, optical fiber interface, full-duplex RS422 serial interface, RapidIO interface, etc.

### 6.7.5 Typical Signal Processor

A typical SAR processing module is shown in Figure 6.39, and its internal structure is shown in Figure 6.40. It uses a mission management module as the communication interface between the signal processor and external equipment. The commercial mezzanine card (CMC) and rear plug board are used to further expand the interface type and to improve extendibility of the processor. CPCI and VME universal buses are used to connect external commercial off-the-shelf products. Dedicated links, such as optical fiber, are used for interboard wideband data transmission. According to application requirements, the module generally uses standard 3 U/6 U size. If necessary, special shapes can be made.

The single processing module consists of four BWDSP100 processing chips. The processing capacity of a single board can reach 72 gigaflops (300 MHz in main frequency)/120 gigaflops (500 MHz in main frequency). The board is a standard 6 U VPX. The power consumption of a single board is less than 60 W. The size is 233.35 mm × 160 mm. The weight of a single processing module is less than 500 g.

A photograph of a typical SAR signal processor is shown in Figure 6.41. Figure 6.42 shows the signal processor composition.

The signal processor uses a standard 6 U VPX architecture, consisting of 17 function modules, including 11 signal processing modules, a power supply module, a data processing module, a beam-forming module, an optical fiber interface module, a mission

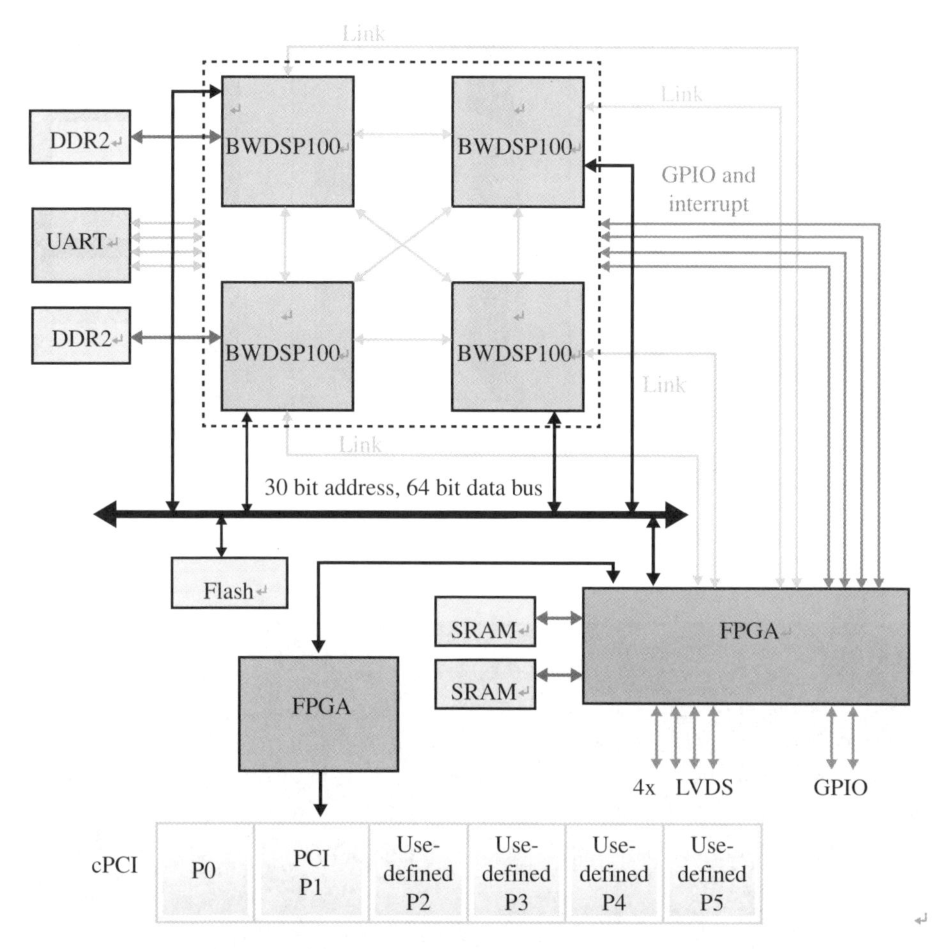


**Figure 6.40** Internal structure of the processing module. GPIO: general purpose input/output; LVDS: low-voltage differential signaling; SRAM: static RAM; UART: universal asynchronous receiver/transmitter. (See color plate section for the color representation of this figure).

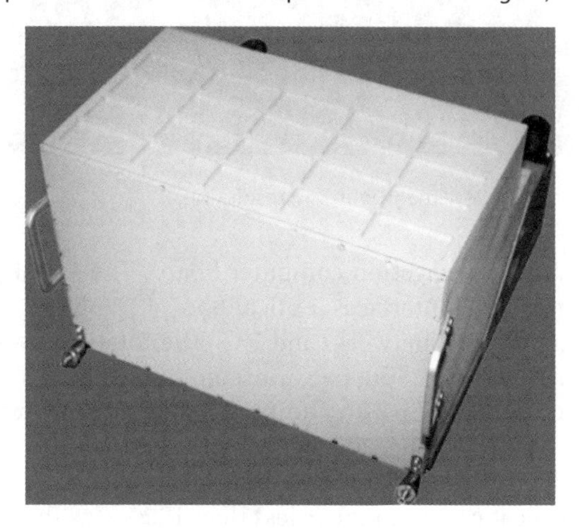

**Figure 6.41** Photograph of a typical SAR signal processor.

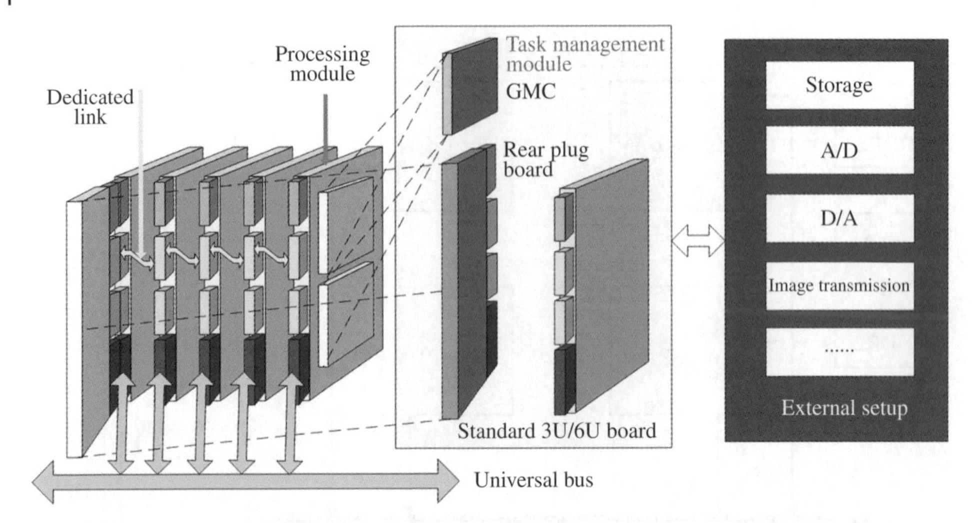


**Figure 6.42** Composition of the signal processor. (See color plate section for the color representation of this figure).

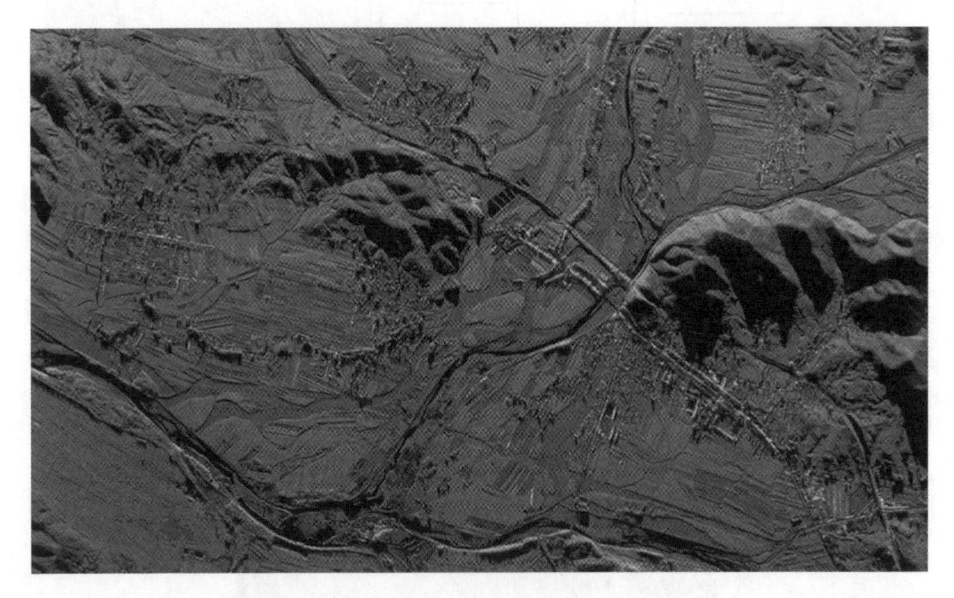

**Figure 6.43** X-band strip-map SAR image with 1 m resolution.

assignment module, and a navigation computer board. The signal processor supports the following external digital interfaces: optical fiber, 429, 1553B, asynchronous serial port, synchronous serial port, network, and $I^2C$. The total bandwidth of the external interface is more than 230 Gbps. The two-way communication speed of an interboard signal processor can reach 40 Gbps. The unit processing capacity can reach 6 teraflops except for FPGA. The outline dimension of the entire unit is 420 mm (width) × 270 mm (height) × 300 mm (depth). The weight is less than 29 kg. The power consumption is less than 1500 W, and the heat consumption is less than 1200 W with liquid cooling.

Figures 6.43 and 6.44 show X-band strip-map SAR images obtained after real-time processing, with resolution of 1 and 0.5 m, respectively. Figure 6.45 shows the real-time image of a slide spotlight SAR with 0.3 m resolution. Figure 6.46 shows real-time image of a strip-map SAR with 0.15 m resolution.

With the progress of technology, signal processors increasingly show the following characteristics: the operation capacity has been significantly improved, and operation

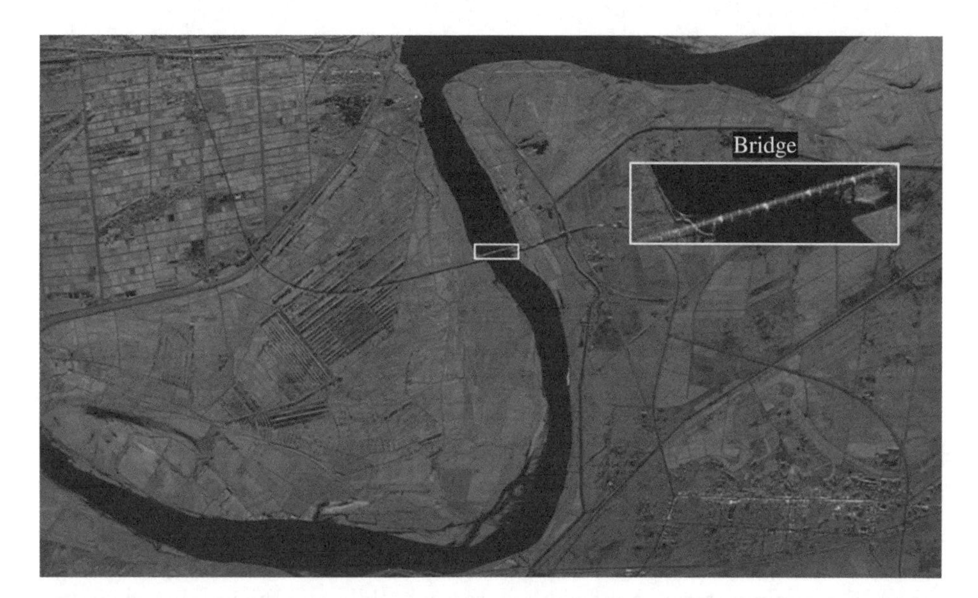


**Figure 6.44** X-band strip-map SAR image with 0.5 m resolution.

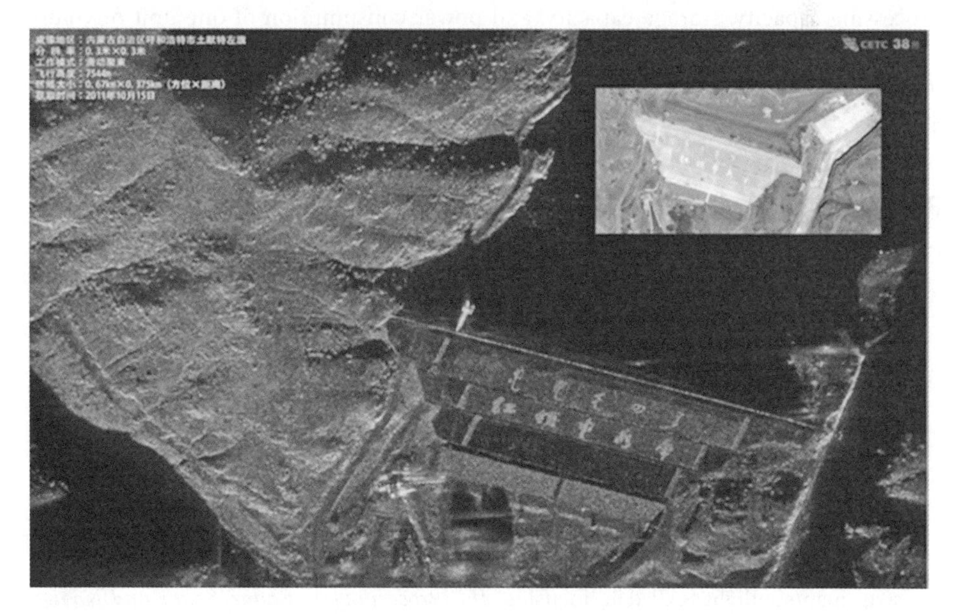


**Figure 6.45** X-band slide spotlight SAR image with 0.3 m resolution.

**Figure 6.46** X-band strip-mode SAR image with 0.15 m resolution.

capacity of a single processing module can reach $10^{12}$ floating-point operations per second. The communication bandwidth increases rapidly, and the interboard communication bandwidth can reach dozens of GB/s. The integration level is improved, and processing capacity, storage capacity, and power consumption of one unit of volume are all significantly increased. As power consumption per unit volume increases, thermal dissipation has been changed from wind cooling to liquid cooling or conduction cooling.

## References


1 Frey, O., Magnard, C., Ruegg, M. et al. (2009). Focusing of airborne synthetic aperture radar data from highly nonlinear flight tracks. *IEEE Transactions on Geoscience and Remote Sensing* 47 (6): 1844–1858.

2 Ulander, L.M.H., Hellsten, H., and Stenström, G. (2003). Synthetic-aperture radar processing using fast factorized bBack-projection. *IEEE Transactions on Aerospace and Electronic Systems* 39 (3): 760–776.

3 Zhu, D. and Zhu, Z. (2007). Range resampling in the polar format algorithm for spotlight SAR image formation using the Chirp-Z transform. *IEEE Transactions on Signal Processing* 55 (3): 1011–1023.

4 Lanari, R., Zoffoli, S., Sansosti, E. et al. (2001). New approach for hybrid strip-map/spotlight SAR data focusing. *IEE Proceedings – Radar, Sonar and Navigation* 148 (6): 363–372.

5 Smith, A.M. (1991). A new approach to range-Doppler SAR processing. *International Journal of Remote Sensing* 12 (2): 235–251.
6 Li, F.K., Held, D.N., and Curlander, J.C. (1985). Doppler parameter estimation for spaceborne synthetic aperture radars. *IEEE Transactions on Geoscience and Remote Sensing* 23 (1): 47–56.
7 C.Y. Chang, J.C. Curlander (1989). Doppler centroid ambiguity estimation for synthetic aperture radars. *Proceedings of the 12th Canadian Symposium on Geoscience and Remote Sensing*, Vancouver, Canada (July 10–14, 1989). Piscataway, NJ: IEEE.
8 Berizzi, F. and Corsini, G. (1996). Autofocusing of inverse synthetic aperture radar images using contrast optimization. *IEEE Transactions on Aerospace and Electronic Systems* 32 (2): 1185–1191.
9 Wahl, D., Eichel, P., Ghiglia, D. et al. (1994). Phase gradient autofocus – a robust tool for high resolution SAR phase correction. *IEEE Transactions on Aerospace and Electronic Systems* 30 (3): 827–835.
10 Cao, F., Zheng, B., and Yuan, J. (2001). The study on DGPS/INS integrated system used for SAR motion compensation. *Chinese Journal of Aeronautics* 22 (2): 121–124.
11 Deng, H. (2014). An accurate compensation method for UWB airborne SAR azimuth space-variant error. *Signal Processing* 30 (2): 221–226.
12 Chan, H.L. and Yeo, T.S. (1998). Noniterative quality phase-gradient autofocus algorithm for spotlight SAR imagery. *IEEE Transactions on Geoscience and Remote Sensing* 36 (5): 1531–1539.
13 Bao, Z., Liao, G., and Wu, R. (1993). Space-time two-dimensional adaptive filtering for phased array airborne radar clutter suppression. *Journal of Electronics* 21 (9): 1–7.
14 Richardson, P.G. (1994). Analysis of the adaptive space time processing technique for airborne radar. *IEE Proceedings – Radar, Sonar and Navigation* 141 (4): 187–195.
15 Zheng, B. (1994). Space-time signal processing for airborne radars. *Modern Radar* 16 (2): 17–27.
16 Deng, H. and Zhang, C. (2009). A real-time signal processing method for airborne three-channel GMTI. *Journal of Electronics and Information Technology* 31 (2): 370–373.
17 Fortmann, T., Bar-Shalom, Y., and Scheffe, M. (1983). Sonar tracking of multiple targets using joint probabilistic data association. *IEEE Journal of Oceanic Engineering* 8 (3): 173–184.
18 Yun, S. (2011). Tracking concatenation algorithm for airbore radar after surveillance intermission. *Radar Science and Technology* 9 (6): 537–541.

# 7

# Image Information Processing System

## 7.1 Overview

Synthetic aperture radar (SAR) image information mainly includes target detection, target change detection, target recognition, multisource SAR image fusion, etc., all parts of the SAR image information processing system.

(1) *Target detection.* Target detection determines whether there is a target of interest, and if so, the target attribution will be extracted. Different users have different definitions of the target. For example, the main objectives of concern for mapping departments are surface features. The forestry department is more concerned about vegetation. The main objectives of the marine sector are sea breezes, waves, and ships. The main objectives of the geological department are rocks and minerals. The main objectives of the military department are military installations and equipments. Different data sources mean different targets, and the methods to extract targets from SAR images are not quite the same, including a variety of target detection algorithms.
(2) *Target change detection.* Target change detection determines the variation of interested targets in multitemporal images. With the accumulation of huge amounts of SAR data, how to extract effective information from the tremendous amount of data rapidly and how to discard redundant intelligence become the challenges in manual interpretation of current SAR images. SAR images are used to detect changes; obtain the changes of surface features and focus on the surface of the target of interest automatically, which can greatly improve the detection efficiency; quickly acquire the monitoring intelligence of the target movement, damage assessment, etc.; and effectively enhance the intelligence support capabilities.
(3) *Target recognition.* Target recognition is based on target detection. Through extraction and analysis of the detected target characteristics, automatic or semiautomatic identification methods are used to judge the target category or specific model. Target recognition is an important requirement for military forces. For example, for maritime ship targets, major target identification can be divided into military ships and civilian vessels. Minor identification can be divided into aircraft carriers, destroyers, cruisers, fishing ships, oil tankers, cargo ships, etc. For even finer identification, the specific model and description of the target can be located. Target recognition is generally supported by a target database. Without the database, it is

*Design Technology of Synthetic Aperture Radar,* First Edition. Jiaguo Lu.

difficult to identify the target. Target recognition methods mainly include template matching, statistical classification, neural networks, etc.

(4) *Multisource SAR image fusion*. Multisource SAR image fusion is intended to solve how to perform target detection, target change detection, and target recognition more effectively under multisource SAR images. When imaging the same scene, SAR images with different phases, different wavebands, and different polarizations can be acquired because of different sensors and platforms. A SAR image from each source is a local indication of the ground feature. The surface features can be comprehensively analyzed by full fusion of the characteristics from various data sources. The fusion method can be classified into pixel-level, feature-level, and decision-level. For every level, there are many kinds of processing algorithms and strategies. The algorithm or strategy to be used is determined by the actual application.

SAR systems are operated with the use of multiband, multipolarization, multiresolution, and multimode. Earth observation in all dimensions has been achieved. A huge amount of image data is produced every day. As a SAR image is quite different from an optical image, information processing is much more difficult than that of conventional optical images, due to influences of speckle noise, shadow, foreshortening, top and bottom inversion, and other geometric features. At present, the SAR image information processing capability is not suitable for development of SAR systems. It cannot meet the information processing requirements of the massive amount of image data obtained by the SAR system in practical applications, so that the potential application performance of the SAR system is not fully reflected.

This chapter begins with the application requirements for SAR image information. The technologies of SAR image target detection, target change detection, target recognition, and multisource SAR image fusion are discussed. Many typical information processing technologies are proposed, and up-to-date information processing results of SAR images are presented.

## 7.2 Target Detection

The goal of SAR image target detection is to define the presence of the target in the image and to separate the target from background clutter. Target detection becomes a challenging task due to the diversity of targets and the complexity of SAR image background clutter. Regarding the detection of highly scattering targets, the constant false alarm rate (CFAR) detection algorithm is most commonly used in SAR image target detection. For non-highly scattering target detection, such as the structure objectives (bridges and airports and so on), CFAR detection is not applicable. These targets should be detected according to other features of the target, such as shape and environment characteristics. The target detection quality has a direct impact on subsequent target identification, classification, and recognition.

### 7.2.1 Highly Scattering Target Detection

CFAR detection is usually used in highly scattering target detection. CFAR detection is a pixel-level detection, which is based on the contrast between target and background.

In practice, the target background tends to be complex, and the fixed threshold cannot be used to detect the target. Instead, the threshold is determined adaptively. CFAR detection completes the purpose of detecting the target pixel through comparison of pixel gray level with a certain threshold. Given a false alarm probability, the detection threshold is specified by the statistical characteristic of clutter.

Assuming $p(x)$ is the probability density function of clutter distribution model and distribution function $F(x) = \int_{-\infty}^{x} p(t)dt$, $F(x)$ is an increasing function in $(-\infty, +\infty)$, then by solving the following equation:

$$1 - P_{fa} = \int_{-\infty}^{I_c} p(t)dt \tag{7.1}$$

the threshold $I_c$ can be obtained. $P_{fa}$ is the known false alarm probability.

CFAR detection is divided into global CFAR and local CFAR. All data is used to calculate the statistical distribution parameters, which is called *global CFAR*; a local window is used to estimate statistical distribution parameters, which is called *local CFAR*. Global CFAR is simple and suitable for target detection in a uniform clutter background. Local CFAR is suitable for target detection in nonuniform background clutter and generally requires three sliding windows: target window, protection window, and background window, as shown in Figure 7.1. The protection window is used to prevent partial pixels of the target from leaking into the background window, which causes the incorrect calculation of the background area statistics ally. A background window is used for background clutter statistics, and its size may be selected empirically. If two targets are approaching too close, this method may lead to missed detection.

To detect targets with different sizes in different directions, the size of the target window needs to be appropriately selected. If the size of a small target is used as the size of the target window, many pixels will be located in the background window for the large

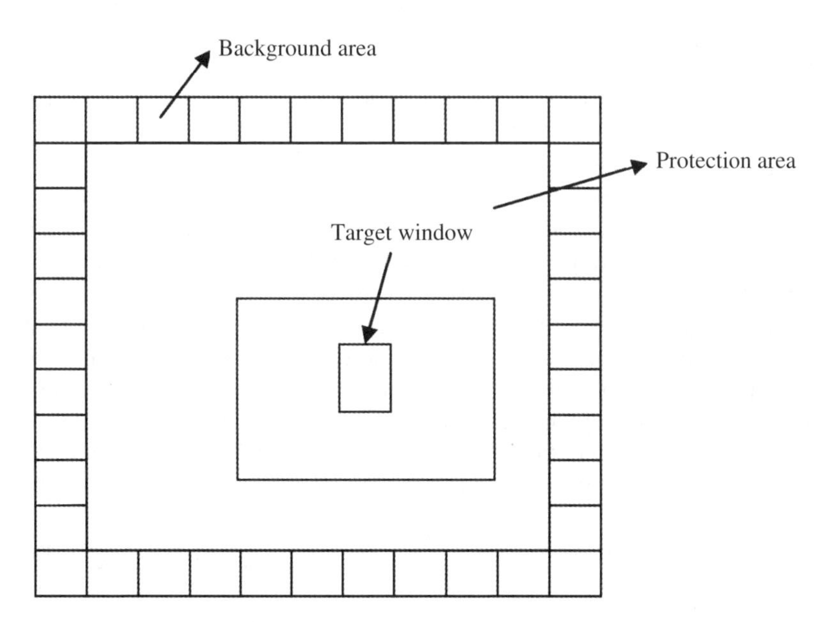


**Figure 7.1** Schematic diagram of CFAR detection.

target, thus increasing the mean and variance of the background window, so that the detection rate decreases. If the size of the large target is used as the size of the target window, then for small targets, the mean is reduced, again reducing the probability of detection. Therefore, generally the target window size is set to be twice the size of the small target, and the protection window size is set to be twice the size of the large target. Background window size = protection window size +2 N, where N = 3 usually.

Different clutter distribution models have different CFAR detection methods. Common clutter distribution models include Gauss distribution, Rayleigh distribution, exponential distribution, Weibull distribution, log-normal distribution, gamma distribution, and K distribution [1]. According to the clutter intensity estimation method, CFAR can be classified into cell average CFAR, greatest of CFAR, least of CFAR, and sort CFAR [2].

One target occupies many resolution units in a SAR image. However, due to the scattering of the target surface itself, in the binary image obtained after CFAR detection, pixels that correspond to the same object are generally not produced in a connected region. So the binary image after detection needs to undertake target pixel clustering.

The clustering of target pixels requires a priori knowledge of target and image, such as target length $L$, target width $W$, and resolution $\Delta A$ of the image. In fact, during the SAR target imaging, not all parts have a highly scattering energy, thus the area of the target pixel $S$ detected by CFAR is less than the actual area of the target, namely

$$N_0 \leq S \leq L \times W/(\Delta A \times \Delta A) \tag{7.2}$$

where $N_0$ is the minimum value of target pixels empirically. In the meantime, the distance of two pixels in the same target area $d$ meets

$$d \leq \sqrt{L^2 + W^2}/\Delta A \tag{7.3}$$

According to the above analysis, the process of target pixel clustering may be established as shown in Figure 7.2. First, the adjacent pixels are connected by morphological "close" operation. Then, the density filter is used to remove the isolated point. Finally, the pixels are clustered according to range threshold, and all pixels whose range is smaller than range threshold are classified as one target. According to area threshold, the isolated noise points and objectives larger than the target size are eliminated.

For offshore ship detection, many research results show that the CFAR detection algorithm alone cannot acquire satisfactory results, which means they usually fluctuate between low detection probability and high false alarm rate. To improve the detection rate and to decrease the false alarm rate, it is necessary to preprocess the marine SAR image before ship target detection. At the same time, the results after the CFAR detection are postprocessed. The ship detection includes five procedures: land mask, preprocessing target enhancement, target detection presegmentation, target identification, and target intelligence extraction, as shown in Figure 7.3.

Ship detection is a typical military application of airborne SAR, with a high-resolution image, which is more useful for small target detection. However, the amplitude of the

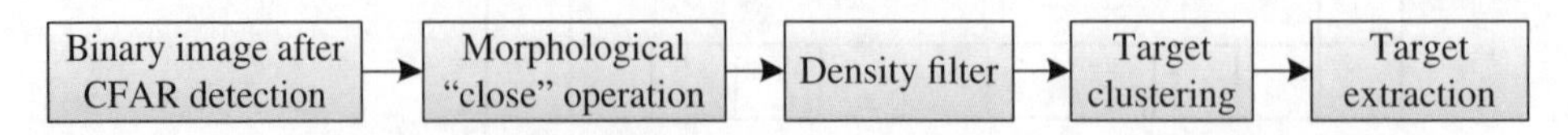


**Figure 7.2** Flowchart of target pixel clustering.

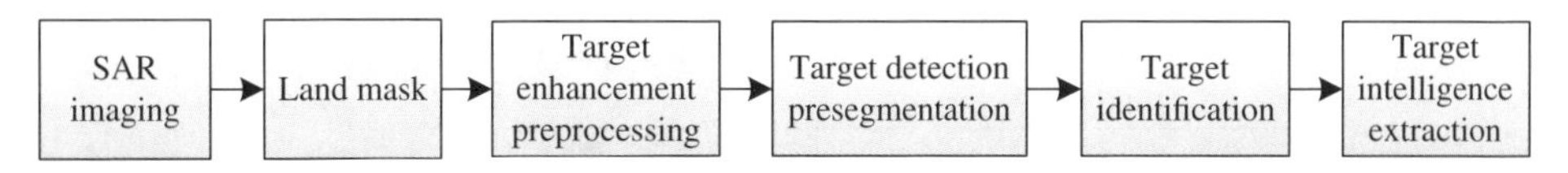


**Figure 7.3** Flowchart of ship detection procedure.

airborne SAR image is not balanced due to aircraft jitter and imaging factors. In addition, the wave is highly scattered. Highly scattering side lobe of a coastal land target will produce heavy blurs in an ocean area. Therefore, the ship detection needs a tradeoff between high detection rate and low false alarm rate, especially for SAR images where land and sea coexist. In a case where the coast target is required to be detected as much as possible, it is likely to cause more false alarms. When reducing the false alarms, an offshore target is likely to be missed. For moving ships, due to the Doppler effect, the ship target is shown in an elongated effect, which is not useful for ship parameter estimation.

Integrated CFAR detection is used for land vehicle detection. First, highlighted pixels in the image are detected by global CFAR, and the pixels of the target area are clustered with local CFAR to remove false alarms, such as canopies and buildings. The process mainly consists of four procedures: global CFAR, local CFAR, target clustering, and target parameter extraction, as shown in Figure 7.4.

Global CFAR is adopted in the runway area to detect aircraft on an airport runway. However, due to the fact that the scattering intensity of different parts of the aircraft are different in a high-resolution image, the pixels of different parts of the aircraft need to be clustered to classify and recognize the whole aircraft. The detection flowchart of aircraft on an airport runway is shown in Figure 7.5.

### 7.2.2 Structure Target Detection

This section focuses on detection of structure objectives, such as bridges and airports.

For bridge detection, the bridge in a low-resolution SAR image shows a highly uniform scattering, in great contrast with the river background. Its characteristics can be summarized as follows: there is a large and legible water area in the SAR image, and the bridge is located across the surface of the water; gray value of the water area is low, and the gray level is evenly distributed (average variance is small), whereas the gray level of

**Figure 7.4** Flowchart of vehicle detection procedure.

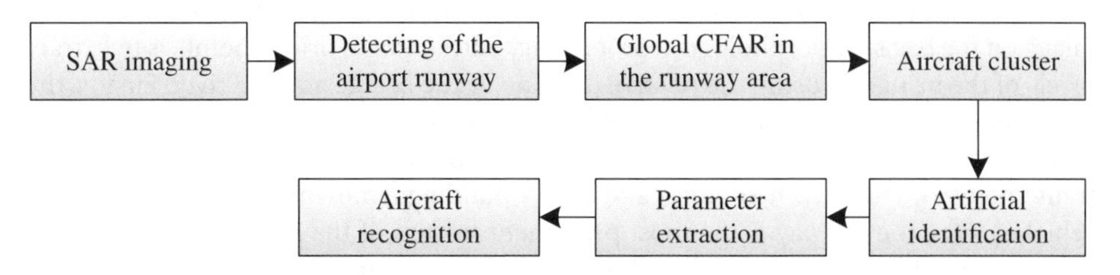


**Figure 7.5** Detection flowchart of aircraft on the runway of the airport.

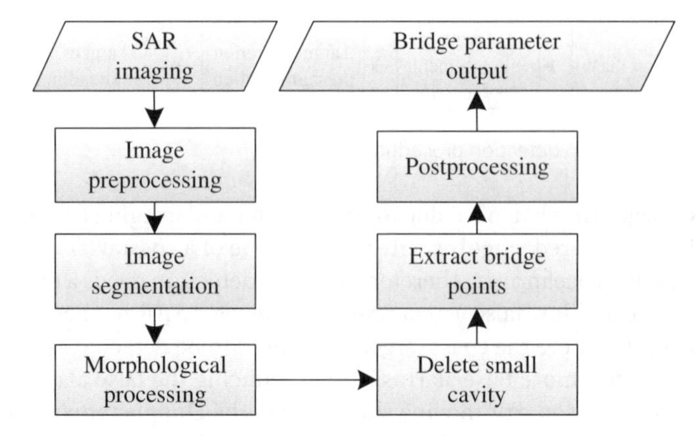


**Figure 7.6** General procedure of SAR bridge detection.

the bridge is close to the land, and the gray value of the bridge is larger than the water area. Both length and width of the bridge are within a certain range, which means the distance between water on both sides of the bridge is also within a certain width. Water is on both sides of the bridge, whereas water is on one side of the riverbank, and land is on the other side. Although the bridge edges do not have very good characteristics of two parallel lines, they will not bend too much.

In airborne SAR data, due to the high resolution, local scattering characteristics of the bridge are more obvious, such as bridge pier scattering, street light scattering, rail scattering, multiple scatterings, etc. These scattering characteristics indicate the bridge does not show continuous highly scattering pixels instead of discontinuous highly scattering points.

Figure 7.6 shows the general procedure for SAR bridge detection. Firstly, the SAR image is preprocessed, mainly including gray scale quantization, gray stretch, speckle noise filtering, and resolution reduction. Gray scale quantization refers to quantizing original SAR amplitude to [0, 255] gray space. Gray level quantization can reduce the data process amount. Gray stretch is to adjust gray scale display of the image to facilitate the image threshold segmentation later. Speckle noise filtering can reduce influence of noise on target detection. The resolution reduction is mainly used in high-resolution airborne SAR images. Different images have different pretreatment requirements. Secondly, the image is segmented. The purpose of image segmentation is to separate the river and the bridge and then to perform such operations as morphological processing, deleting small cavities, and extracting bridge points. The purpose of morphological processing is to connect segments of the bridge to remove small burrs and isolated noise. Deleting small cavities is to continue deleting the small closed sections of the image on the basis of morphological processing. Extraction of bridge points is to extract pixels of the bridge by using the relative position of the bridge and the river. Finally, the bridge parameters are extracted by using the bridge target points.

For airport detection, an airport is mainly composed of the following parts: airport traffic network (runway, taxi lanes, and road), buildings, moving targets (aircraft and vehicles), etc. The runway is the most prominent feature of the airport, which is characterized by a long and thin black rectangular strip in the image. Obviously, the airport detection should start by detecting the runway. Finding a runway means finding the

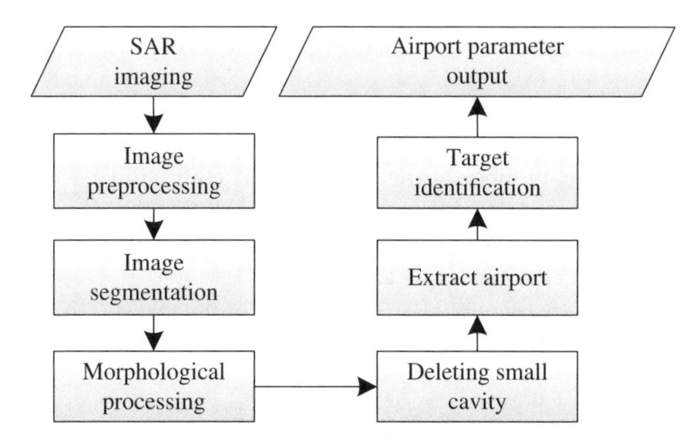


**Figure 7.7** Basic procedure of airport detection.

airport. However, there are the following factors that increase the difficulty of runway extraction: influence of speckle, inhomogeneity of the gray surface, and the interference of other targets.

In summary, the airport targets generally have the following characteristics: the runway length is in a certain range, whose mathematical expressions are $l_1 \leq L \leq l_2$, where $L$ is length of runway, and $l_1$ and $l_2$ are constants ($l_1$ is usually 900 m and $l_2$ is usually 4200 m). The width of the runway is in a certain range, whose mathematical expression is $w_1 \leq W \leq w_2$, where $W$ is the width of runway, and $w_1$ and $w_2$ are the constants (usually $w_1$ is 18 m and $w_2$ is 60 m). The length of the runway is far larger than the width of the runway. The average gray value of the runway surface is low, and gray level is distributed uniformly, which is a long and narrow dark area in the image. If there are several runways, they are shown in cross shape or in parallel. The airport width is at least four times the runway width. The length-to-width ratio of the airport area meets a certain threshold range, generally bigger than three.

Figure 7.7 shows the basic procedure of airport detection in SAR image. Similar to the bridge detection procedure, it includes image preprocessing, image segmentation, morphological processing, deleting small cavities, and target detection and identification.

### 7.2.3 Target Parameter Extraction

In the process of target detection and recognition, it is very important to extract target parameters, which guide the development of image interpretation. The main target parameters include area, coordinates of center of gravity, direction angle of principal axis, aspect ratio, average gray scale, etc. Area, coordinates of center of gravity, and direction angle of principal axis are calculated using the concept of spatial moment in mechanics.

For discrete image $f(i,j)$, $p + q$ moment is defined as

$$m_{pq} = \sum_i \sum_j i^p j^q f(i,j) \tag{7.4}$$

$p + q$ central moment is calculated by using the gravity center $(\bar{i}, \bar{j})$ as the origin, which is defined as

$$u_{pq} = \sum_i \sum_j (i - \bar{i})^p (j - \bar{j})^q f(i,j) \tag{7.5}$$

(1) *Area.* The total pixel number of the target area is commonly used to estimate the target area. However, due to the impact of the CFAR threshold, not all target pixels can be extracted. Therefore, one better alternative is to select a minimum enclosing rectangle of the target as the target area.
(2) *Location of the gravity center.* The gravity center of the target $(\bar{i}, \bar{j})$ is the ratio of the first- to zero-order moment, that is

$$\bar{i} = \frac{m_{10}}{m_{00}}, \quad \bar{j} = \frac{m_{01}}{m_{00}} \tag{7.6}$$

(3) *Direction angle of principal axis.* The principal axis is the longest, which goes through the center of gravity of the target. The angle between the principal axis of the target and the $Y$ axis (i.e. the north) is the direction angle of the principal axis:

$$\theta = \frac{1}{2} \arctan\left(\frac{2u_{11}}{u_{20} - u_{02}}\right) \tag{7.7}$$

where $u_{11}$ is the product mean of the target pixel coordinates relative to center coordinates, $u_{20}$ is the square mean of the $x$ coordinate of the target pixels relative to center coordinates, and $u_{02}$ is square mean of the $y$ coordinate of the target pixels relative to center coordinates.

Then the direction angle of the target is either $\theta$ or $\theta + \pi$.
(4) *Aspect ratio.* Aspect ratio is an important parameter of the target. It is defined as the ratio of length to width of the minimum enclosing rectangle determined by the principal axis.
(5) *Velocity extraction.* Whenever there is ship wake, it may be used to estimate the ship velocity. The main characteristics of a ship wake are listed in Table 7.1.

There are three methods to estimate the ship velocity by using the ship wake.

Firstly, Doppler shift during the moving target imaging will allow the moving ship to deviate from the actual location, which makes the ship deviate from its wake location in the SAR image. Assume that the angle between the moving direction of the ship and azimuth direction is $\phi$. The velocity component along range direction $V_{ship} \cos\phi$ affects Doppler shift, which finally leads to drift off the ship target in x azimuth.

In case the deviation is known, the ship velocity can be calculated via the deviation:

$$V_{ship} = \frac{dV_{sat}}{R\cos\phi} \tag{7.8}$$

where $V_{ship}$ is the ship velocity, $d$ is the azimuth deviation of the ship, $V_{sat}$ is the speed of the platform, $R$ is the distance between the ship and the satellite, and $\varphi$ is the angle between ship velocity vector and range direction. From Eq. (7.8), it can be seen that when the ship sails along the range direction toward the radar, the ship will be imaged along

**Table 7.1** Classification and characteristics of ship speed.

| Wake characteristics | | Display in SAR image | Wake imaging mechanism | Wake angle |
|---|---|---|---|---|
| Surface wave | Prague wave | Narrow V-type bright line | Prague effect | Smaller than 10° |
| | Kelvin wake | Kelvin arm | Tilt and hydrodynamic modulation | About 39° |
| Turbulence or vortex wake | | Dark band (with bright edge) | Flow and inhibition | Close to the axis |
| Internal wave from ship | | Internal wave wake | Flow field effect | Depending on the velocity |

the forward azimuth direction. When the ship sails along the range direction against the radar, the ship will be imaged along the backward azimuth direction.

Second, when the sea surface is calm, a narrow V-shaped wake may be detected in the SAR image. The ship wake can be estimated by using the angle of two narrow V-shaped wakes:

$$V_{ship} = \frac{\sin \varphi}{2 \tan \alpha} \sqrt{\frac{\lambda g}{4\pi \sin \theta}} \tag{7.9}$$

where $\lambda$ is the radar wavelength, $g$ is acceleration of gravity, $\theta$ is the radar incident angle, $\phi$ is the angle between the radar side-looking direction and the ship course, and $\alpha$ is the half-angle of a narrow V-shaped wake.

Third, when the wavelength of the tail wave is measured, the relationship of the phase velocity of the tail wavelength equal to the velocity of the ship can be used to estimate the ship speed:

$$V_{ship} = \sqrt{\frac{\sqrt{3} g \lambda_w}{4\pi}} \tag{7.10}$$

where $\lambda_w$ is the wavelength of tail wave and g is the acceleration of gravity.

### 7.2.4 Typical Examples

(1) *Ship detection.* High-resolution airborne SAR data in this section is used in a ship detection experiment. Figure 7.8 shows detection results of three SAR images with 2 m image resolution. Figure 7.8a and b are the ocean area, while Figure 7.8c is the image of ocean and land, and Figure 7.8d is a local zoom-in of Figure 7.8c. Table 7.2 is the finding of image detection. For detection of Figure 7.8c, under the condition of ensuring sufficient detection rate, it is appropriate to relax control of the false alarm rate. In the experiment, the average detection rate is more than 90%, and the false alarm rate is less than 10%.

(2) *Vehicle detection.* In contrast with marine target detection, the land target detection is more complicated, because of the impact of complex background. Figure 7.9

**Table 7.2** Detection parameters of airborne SAR ship.

| No. | Actual target number | Detected target number/detection rate | False alarm data/false alarm rate |
|---|---|---|---|
| Image 1 | 3 | 3/100% | 0/0 |
| Image 2 | 10 | 10/100% | 0/0 |
| Image 3 | 91 | 87/95.6% | 11/11.2% |
| Total | 104 | 100/96.15% | 11/9.9% |

**Table 7.3** List of MSTAR target attributes (unit: pixel).

| Serial number | Target category | Central position (row, column) | Target length | Target width | Aspect ratio | Azimuth angle (°) |
|---|---|---|---|---|---|---|
| #1 | Tank armor | (247, 1269) | 29 | 14 | 2.07 | 73.74 |
| #2 | Tank armor | (597, 669) | 26 | 11 | 2.36 | 122.91 |
| #3 | Tank armor | (1353, 1274) | 27 | 14 | 1.93 | 77.47 |
| #4 | Tank armor | (1516, 198) | 32 | 12 | 2.67 | 157.07 |

shows an MSTAR image, where there are houses, grasslands, forests, and separate trees. The statistical distribution of a whole image is in gamma distribution. All four tanks were detected, as shown in Table 7.3, and the false alarm rate was 0. Figure 7.10 shows the image of a Chinese airborne SAR with a resolution of 0.5 m for vehicle target detection. There are 17 large trucks and 10 small vehicles in the image. The large tracks are the detection target. It can be seen that 16 trucks were detected and other two trucks were false alarms. The detection rate was 94%, and the false alarm rate was about 11%.

(3) *Bridge detection.* The bridge target detection results of spaceborne (Radarsat 1) and airborne SAR are shown in Figures 7.11–7.13 and listed in Tables 7.4 and 7.5, where the bridge target detection rate is more than 90%.
(4) *Airport detection.* The airport detection experiments are performed by using four high-resolution airborne SAR images. Experimental results are shown in Figure 7.14. Airport targets can be extracted, and the detection rate reaches 100%. In addition to the image size, the other key factor affecting the detection rate is the number of dark areas in the image. The more dark areas are, the slower the detection speed is.

## 7.3 Target Change Detection

Change detection uses multiple images obtained from one area at different times to determine the characteristics and process of the surface feature change. It identifies the

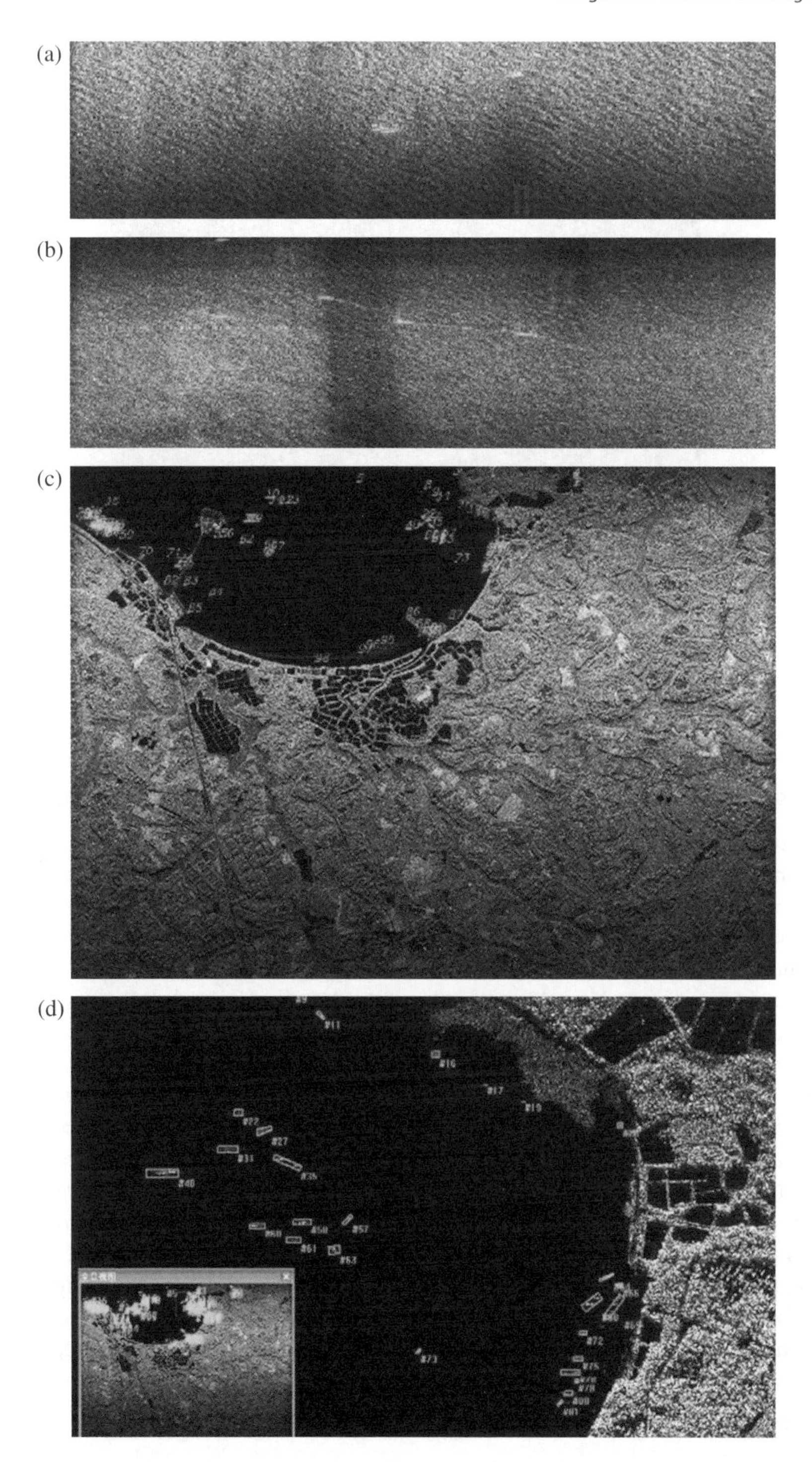

**Figure 7.8** Ship detection finding of airborne SAR image. (a) Image 1, resolution: 2 m, column × row: 4096 × 864; (b) image 2, resolution: 2 m, column × row: 5462 × 3936; (c) image 3, resolution: 2 m, column × row: 5419 × 1512; (d) local zoom-in of image 3. (See color plate section for the color representation of this figure).

**Figure 7.9** MSTAR detection results. (See color plate section for the color representation of this figure).

change of the surface feature type and the internal conditions and states mainly through extracting and analyzing the difference of spectral characteristics and spatial structures of the images. The main changes are reflected in the gray value or the local texture. Based on this, variations of shape, location, quantity, and other attributes of the interested areas are detected.

A SAR image change detection procedure may be summarized as three steps: image preprocessing, acquisition of different images, and segmentation of difference image [3], as shown in Figure 7.15. In application, the image obtained in a historical moment is used as the reference image, and the acquired real-time image is used as the image to be detected. After registration of the reference image and the image to be detected, the initial difference image is generated. Then pixels in the initial difference image are judged and identified automatically, and classified difference image is generated for the image interpreter to judge. The usual classified difference image is a binary image, where 1 refers to changed pixels and 0 refers to unchanged pixels, namely, white represents change, black indicates no change. Sometimes the classified difference image is represented by a ternary image, where 1 represents pixels with increased target (blue), −1 represents pixels with decreased target (red), and 0 represents unchanged pixels (black).

### 7.3.1 Preprocessing

Image preprocessing mainly includes image registration, speckle noise filtering, radiometric correction, and geometric correction. The image registration is a

**Figure 7.10** Vehicle detection results of airborne SAR image. (See color plate section for the color representation of this figure).

prerequisite for SAR image change detection, which ensures that the pixel size and geographical position of two images are consistent. The registration accuracy is generally required within 1 pixel. Many registration algorithms are very mature and unsupervised and can be used directly in change detection steps, such as the correlation coefficient method, coherent coefficient method, phase correlation method based on fast Fourier transform, and matching algorithm based on invariant feature. Speckle noise seriously affects the quality of the SAR image and the later application. Speckle noise filter can effectively suppress speckle and improve image quality. After many years of development, there are a variety of adaptive filtering algorithms, such as the mean filter, the median filter, and the neighborhood model-based filter. Filtering damages details of an image to a certain degree, though it reduces the impact of noise. In spite of this, speckle noise filtering is still an important step in the process of SAR image change detection.

Radiometric calibration includes relative radiometric correction and absolute radiometric calibration. Relative radiometric correction refers to the relative normalization of intensity in two images. Absolute radiometric calibration transforms amplitude or intensity data to backscattering coefficient or backscattering cross-section area by using calibration parameters. When the difference image is generated by ratio treatment, radiation correction can be neglected. Formal SAR image geometric correction includes range conversion, direction correction, and terrain correction. Image

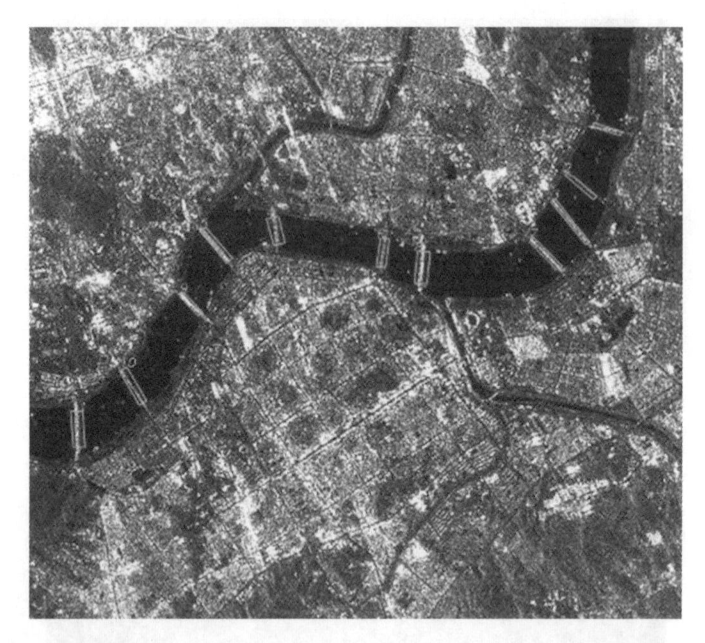

**Figure 7.11** Spaceborne SAR bridge image (resolution about 10 m). (See color plate section for the color representation of this figure).

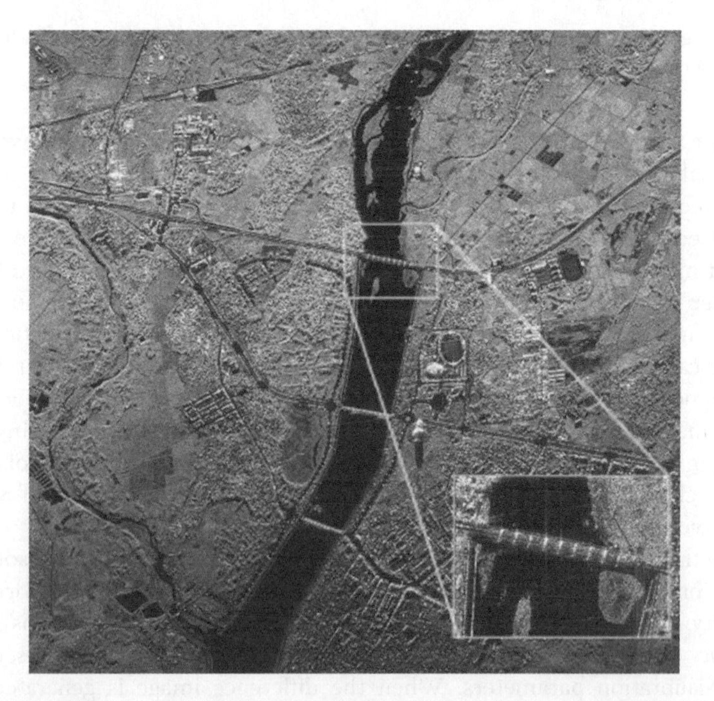

**Figure 7.12** 1 m resolution airborne SAR image of bridges. (See color plate section for the color representation of this figure).

**Figure 7.13** Finding of bridge target detection.

**Table 7.4** Parameters of bridge target, spaceborne (unit: pixel).

| Serial number | Target category | Central position (row, column) | Target length | Target width | Aspect ratio | Azimuth angle (°) |
|---|---|---|---|---|---|---|
| #1 | Bridge | (121, 706) | 30 | 5 | 6.00 | 169.22 |
| #2 | Bridge | (189, 674) | 41 | 6 | 6.83 | 140.91 |
| #3 | Bridge | (233, 661) | 61 | 6 | 10.17 | 128.93 |
| #4 | Bridge | (244, 302) | 29 | 11 | 2.64 | 107.10 |
| #5 | Bridge | (261, 229) | 44 | 9 | 4.89 | 129.40 |
| #6 | Bridge | (270, 433) | 31 | 7 | 4.43 | 81.03 |
| #7 | Bridge | (272, 631) | 43 | 8 | 5.38 | 135.00 |
| #8 | Bridge | (288, 481) | 34 | 10 | 3.40 | 83.29 |
| #9 | Bridge | (332, 199) | 26 | 7 | 3.71 | 135.00 |
| #10 | Bridge | (429, 127) | 43 | 10 | 4.30 | 118.61 |
| #11 | Bridge | (481, 58) | 39 | 11 | 3.55 | 93.18 |

**Table 7.5** Parameters of bridge target, airborne (unit: pixel).

| Serial number | Target category | Central position (row, column) | Target length | Target width | Aspect ratio | Azimuth angle (°) |
|---|---|---|---|---|---|---|
| #1 | Bridge | (1027, 1625) | 167 | 17 | 9.82 | 168.69 |
| #2 | Bridge | (1713, 1541) | 202 | 18 | 11.22 | 170.13 |
| #3 | Bridge | (2213, 1335) | 167 | 18 | 9.28 | 161.57 |

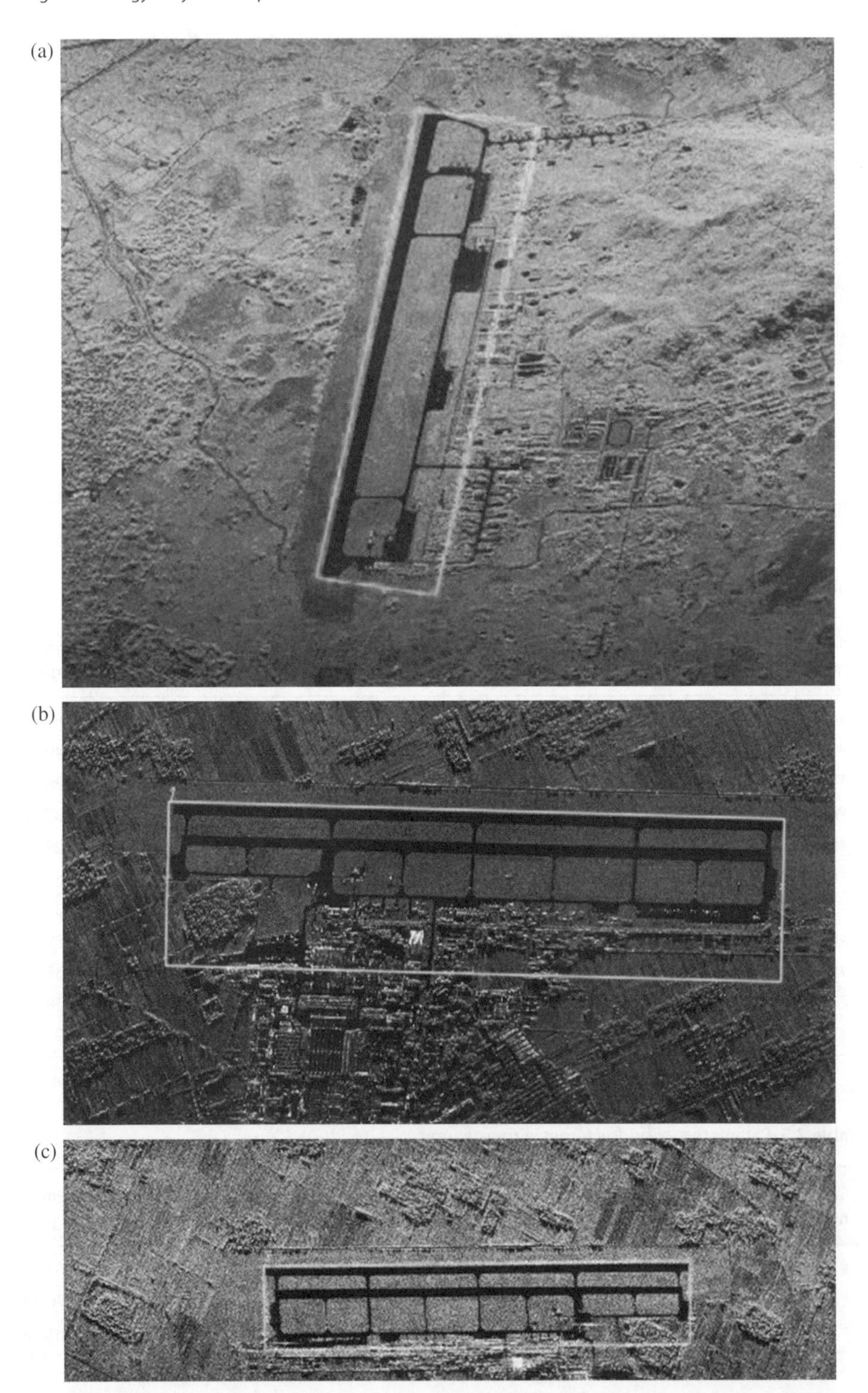


**Figure 7.14** Airport detection finding from airborne SAR images. (a) 2 m resolution; (b) 1 m resolution; (c) 0.6 m resolution; (d) 1 m resolution. (See color plate section for the color representation of this figure).

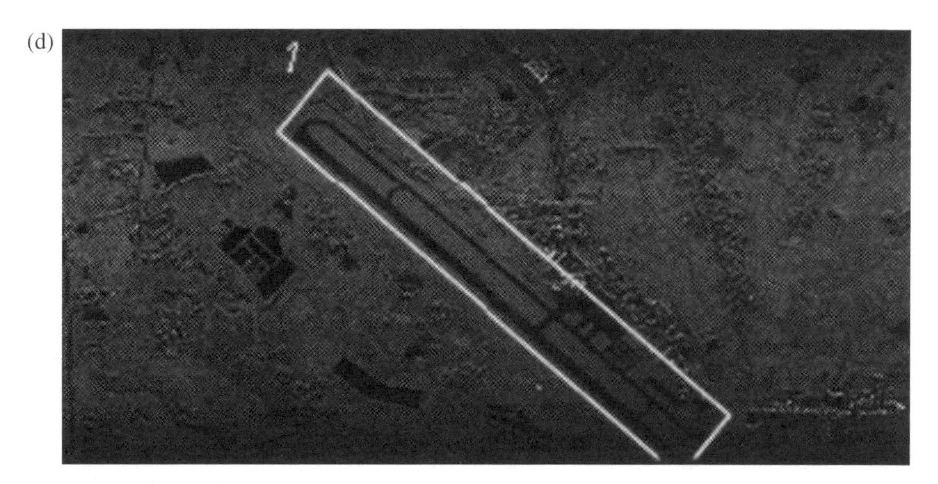


Figure 7.14 (*Continued*)

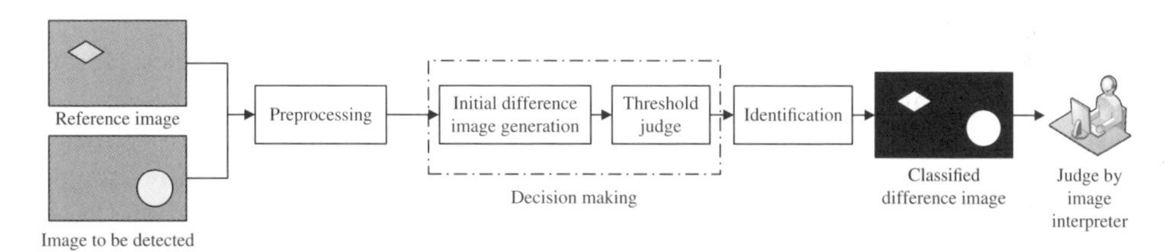


Figure 7.15 Change detection procedure. (See color plate section for the color representation of this figure).

processing includes resampling, interpolation, rotation, and mirroring. Resampling and interpolation are not useful for maintaining the characteristics of original data. As a result, geometric correction can be ignored in image preprocessing. Geometric correction processing is performed on the detection results, after the change detection findings are obtained.

### 7.3.2 Difference Image Acquisition

SAR difference image acquisition is a very important step in research of unsupervised change detection. The main approaches to obtain the difference image can be summarized as logarithmic ratio-based difference image acquisition, correlation-based difference image acquisition, feature transform-based difference image acquisition, and multichannel-based difference image acquisition. The different acquisition approaches affect the choice of the segmentation method.

The most direct approaches to obtain a difference SAR image are the difference method and the ratio method. The difference method generally requires that bitemporal SAR images are first relatively calibrated and absolutely calibrated, and then a difference image is acquired by subtraction. Otherwise the change information will be submerged by noise. The ratio method does not require calibrating bitemporal images, and a difference image can be directly obtained by the ratio of magnitude or intensity. The ratio processing not only can eliminate the influence of multiplicative noise to a large extent and reduce extra error introduced by calibration but also can highlight the relative change region of the SAR image.

The extraction of difference image based on correlation is also an important method of change detection. The main indicators of correlating two SAR images include the coherence coefficient and the correlation coefficient. The correlation coefficient of multitemporal SAR images is an important indicator of surface feature change detection. The coherence coefficient is calculated based on single-look complex data. Coherence is used to investigate the variation of SAR images, which is limited because the coherence coefficient is affected by the interference baseline.

From interference theory, the interference baseline distance is one main factor that overwhelms coherence. The greater the baseline distance is, the lower the coherence is. When baseline distance exceeds the critical baseline distance, two images will be completely incoherent. Therefore, use of the correlation coefficient should be limited to detecting and analyzing the change. In contrast, the correlation coefficient is more suitable for change detection of SAR images than the coherent coefficient. The correlation coefficient ignores the phase information of the echo and reflects similarity of the local spatial texture.

In essence, the difference of the SAR image is the difference of characteristics. How to better explore characteristics of SAR images has become the focus of many researchers in studying change detection. In addition to such common features as intensity, mean, variance, and texture feature, features obtained by appropriate transform have brought innovation to the study of change detection, such as principal component analysis (PCA) transform, independent component analysis (ICA) transform, discrete wavelet transform, and discrete cosine transform. Features acquired by transformation can better characterize changes of SAR images. The typical process for such change detection is to extract the feature vector first, and then perform spatial clustering processing on the feature vector.

For multichannel (such as multiband or multipolarization) SAR image data, the data from each channel needs to be fused so that variation can be fully exploited. There are a variety of fusion methods. In addition to the PCA transform method previously introduced, there are ICA, statistical feature fusion, polarization likelihood ratio, polarization characteristic decomposition, etc.

### 7.3.3 Difference Image Segmentation

Threshold segmentation of the difference image is a common method to obtain the change image. There are a lot of automatic threshold segmentation algorithms, such as the maximum classes square error (Otsu) algorithm, minimum error criterion-based segmentation algorithm, the K&I threshold segmentation algorithm, the CFAR threshold segmentation algorithm, the spatial clustering segmentation algorithm, the Markov random field (MRF) statistical segmentation algorithm, etc. Among them, only the MRF statistical segmentation algorithm makes full use of the neighborhood information between pixels. Compared with other algorithms, this can acquire more accurate change detection results.

If a pixel is marked as changed or unchanged area, then pixels around it are likely to be the same. Therefore, the use of neighborhood information will produce more reliable and more accurate change detection results. MRF is used to define the dependence of pixels, and two Gauss functions are used to describe statistical properties of the pixel intensity in the changed region and unchanged region.

$C = \{C_l, 1 \leq l \leq L\}$ is defined as the pixel class identifier, which represents the class of corresponding pixel in the difference image, where $C_l = \{C_l(i,j), 1 \leq i \leq I, 1 \leq j \leq J\}$ and $C_l(i,j) \in \{\omega_c, \omega_n\}$. To solve the class identifier of pixel $C_l(i,j)$, according to difference image $X_D$, maximum posteriori estimation of Bayesian theory is used to determine the class identifier of each pixel in the difference image. That is, posterior probability of $C_l$ to meet

$$\begin{aligned} C_k &= \arg\max_{C_l \in C}\{P(C_l/C_D)\} \\ &= \arg\max_{C_l \in C}\{P(C_l)p(X_D/X_l)\} \end{aligned} \tag{7.11}$$

Solving the above maximum posteriori probability is equivalent to solving the minimum value of the energy function:

$$U(C_l \mid X_D) = U_{data}(X_D \mid C_l) + U_{context}(C_l) \tag{7.12}$$

where $U_{data}$ is the likelihood energy associated with statistical features, and $U_{context}$ is the neighborhood energy that represents the category relationship between pixels. The classification of the difference image is realized by finding the $C_l$ value that minimizes the energy function. The energy minimization process can be realized by the maximum flow (minimum cut) algorithm of network flow theory [4].

### 7.3.4 Artificial Auxiliary Intelligence Analysis

When the difference image is acquired, intelligence personnel can quickly perform intelligence analysis on key targets, including target identification and recognition, comprehensive information analysis, etc. Artificial auxiliary approaches mainly include:

(1) *Target identification and recognition.* Based on previous target detection or change detection, the critical area is located. However, it is inevitable to have false alarms and missing detection. The process of artificial identification is to eliminate false alarms and to keep the target area of interest. After identification, the target recognition is followed. It is performed by using a database of the targets. In the absence of a target database, target recognition relies on the experience of intelligence personnel. In the process of target identification and recognition, the computer provides such functions as target feature calculation, feature matching, parameter estimation, and display.

(2) *Comprehensive information analysis.* On the basis of key target analysis, it is also required to consider the situation around the target, such as traffic network, water network, landforms, vegetation cover, etc. These are dependent upon the analysis of intelligence personnel and require that intelligence personnel be globally aware. In the process of comprehensive information analysis, the basic geographic information provides important support to ensure the geographical attributes of the intelligence.

### 7.3.5 Damage Assessment

Usually, after a target is hit, its geometry features and texture features will change greatly. Damage assessment is based on change detection findings to further analyze the geometry features and texture features of the target before and after the attack.

Normally, intelligence analysts are involved to improve the quality of damage assessment.

(1) *Building damage assessment*. A building is an important facility used to hide personnel and material and to survive the potential battle. Building damage is mainly assessed through geometry change caused by damage. The building damage is generally divided into four hierarchies:

*Serious damage*. According to geometry features assessment, if there is major change in the geometry features of the target, the target can be considered seriously damaged. If geometry features are basically the same, but texture features show a major change, which means the internal building structure is seriously damaged, the target may also be considered seriously damaged.

*Moderate damage*. According to the texture features assessment, if there is no change in geometry characteristics of the target, but only the texture characteristics change, it can be considered that the target is moderately damaged.

*Minor damage*. The change of geometry and texture features of the target is small.

*No damage*. No change occurs.

(2) *Airport damage assessment*. Damage assessment of the airport runway mainly considers existence of a rectangular area with a certain length and width that meets requirements for aircraft to take off. Both length and width of the rectangular area need to be converted into the length and width in pixels according to resolution of the SAR image. The image of the damaged runway can be acquired according to the runway target recognition image and change detection image. Then, given the length and width of the runway required by the aircraft to take off, it can be further determined whether there is a rectangular area.

The airport damage is generally divided into four hierarchies:

*Serious damage*. It is not possible to take off. Both width and length of airport runway without damage are smaller than the minimum takeoff runway ranges required by the aircraft.

*Moderate damage*. Taking off is possible in local areas. The airport runway is heavily damaged. However, there are still some unaffected local areas whose width and length meet the minimum runway range required for takeoff.

*Minor damage*. Most areas can be used for takeoff. The damage of the airport runway is minor, and most regions are suitable for aircraft to take off.

*No damage*. The airport runway is basically not damaged, i.e. no change occurs.

(3) *Bridge damage assessment*. A bridge, somewhat different from an aircraft, belongs to nonmoving targets. Therefore, a bridge damage assessment is only needed to first identify the bridge before damage and to extract the damage information via change detection findings, and finally, to determine the damage level according to damage index.

The bridge damage is generally divided into three hierarchies:

*Serious damage*. No entry. The bridge is damaged in transverse fracture or cross section.

*Minor damage*. It is possible to pass through locally. The width of the local area not affected by the damage meets the minimum width of the vehicle.

*No damage or basically no damage*. No change occurs.

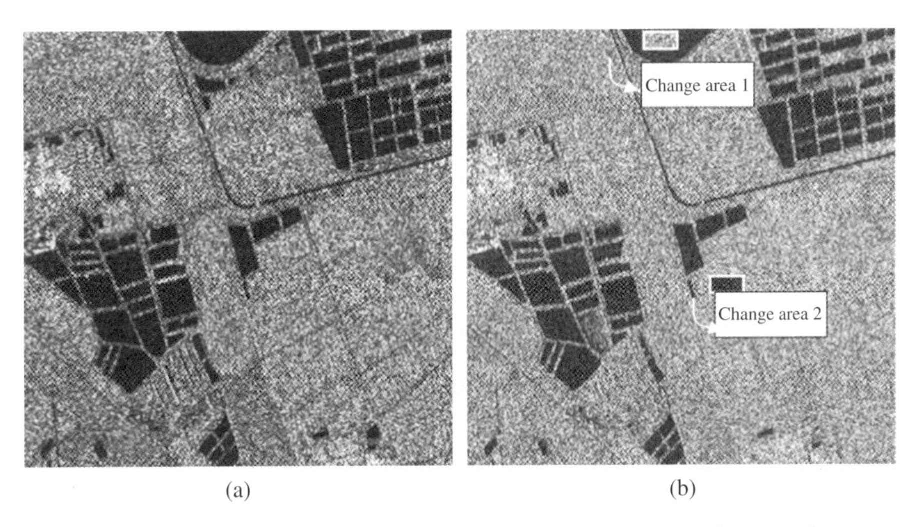


(a) (b)

**Figure 7.16** Bitemporal SAR data of one suburb (512 × 512). (a) June 25, 2005; (b) September 29, 2005. (See color plate section for the color representation of this figure).

## 7.3.6 Typical Examples

### 7.3.6.1 Detection of Mutual Change of Land and Water

Change detection is mainly used in floods, water area reduction, disaster assessment, etc. The experiment on water and land mutual change detection uses a multitemporal RADARSAT-1 SAR image of one suburb as the data, and image resolution is 8 m.

As shown in Figure 7.16, the data interval of the bitemporal is three months. During three months, in addition to the water and land area changes of the image itself, two change regions are artificially simulated to better analyze the effect of the change detection method, as shown in Figure 7.16b, where both land and water data inside

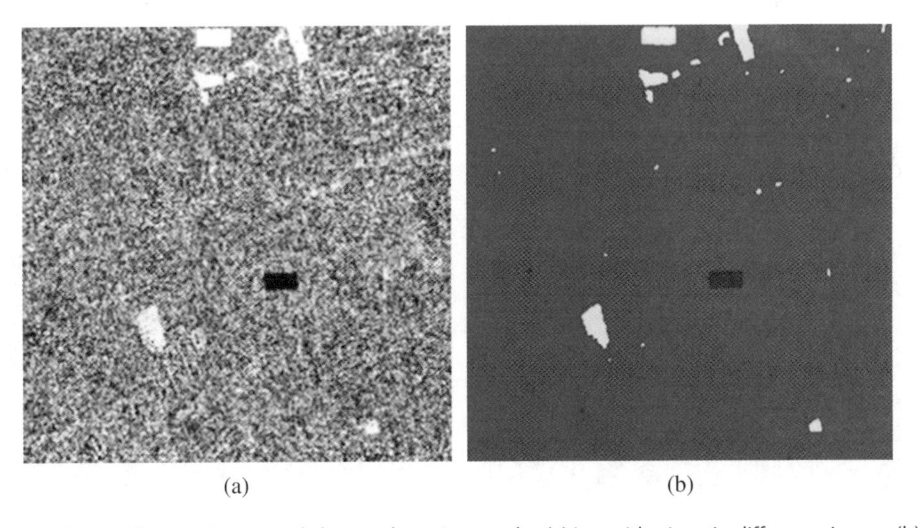

(a) (b)

**Figure 7.17** Difference image and change detection results. (a) Logarithmic ratio difference image; (b) detection results of MRF segment change. (See color plate section for the color representation of this figure).

**Table 7.6** Statistics of change detection findings (Figures 7.16 and 7.17).

| Area | Actual changed pixels | Detected changed pixels | Correctly detected pixels | Detection rate | False alarm rate |
|---|---|---|---|---|---|
| Changed area 1 | 800 | 883 | 799 | 99.88% | 9.51% |
| Changed area 2 | 800 | 734 | 728 | 91.00% | 0.82% |

two equal-sized boxes are exchanged. The data size is 40 rows × 20 columns, and the total number of pixels is 800. Figure 7.17a is the difference image, where the light areas indicate the enhanced region, and dark area indicates the weakened region. The statistical distribution of the difference image accords with the mixed Gaussian model. Figure 7.17b shows detection results. It can be seen from the findings of the two test areas that the detection rate is above 90% and false alarm rate is better than 10% (Table 7.6).

#### 7.3.6.2 Change Detection of Vegetation Growth

The different vegetation growth and soil moisture content will lead to major changes in the backscattering coefficient. Figure 7.18 shows a RADARSAT-1 bitemporal image of one suburb, where the time interval is about 80 days. During this period, the surface crops have changed greatly. Both enhanced and weakened regions can be extracted through detecting. Figure 7.19a shows a logarithmic ratio difference image. Figure 7.19b shows MRF segmentation change detection results. Figure 7.20 is real change detection result of a weakened region artificially identified. The detected assessment parameters for this area are listed in Table 7.7. It can be seen that detection rate has reached 95% with false alarm rate only 5.8%.

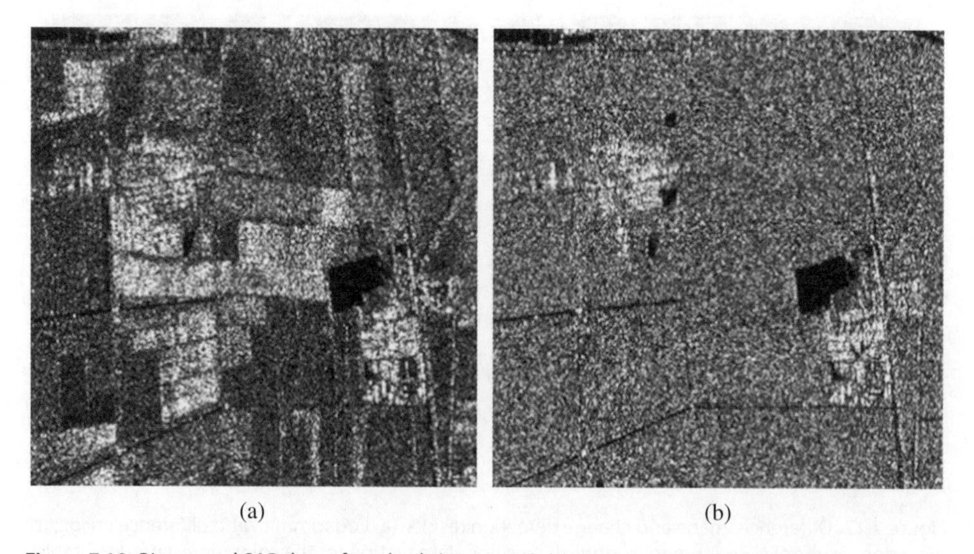

**Figure 7.18** Bitemporal SAR data of a suburb (400 × 400). (a) June 25, 2005; (b) September 5, 2005.

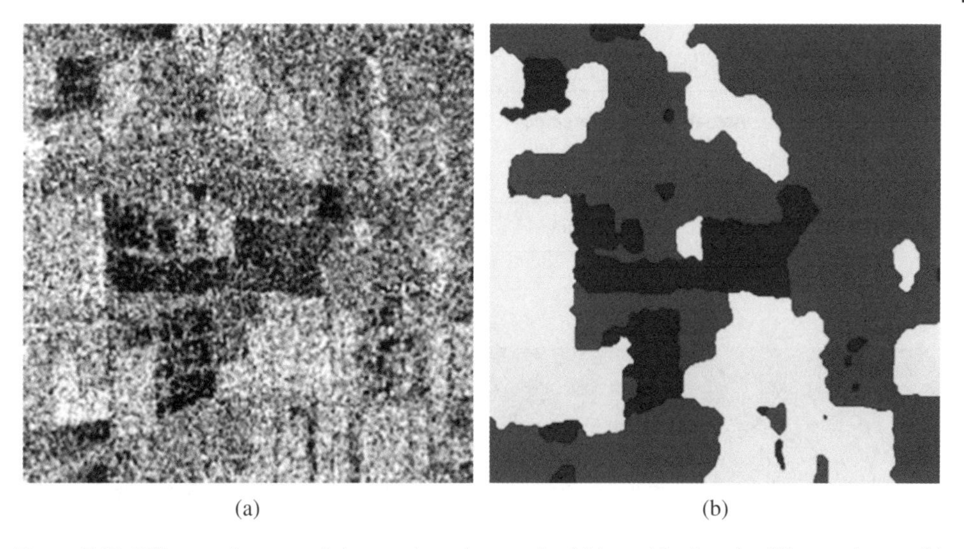

(a) (b)

**Figure 7.19** Difference image and change detection results. (a) Logarithmic ratio difference image; (b) detection results of MRF segmentation change. (See color plate section for the color representation of this figure).

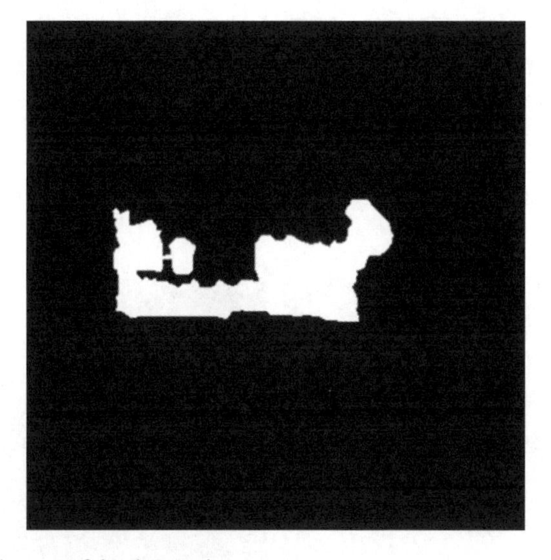

**Figure 7.20** Actual changes of the detected area.

#### 7.3.6.3 Change Detection of Urban Buildings

The assessment of city expansion, illegal construction, and collapse disasters is an important application of SAR image change detection. Bitemporal SAR images are used to detect the change of urban construction, as shown in Figures 7.21 and 7.22. Figure 7.23 shows the actual change of the assessed area. Table 7.8 lists the statistical results. As a whole, changes can be found on the corresponding location where they are detected in the image. This means that the change location is accurate. The pixel-level detection accuracy reaches 90%, with false alarm rate less than 10%.

**Table 7.7** Statistics of change detection results (Figures 7.18–7.20).

| Area | Actual changed pixels | Detected change pixels | Correctly detected pixels | Detection rate | False alarm rate |
|---|---|---|---|---|---|
| Detected area | 11,201 | 11,300 | 10,645 | 95.04% | 5.80% |

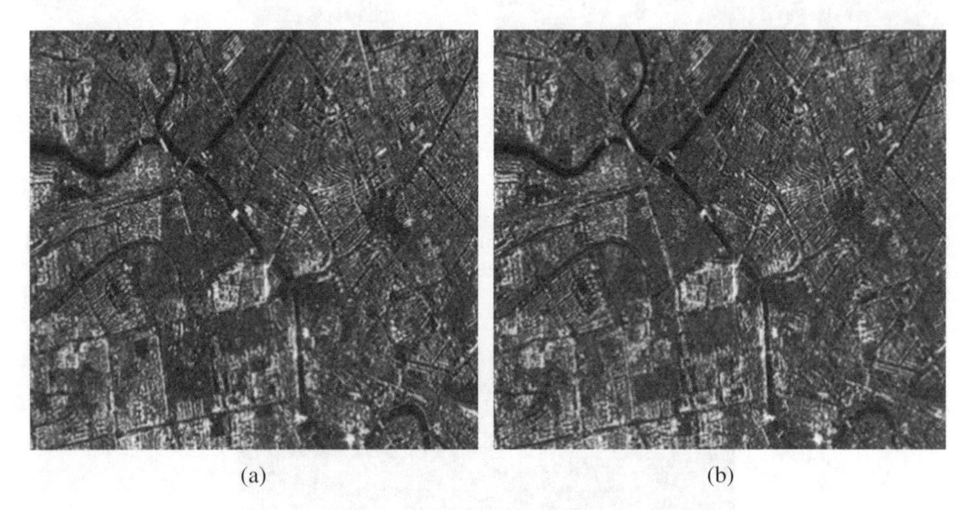

**Figure 7.21** Bitemporal SAR data (944 × 862). (a) April 14, 2005; (b) September 5, 2005.

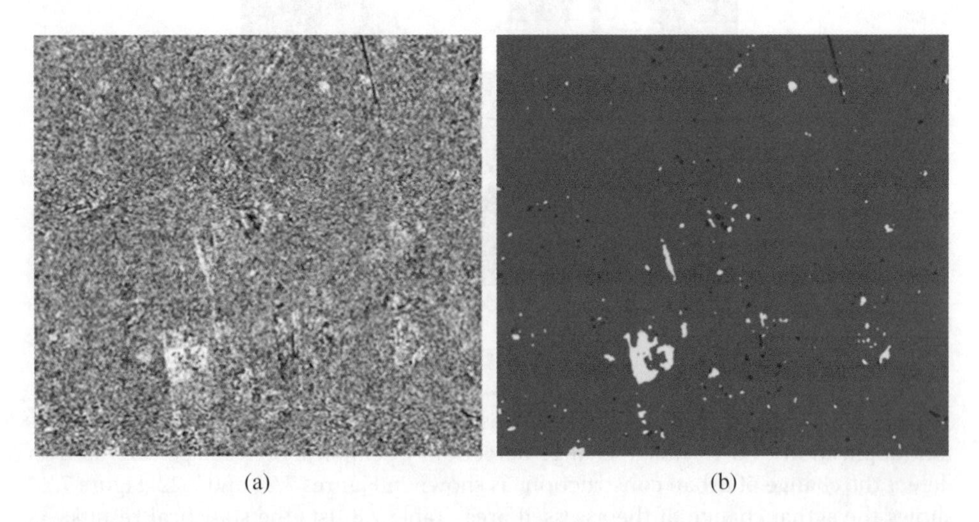

**Figure 7.22** Difference image and change detection results. (a) Logarithmic ratio difference image; (b) detection results of MRF segment change. (See color plate section for the color representation of this figure).

**Figure 7.23** Actual changes of the detected area.

**Table 7.8** Statistics of change detection results (Figures 7.21–7.23).

| | Actual change pixels | Detected change pixels | Correctly detected pixels | Detection rate | False alarm rate |
|---|---|---|---|---|---|
| Detected area | 5483 | 5373 | 4934 | 90.00% | 8.14% |

#### 7.3.6.4 Airport Change Detection

Figure 7.24 shows the change detection result of one area in one airport by repeat navigation. Figures 7.25 and 7.26 show local enlarged detail. It can be seen that Figure 7.25 has one small plane missing. In Figure 7.26, the upper corner of the airport lawn moved, and a new car appears on the runway. Figure 7.27 shows change detection results of the other area in the airport by repeat navigation, where the corner reflector array is the target. The change of angle reflector can be clearly seen from the detection image.

## 7.4 Target Recognition

The target recognition of the SAR image selects and extracts effective features of SAR image targets. The target in a SAR image is highly variable due to the influence of the target characteristics, SAR parameters, and environmental factors. That is, the moving parts of the target and changes to the target posture, sensor parameters, and background will cause significant change of the target SAR image characteristics. Obviously, this will make it difficult for extracting the target features and differences to describe target variability, resulting in a decline in system identification performance. Therefore, it is required to choose different characteristics for different goals and environments.

Generally, texture features reflect regional characteristics of the target, used to classify different terrains in SAR images. Polarization characteristics are mainly used in full polarization data. The peak feature of the target shows distribution in local extreme points of the target. The shadow feature can effectively reflect target geometry.

Feature-based target recognition can be roughly classified into the template matching-based method [4] and the statistical model-based method [5]. In the template

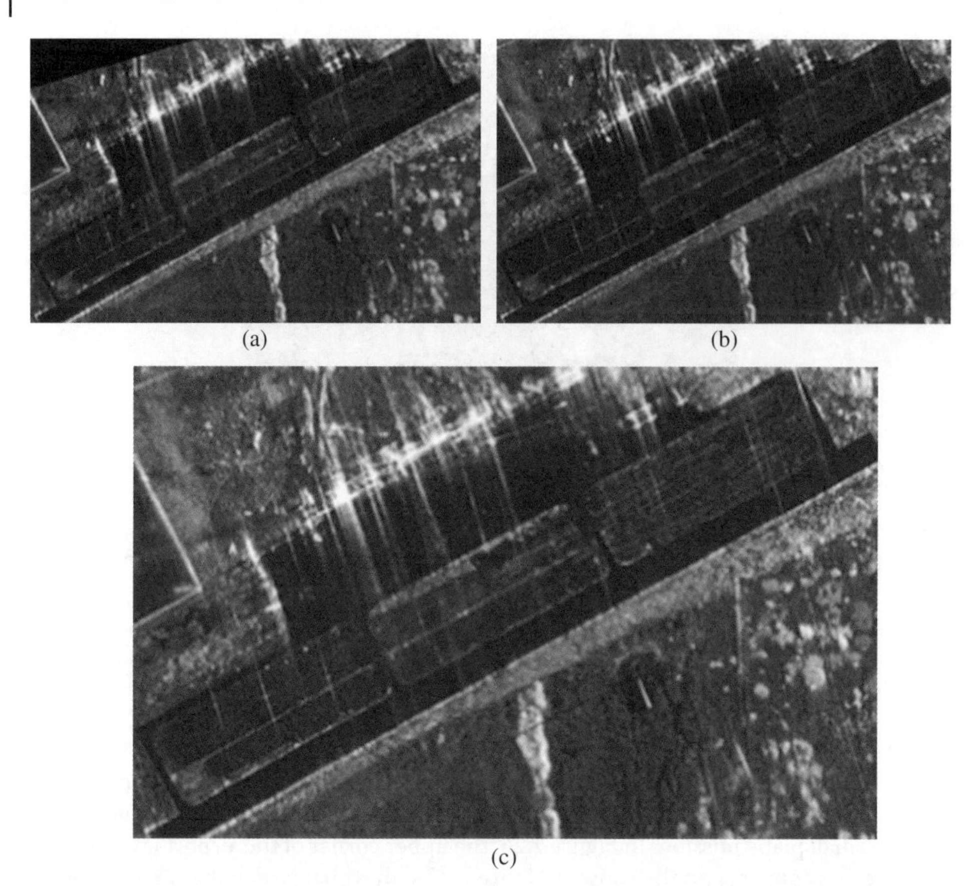

Figure 7.24 Change detection results of an airport. (a) Before change; (b) after change; (c) change detection results.

matching-based method, detected targets with the same characteristics are matched with template objectives in the prior knowledge database according to certain criteria. This method usually needs a large number of templates, and its recognition performance lacks stability in the presence of occlusion, camouflage, and deformation. In the statistical model-based method, the model of the target or classifier is firstly constructed by using prior knowledge and sample data. Then the target or classifier model is used to realize target recognition. Compared with the former, this method does not require the support of massive template data and has certain robustness in the presence of occlusion, camouflage, and local deformation. However, this method is difficult to realize. Intense training and testing are needed before operation.

### 7.4.1 Template Matching Recognition

The template-based target recognition process is shown in Figure 7.28. It includes three parts: target template library establishment, target azimuth angle estimation, and target recognition. The amplitude feature is used as the target feature. The recognition is realized by calculating the correlation coefficient. To control the number of templates

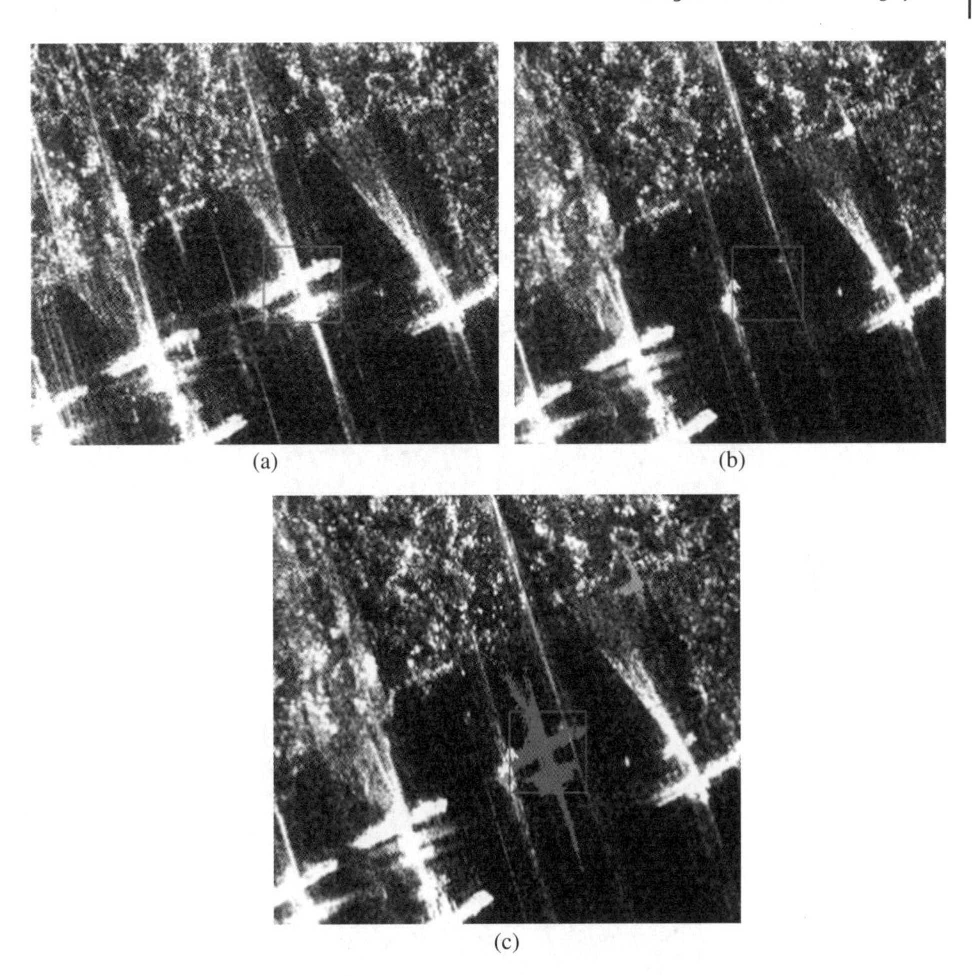


**Figure 7.25** Part 1: One plane disappears on the tarmac (see box). (a) Before change; (b) after change; (c) change detection results. (See color plate section for the color representation of this figure).

and to improve matching efficiency, on one hand, the method of establishing an average template with a certain angle is used. On the other hand, target azimuth angle is estimated so that the search is performed in a certain local azimuth angle during template matching.

The pretreatment of the sample is needed before the template is produced and matched. It is an important part of radar target recognition. Its main mission is to minimize influences of various uncertain factors on the performance of target recognition, such as denoising, amplitude compensation, energy normalization, translation, target segmentation, and azimuth angle estimation. In the following text, MSTAR data [6, 7] is used as an example to illustrate the template-based target recognition process.

#### 7.4.1.1 Target Segmentation Preprocessing

Target segmentation on the slice target of interest is an essential step to extract the target feature. Due to the influence of speckle noise, it is difficult to obtain satisfactory results by only using information provided by a single pixel to segment the SAR image.

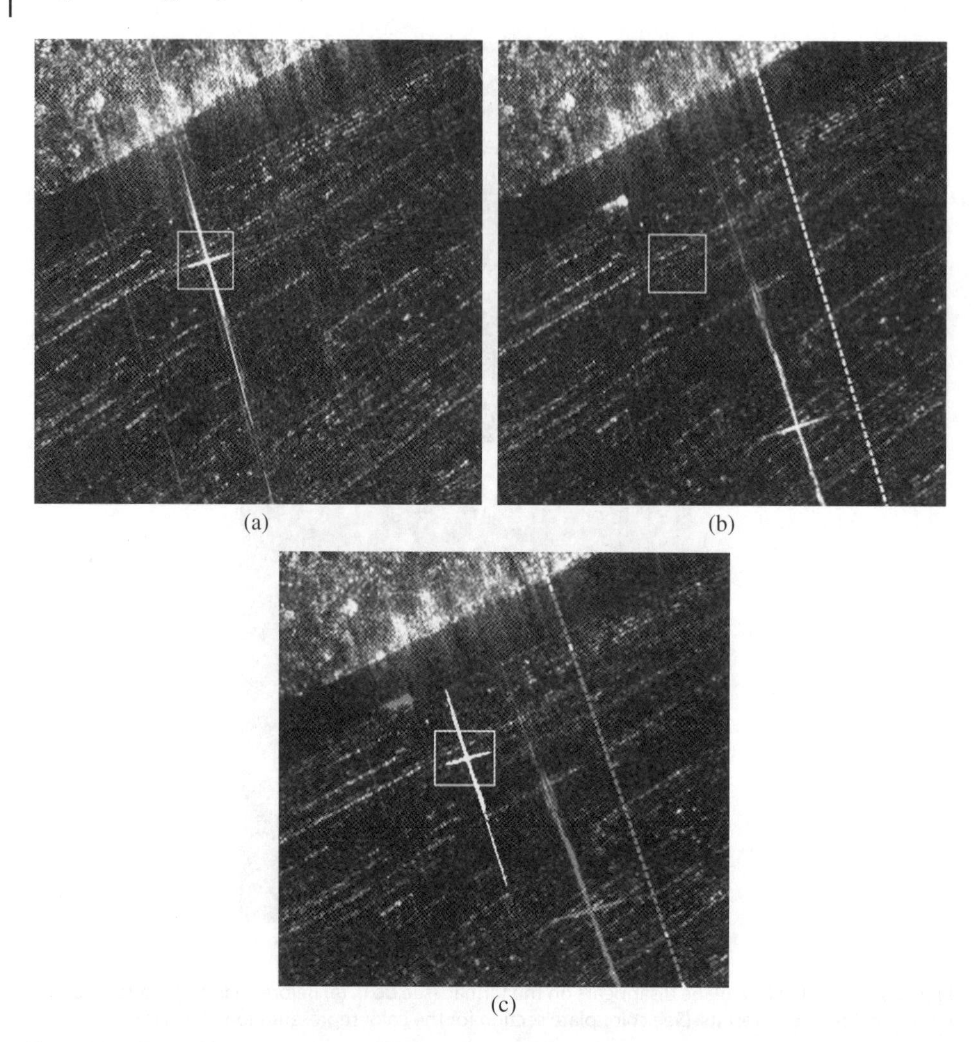

**Figure 7.26** Part 2: The upper corner of the airport lawn moved, and a new vehicle appears on the runway. (a) Before change; (b) after change; (c) change detection results. (See color plate section for the color representation of this figure).

A segmentation algorithm based on the MRF model makes full use of neighborhood information of pixels and effectively reduces the effect of speckle noise. Through iterative image segmentation, it can achieve more accurate results [8].

To ensure accuracy and speed of segmentation, it is best to provide initial segmentation for the algorithm. There is usually only one target in each slice target. Although the brightness of each image may vary significantly, proportion of the target and shadow region does not change much. The image is divided into three parts: shadow, target, and background:

$$x = \begin{cases} \text{shadow} & y \leq T_s \\ \text{target} & y \geq T_t \\ \text{background} & \text{else} \end{cases} \tag{7.13}$$

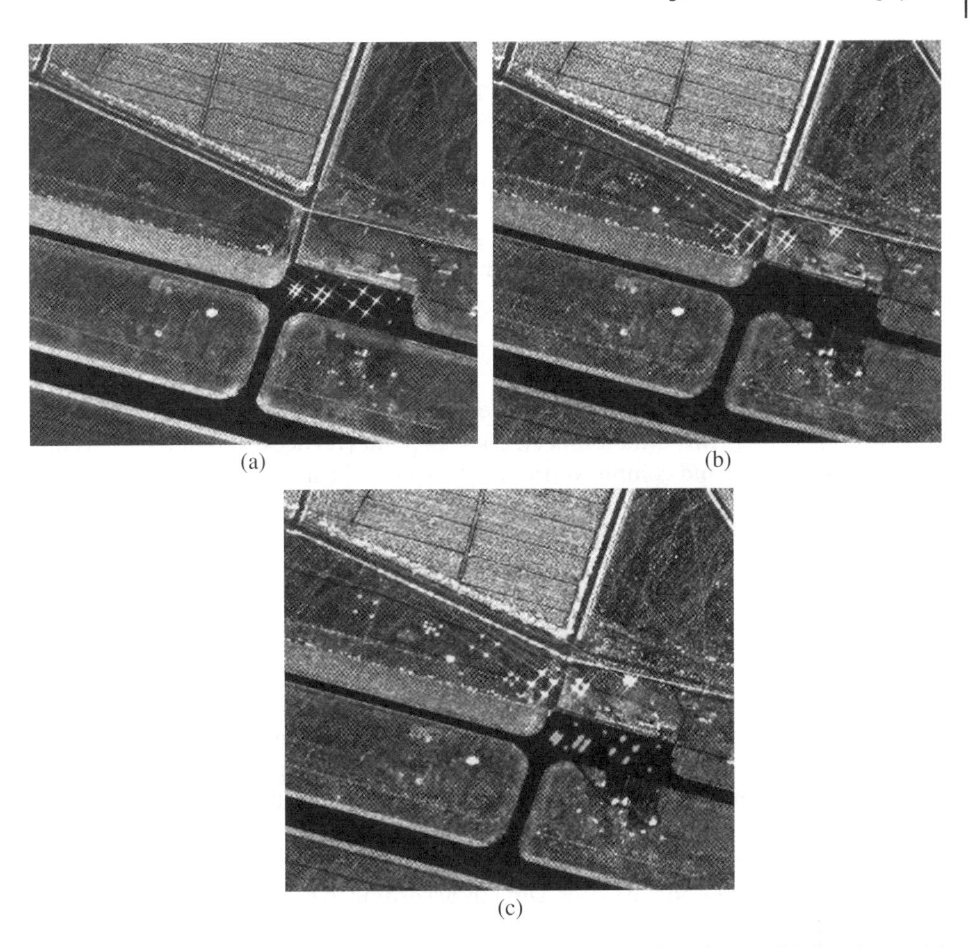


**Figure 7.27** The angle reflector change detection of an airport. (a) Before change; (b) after change; (c) change detection results. (See color plate section for the color representation of this figure).

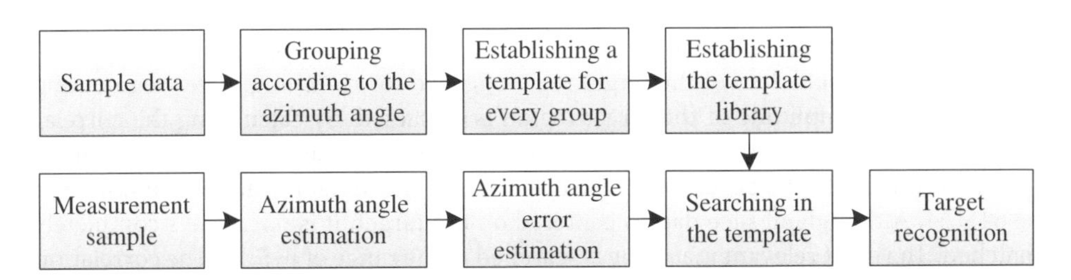


**Figure 7.28** Flowchart of template-based target recognition technology.

where $x$ is the image after segmentation, and $y$ is the amplitude/intensity of the raw image pixel. Through testing, threshold $T_s$ and $T_t$ are generally selected to meet

$$\sum_{t=0}^{T_s} p(t) = 0.03 \tag{7.14}$$

$$1 - \sum_{t=0}^{Tt} p(t) = 0.02 \tag{7.15}$$

where $p(t)$ is the histogram of $y$. Normalizing the histogram gives $\sum_{t=0}^{T_{\max}} p(t) = 1$. The selected threshold $T_s$ and $T_t$ make the shadow region account for 3% of the slice image, and the target area accounts for 2% of the slice image. After the initial segmentation, the mean and variance of each class are estimated. Then MRF iterative segmentation is performed by using graph-cuts algorithm.

#### 7.4.1.2 Peak Feature Extraction

In the SAR image, two kinds of peak features can be defined according to the nature of the scattering center: two-dimensional (2D) peak point (vertices) and one-dimensional (1D) peak point (row and column vertices). The row and column vertices are row local maximum and column local maximum of the image in the target region, respectively. They both are 1D local maximum, while the vertex is 2D local maximum of the image in the target region. The peak feature of the target can be defined by pixels in its neighborhood:

$$p_{ij} = \begin{cases} 1, & \min(a_{ij} - a_{mn}) > \sigma, a_{mn} \in \mathrm{U}(a_{ij}) \\ 0, & \text{else} \end{cases} \tag{7.16}$$

where $i$ is the row number, $j$ is the column number, $\alpha_{ij}$ is current pixel, $\mathrm{U}(\alpha_{ij})$ is the local neighborhood of $\alpha_{ij}$, $\sigma$ is the standard deviation of the target neighborhood pixel, $p_{ij} = 1$ indicates that current pixel is the peak pixel, and $p_{ij} = 0$ indicates that the current pixel is not the peak.

However, in the region of interest (ROI), there is not only the target itself but also background clutter around the target. Directly according to definition of the target peak feature, feature extraction will introduce a large number of false peak points due to interference of background clutter. Therefore, during peak feature extraction, it is necessary to segment the target firstly and extract the peak in the target region.

#### 7.4.1.3 Building Target Template Library

The correlation coefficient of the target region is used for matching. After segmenting the slice target, amplitude of the target region is extracted. By calculating the correlation coefficient between them and the target area of the template, the target category is then determined. In the experiment, the calculation related data window is assumed to be $64 \times 64$. Although all slice data is centered on the target, it is not always completely matched. The most relevant matching is searched in the range of $[-5, 5]$. The correlation coefficient is calculated as

$$\rho = \frac{E((x - \bar{x}) \cdot (y - \bar{y}))}{\sqrt{D(x)} \cdot \sqrt{D(y)}} \tag{7.17}$$

where $x$ and $y$ are the amplitude of the pixel in two windows, $\bar{x}$ and $\bar{y}$ are the average of amplitudes, and $D(x)$ and $D(y)$ are the variance. The region correlation coefficient method is used to match by making full use of low-frequency amplitude and shape information of the target, and the stability of target recognition can be ensured.

To reduce the number of templates and improve search efficiency, a template library is usually established according to a certain azimuth angle interval. For example, if an average template is established every 10°, then a total of 36 templates can be available. The process of the template establishment includes target registration, amplitude normalization, and target average.

Target registration is based on the correlation coefficient. Suppose that there are $n$ samples $f_i(x, y), i = 1, 2, \ldots, n$ in the angle interval. One sample is selected as a standard sample, and the other $n - 1$ samples are registered to the position with the largest correlation coefficient. Here a standard sample can be selected randomly or can be selected close to the center angle. The initial position of registration is the center of the target region after MRF segmentation. Registration accuracy can be achieved by the correlation coefficient matching in a certain range.

Amplitude normalization is used to reduce the influence of amplitude-frequency characteristics of radar transmitter and receiver. Here the power normalization is used, namely

$$f_i(x, y) = \frac{f_i(x, y)}{\sum_{x,y} f_i(x, y)} \tag{7.18}$$

where $f_i(x, y)$ is the amplitude of the pixel at $(x, y)$.

According to registration position, the images around the target are averagely extracted, which means that the target angle template can be obtained. The formula for calculating the average template is

$$\bar{f} = \frac{1}{n} \sum_i f_i(x, y) \tag{7.19}$$

#### 7.4.1.4 Estimation of Target Azimuth Angle

The target in a SAR image is not only affected by various imaging parameters but also by the azimuth angle of the target principal axis relative to the radar beam. The target image displays differently at different azimuth angles. In SAR image recognition, template matching is an important recognition strategy. Usually, it needs to search for matching templates in different azimuth angles. Therefore, if the target azimuth angle of a SAR image is estimated within a certain range, the efficiency of searching and matching will be greatly improved.

An estimation algorithm robust azimuth angle is introduced here [9]. The method is based on the MRF segmented target, and the shadow and background image. According to the relationship between azimuth angle and dispersion coefficient of the target and shadow, the azimuth angle estimation interval (0°, 180°) is divided into the following three sections: (0°, 20°) and (160°, 180°), (20°, 60°) and (120°, 160°), and (60°, 120°). After that, the coordinate axis projection estimation, the target edge estimation fitting, and the improved edge fitting are applied in three intervals.

Define two dispersion coefficients as

$$R_1 = \frac{\text{std}(X_1)}{\text{std}(Y_1)} \tag{7.20}$$

$$R_2 = \frac{\text{std}(X_2)}{\text{std}(Y_2)} \tag{7.21}$$

where $R_1$ is the dispersion coefficient of the target, $R_2$ is the joint dispersion coefficient of the joint region in the target and the shadow, std(·) represents standard deviation, $X_1$ and $Y_1$ are the vectors generated by the $x$ and $y$ coordinates of the target region pixels, and $X_2$ and $Y_2$ are the vectors generated by the $x$ and $y$ coordinates of pixels in the joint region of the target and the shadow. These two dispersion coefficients have common properties: when $R_1 < 0.5$, the target azimuth range is (60°, 90°). When $R_2 > 2.5$, the target azimuth range is (0°, 20°) and (160°, 180°). The above properties will be helpful to estimate the azimuth angle of different intervals by using different estimation methods.

There are some drawbacks to the above azimuth angles estimation method, which cannot estimate the azimuth angle in the range of 0~180° for all images. However, it performs well in certain azimuth angle. The flowchart of azimuth angle estimation according to the relationship between the azimuth angle and the dispersion index is shown in Figure 7.29. The process can ensure that the azimuth angle estimation error is restricted to a small value.

### 7.4.2 Statistical Pattern Recognition

Without considering the estimation error of the target azimuth angle, a support vector machine (SVM) classifier based on PCA can be used to classify the target. The conventional identification process can be summarized as: firstly, the target region is extracted

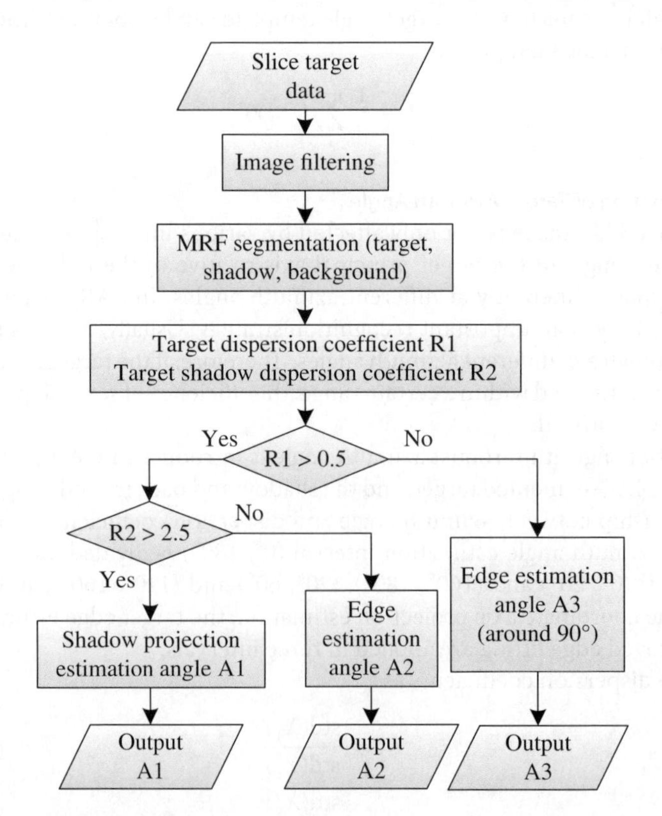


**Figure 7.29** Flowchart of azimuth angle estimation.

from the slice target in detection phase. Then the azimuth angle of the target relative to the radar beam is estimated. In the template library, the template or classifier close to the estimated azimuth is found. The final classification result is obtained by matching or classification. In this process, the estimation accuracy of the azimuth angle is required to be high. Otherwise, it will not find the correct matching template or classifier, and a wrong identification result is released. The matching process will be time-consuming. Because it is difficult to achieve 100% accuracy in target azimuth angle estimation, normally, only error within a certain scope is guaranteed, and there is a 180° ambiguity. Therefore, it is necessary to change the idea of classifying templates or classifier according to a small azimuth angle interval and to develop a better practical classifier.

SVM is a new pattern recognition technology developed in the early 1990s [10]. It is based on the Vapnik–Chervonenkis dimension theory and minimal structural risk principle. It is suitable for classification problems with fewer training samples. It has the advantages of simple structure, global optimization, good generalization ability, and low computational complexity. It also has strong capability to deal with high-dimensional samples. SVM mainly includes support vector classification and support vector regression.

Due to the advantages for solving problems of small samples, nonlinear, high dimension, and uniqueness of solution, the SVM classifier can be widely used in the field of recognition and classification, such as face recognition, voice recognition, and handwriting recognition. SAR image target recognition and classification is the critical component of SAR image interpretation. It is helpful to improve target recognition and classification rate by introducing the SVM classifier into SAR target recognition and classification.

In statistical pattern recognition, feature selection for SVM classification is mostly used, and an appropriate classification model is constructed. As a statistical feature, the PCA feature has been effectively applied in SVM. The analysis method of the SVM classifier based on PCA feature is introduced below, and the SVM classifier is discussed without considering the error of target azimuth angle estimation.

#### 7.4.2.1 Technical Process

Figure 7.30 shows the target recognition process based on SVM classifier. The preprocessing includes target segmentation, peak point extraction, and amplitude normalization. The purpose of target segmentation is to extract the target area in ROI slice and the peak point. Target segmentation can use MRF-based random field segmentation, which results in better segmentation of the target area. The peak point is an important feature of the target, which represents local maximum value point. The aim of extracting peak point is to extract the center of the target, which is convenient for target alignment. Amplitude normalization is used to prevent power differences in the SAR system.

After the above preprocessing, the training samples are grouped in equal angle intervals according to true azimuth angle. Each group is aligned according to target peak point. $64 \times 64$ slice images around the peak are extracted, and PCA transform is applied on this slice data. The corresponding PCA features are extracted, and feature vector matrix is preserved.

On the basis of PCA feature extraction, the SVM classifier is established. The classifier is built in a one-to-one approach and classified according to the voting rules. The

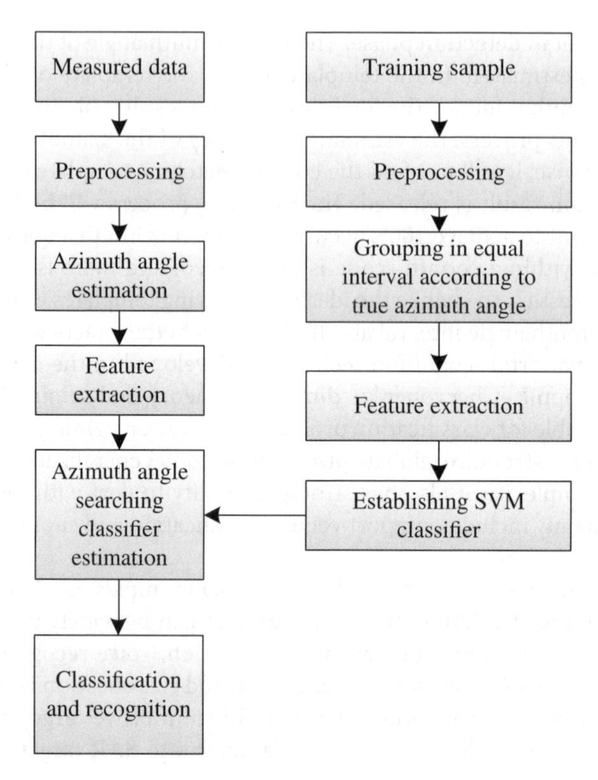


**Figure 7.30** Target recognition process based on SVM classifier.

parameter selection of the classifier has an important influence on classification accuracy. Cross validation is a common method to obtain parameters of the classifier model. When parameters are obtained by cross validation, high accuracy and a small number of support vectors are ensured as much as possible.

In the target testing phase, appropriate classifier is selected according to the estimated azimuth angle.

#### 7.4.2.2 PCA Feature Extraction

The PCA, also known as *K-L transform*, is a transformation based on the statistical properties of random variables. The purpose is to explain the variance–covariance structure of a set of variables by variable linear variables combinations. For $m$ random variables $X = [X_1, X_2, \ldots, X_m]$, although $m$ components can reproduce variability in the whole system, most of variability can be explained by only a few $k$ components. In this case, information contained in the $k$ principal components is (almost) as much as the original $m$ variable. Thus, these $k$ principal components can be used to represent initial $m$ variables.

For an $N \times N$ image $f(x, y)$, the image sample collection is

$$\{f_1(x, y), \quad f_2(x, y), \ldots, \quad f_M(x, y)\} \tag{7.22}$$

where $M$ is the number of samples of the $N \times N$ image.

Equation (7.22) is a statistical integrity. Its properties depend on sample characteristics of the image. Arrange any image $f_i(x, y)$ by column to generate an $N^2$-column vector.

The entire image set becomes $M$ column vector, and each column vector has $n = N^2$ elements. Consider these $M$ random vectors as a random variable $X$, and there are $M$ $x_i$ in $X$. The covariance matrix of variables is

$$C = E\{(X - \mu)(X - \mu)^T\} \text{ OR } C = \frac{1}{M}\sum_{i=1}^{M}(x_i - \mu)(x_i - \mu)^T \tag{7.23}$$

where $\mu$ is the average image vector of the sample.

It is difficult to directly calculate eigenvalue and orthonormal eigenvector of the $N^2 \times N^2$ matrix $C$. The following theorem is herein introduced.

Singular value decomposition (SVD) theorem: assume $A$ is an $n \times r$ matrix with a rank of $r$, and there are two orthogonal matrices:

$$U = [u_1, u_2, \dots, u_r] \in R^{n\times r} \quad U^T U = I \tag{7.24}$$

$$V = [v_1, v_2, \dots, v_r] \in R^{n\times r} \quad V^T V = I \tag{7.25}$$

and the diagonal matrix is

$$\Lambda = diag[\lambda_1, \lambda_2, \dots, \lambda_r] \in R^{n\times r}, \tag{7.26}$$

and

$$\lambda_1 \geq \lambda_2 \geq \dots \geq \lambda_r \tag{7.27}$$

meets $A = \mathrm{U}\Lambda^{1/2}\mathrm{V}^{\mathrm{T}}$, where $\lambda_i$ is the nonzero eigenvalue of matrix $\mathrm{AA}^{\mathrm{T}}$ and $\mathrm{A}^{\mathrm{T}}\mathrm{A}$, and $u_i$ and $v_i$ are the eigenvectors corresponding to $\lambda_i$ of matrix $\mathrm{AA}^{\mathrm{T}}$ and $\mathrm{A}^{\mathrm{T}}\mathrm{A}$, respectively. The above decomposition is called the SVD of the matrix, and $\sqrt{\lambda_i}$ is the singular value of $A$.

Based on the above theorem, it is inferred that

$$U = AV\Lambda^{1/2} \tag{7.28}$$

$C$ can be given by

$$C = \frac{1}{M}\sum_{i=1}^{M}(x_i - \mu)(x_i - \mu)^T = \frac{1}{M}XX^T \tag{7.29}$$

where $X = [x_1 - \mu, x_2 - \mu, \dots, x_M - \mu]$. Knowing the structure matrix $R = XX^T \in R^{M\times M}$, it is easy to calculate eigenvalue $\lambda_i$ and its corresponding orthonormal eigenvector $v_i$. Then the orthonormal eigenvector $u_i$ is

$$u_i = \frac{1}{\sqrt{\lambda_i}}Xv_i \tag{7.30}$$

This is the eigenvector of the image, which is indirectly obtained by computing eigenvalue and eigenvector of the low-dimensional matrix R.

Sort the eigenvalue from big to small $\lambda_1 \geq \lambda_2 \geq \cdots \geq \lambda_r$, and the corresponding eigenvector is $u_i$. In this way, every image can be projected to a subspace composed of $u_1, u_2, \dots, u_r$. Each image corresponds to one point in the subspace. Similarly, any point in the subspace corresponds to an image. For any image, a group of coordinate coefficients can be obtained by projecting its subspace. This group of coefficients is the position of the image in the subspace, which can be used as the basis of target recognition. Thus, for any group of samples $f$, the coefficient vector is represented as

$$y = U^T f \tag{7.31}$$

As PCA transform has compressed the image data, the largest $k$ eigenvectors are chosen to have

$$\sum_{i=1}^{k} \lambda_i / \sum_{i=1}^{M} \lambda_i \geq \alpha \tag{7.32}$$

where $\alpha$ is the proportion of total energy.

### 7.4.3 Typical Examples

Template-based matching recognition is performed by using three kinds of data, T72-132, BMP2-9563, and BTR70-C71 of MSTAR under 17°, as the template data [11] and all types of data under 15° as test data. If the template is established at an interval of 10°, this three-kind data has a total of 108 matching templates. First, the target azimuth angle is estimated, and the template is searched within a certain error range and is matched according to correlation coefficients. Table 7.9 lists the recognition rate at different azimuth search ranges. It can be seen that, with the expansion of search scope, the accuracy increases. Due to influence of the azimuth angle estimation accuracy, when search scope is larger than 20°, the recognition rate basically does not change.

In the above experiments, only three kinds of target templates are used. Subsequently, all the seven kinds of targets are used to make the template at an interval of 10° azimuth angle, leading to a total of 252 templates. This template library contains all types of targets. By using the test data, the target recognition is classified into three types of confusion matrixes as listed in Table 7.10, where the angle search scope is [−20°, 20°].

## 7.5 Multisource SAR Image Fusion

The multisource SAR image fusion is an important branch of image fusion that mainly solves the problem of image information redundancy and complementarity of one area with different time intervals, different bands, and different polarizations.

**Table 7.9** Recognition rate % in different azimuth angles based on three kinds of target templates.

| Type | 5° | 10° | 20° | 35° |
|---|---|---|---|---|
| T72-132 | 92.36 | 95.41 | 98.98 | 98.98 |
| T72-812 | 84.10 | 90.77 | 92.82 | 92.31 |
| T72-S7 | 91.10 | 93.72 | 93.72 | 93.72 |
| BMP2-9563 | 90.26 | 98.46 | 98.97 | 98.97 |
| BMP2-9566 | 88.27 | 92.35 | 92.35 | 92.35 |
| BMP2-C21 | 89.80 | 93.88 | 95.92 | 95.41 |
| BTR70-C71 | 96.43 | 97.45 | 98.98 | 98.98 |

**Table 7.10** Target recognition confusion matrixes in the template library of seven types of targets (azimuth angle error [−20°, 20°]).

| Type | T72 | BMP2 | BTR | Recognition rate (%) |
|---|---|---|---|---|
| T72-132 | 194 | 0 | 2 | 98.98 |
| T72-812 | 195 | 1 | 0 | 99.49 |
| T72-S7 | 190 | 0 | 1 | 99.48 |
| BMP2-9563 | 1 | 194 | 0 | 99.49 |
| BMP2-9566 | 0 | 196 | 0 | 100 |
| BMP2-C21 | 3 | 193 | 0 | 98.47 |
| BTR70-C71 | 1 | 1 | 194 | 98.98 |

(1) SAR images at different times reflect variation of surface features and targets. Therefore, the change can be detected by fusion. Multitemporal image change detection is widely used in land usage, urban change, target monitoring, etc.
(2) The SAR image features are quite different in both high- and low-frequency bands. The vegetation penetration in high frequency is weak, and the SAR image provides the surface image, which is close to the visual effect of the optical image. The vegetation penetration in low frequencies is strong and the SAR image is dark, which displays the strong reflection objects and the scene image of objectives under vegetation. If both high- and low-frequency SAR images are fused, a three-dimensional (3D) image of the scene can be offered that is very useful to detect hidden targets.
(3) The multipolarization images reflect the scattering characteristics of the same ground objects in different polarization receiving and transmitting states. By means of polarization decomposition, the polarization characteristics are extracted, which is beneficial to analyzing structural characteristics of the target, such as single scattering, second-order scattering, or volume scattering. Multipolarization data fusion is conducive to classification of objects, artificial target detection and recognition, etc.

### 7.5.1 Image Fusion Method

Image fusion is generally divided into three levels, namely, pixel-level fusion, feature-level fusion, and decision-level fusion. Each level has different processing approaches.

(1) Pixel-level fusion is the direct fusion of the images originally collected, which is the lowest level of fusion. It requires accurate registration of the image (generally better than 1 pixel). The advantage of pixel-level fusion is that it preserves original data and detailed information as much as possible. The disadvantage is the large amount of image data, low efficiency, low stability, and anti-interference. Pixel-level fusion is often used for multisource image fusion and image enhancement and understanding. Common methods are intensity hue saturation (HIS) transform, PCA transform, high-pass filter (HPF) and wavelet transform fusion algorithm, etc. There is more polarization decomposing fusing for multipolarization data.

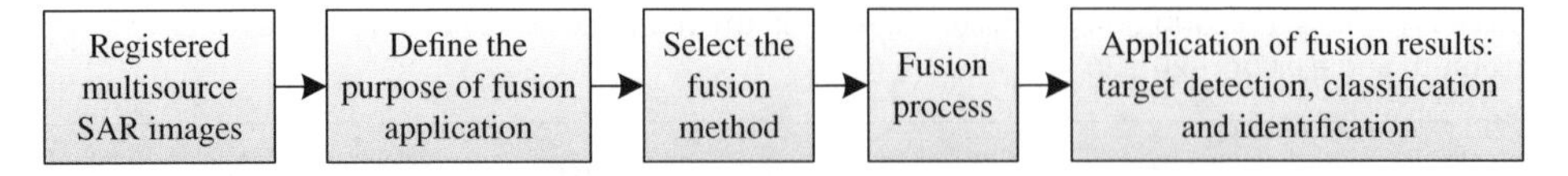


**Figure 7.31** General procedures of multisource SAR image fusion.

(2) Feature-level fusion belongs to the second level of fusion. It requires each image source to complete the target detection and feature extraction alone, and complete fusion is based on the characteristic vector. The advantage of feature-level is finalization of information compression, beneficial to real-time processing. In addition, because the extracted features are directly related to decision analysis, the fusion result can give the maximum characteristic information needed for decision analysis. Feature-level fusion is mainly used for image change detection, classification, recognition, etc. The major means of feature-level fusion include cluster analysis, Bayes estimation, entropy method, weighted-average method, voting method, neural network method, etc.
(3) Decision-level fusion is the highest level of fusion, which requires each image source to complete the decision making alone, and all results are fused. It coordinates according to certain criteria and credibility of each image decision to make the optimum decision. The main advantages of decision-level fusion are good real-time performance, high efficiency, fault tolerance, and openness. However, the independent decision making of each image requires a large amount of computation and artificial decision making before fusion. Decision-level data fusion methods mainly include Bayes estimation method, expert system, neural network method, fuzzy set theory, reliability theory, logic template method, etc.

The general procedures of multisource SAR image fusion are shown in Figure 7.31.The information of the same object in different images is correlated after accurate registration of the multisource SAR images. The purpose of fusion application is determined, such as target enhancement, target detection, target classification, and identification. With different purposes, the fusion approach adopted is different as well.

### 7.5.2 Fusion Effect Evaluation

The image fusion effect evaluation is normally divided into two approaches: qualitative assessment and quantitative assessment. Qualitative assessment mainly includes visual comparison and algorithm operation time efficiency comparison. Quantitative evaluation uses image parameters to assess, such as information entropy, average gradient, correlation coefficient, standard deviation, edge preservation, etc. However, from the perspective of the SAR image application, image fusion is only one of the possible approaches. The fusion effect should improve application value. The fusion evaluation should be reflected in the application effects, such as detection rate, false alarm rate or recognition rate, etc.

### 7.5.3 Typical Examples

This section presents fusion results of multiband and multipolarization images.

#### 7.5.3.1 Target Detection of Multiband Vegetation Penetration

SAR with different frequencies has different penetration capability. Low-frequency electromagnetic wave has a powerful penetration to ground targets, so that surface features and special features that cannot be detected by high-frequency SAR data can be acquired. This section and the next five sections introduce the use of the CFAR detection algorithm to detect the ground target from P-, X-, L-, and Ku-band images. Detection results are compared to analyze the hidden target detection capability at different bands.

It is known that there are a total of seven trucks under a roadside trees shelter, and there are three groups of large corner reflectors and a group of small corner reflectors, as shown in Figure 7.32. Corner reflectors and trucks are circled. The arrangement of seven trucks, three groups of large corner reflectors, and one group of small corner reflectors is shown in Figure 7.33.

Detection results obtained by CFAR are shown in Figure 7.34. Figure 7.34a shows Ku-band detection results, where points 2, 3, 4, and 5 are corner reflectors, and other points are the detected targets. Figure 7.34b shows X-band test results, where points 2, 3, 4, and 5 are corner reflectors, and other points are the detected targets. Figure 7.34c shows the L-band detection results, where points 2 and 3 are corner reflectors, and the remaining points are the detected targets. Figure 7.34d shows test results of multiband data, where points 2–9 and 19–21 are corner reflectors, and points 11 and 12 are false targets. Figure 7.34e shows P-band detection results, where point 2 is a corner reflector, and other points are the detected targets.

It can be seen from Figure 7.34 that the truck detection at P band has the best result with accurate positioning and clear detection, and interference of false targets is at the minimum. But the detection of corner reflectors is not good enough, and only one corner reflector of a large corner reflector group is recognized and located. The other corner reflectors in the same group are abandoned because they are not clear enough. All seven trucks at L band are also completely detected even though they are sheltered by street trees, but detection results are not as clear as those at P band. And there are quite a number of interference points. Its corner reflector detection is better than that at P band, with a total of two corner reflectors accurately detected and located. At X and Ku band, only two trucks are detected, but the corner reflector detection is better. Multiband data produced by fusion of Ku, X, L bands has detected all seven trucks, as shown in Figure 7.35.

The experimental results have clearly shown that multiband data has more abundant information than the single-band data, so it has better detection capability for both hidden and exposed targets.

#### 7.5.3.2 Target Detection of Multiband Grassland Background

The multiband SAR data is at the L, C, and X band. There are four vans and three telegraph poles in the experimental region, and the background is grassland. There is a major difference in the intensity response of the target in different wavebands, as shown in Figures 7.36 and 7.37. In the L-band image, four vehicles and three telegraph poles are capable to highlight the background with a high signal-to-noise ratio (SNR). In the C- and X-band images, the telegraph poles are lost in the background and vans with top roofs do not have high scattering but have obvious features of shadow. The SNR of vehicles is listed in Table 7.11. Obviously the target in the background of the grass, in L band, is even more prominent.

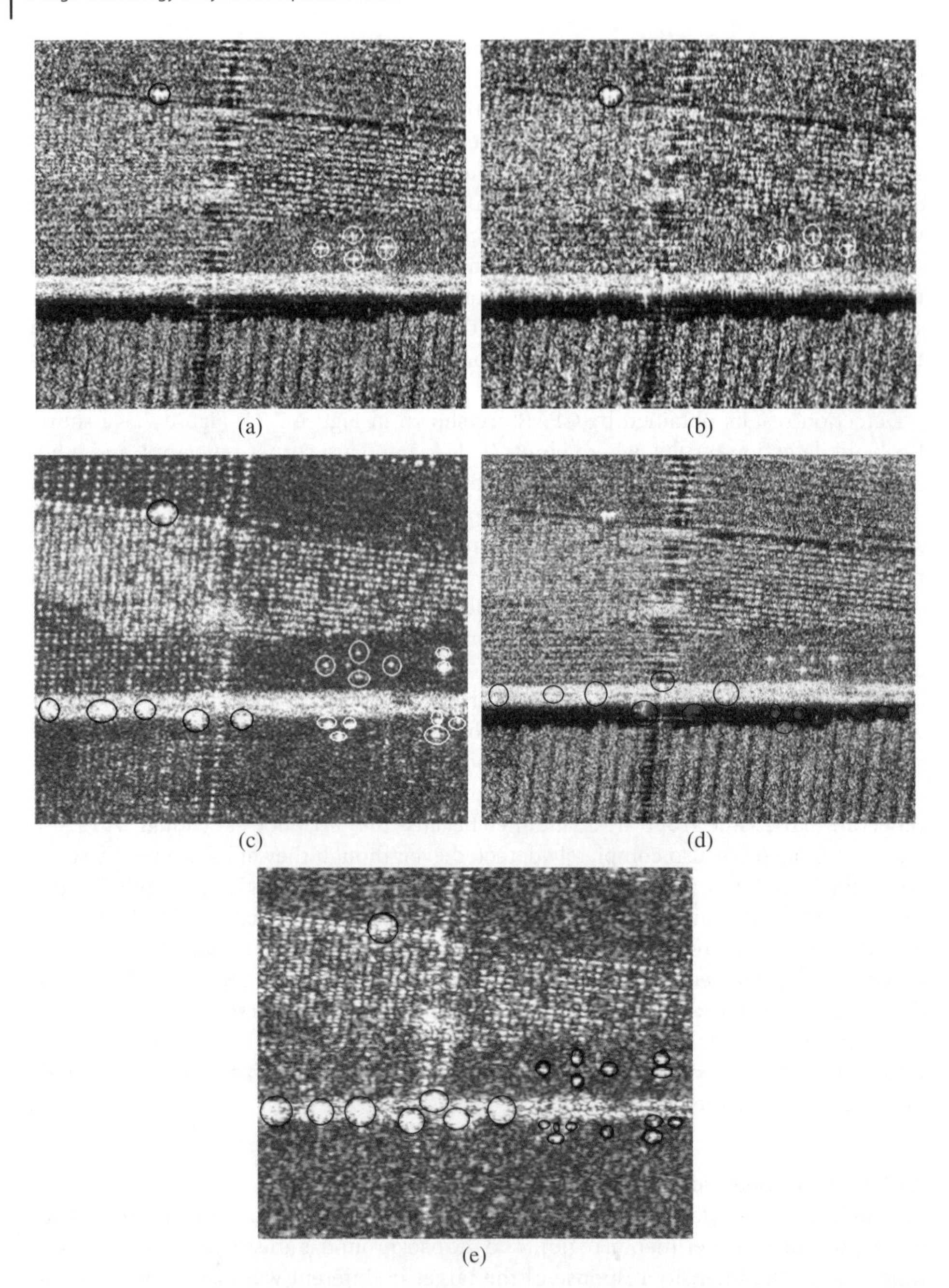


**Figure 7.32** Experimental images used for hidden target detection. (a) Ku band; (b) X band; (c) L band; (d) Ku, X, L triband fusion multiband data; (e) P band. (See color plate section for the color representation of this figure).

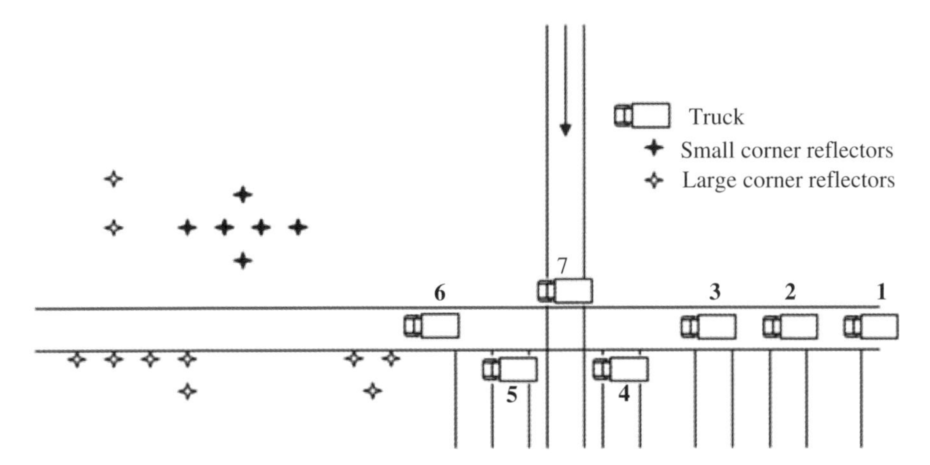


**Figure 7.33** Targets to be detected.

The reason why the scattering intensity of the poles on different wavelength images is different is that under the background condition of the grass, a large amount of diffuse reflection occurs in the short wave. The long wave has strong specular reflection in the soil layer, after it penetrates the surface grass layer, and a dihedral angle is formed after the secondary scattering by the pole, as shown in Figure 7.38, and all the pixels reflected by the dihedral angle are condensed at one point, which is the bottom of the utility pole. When the background of the pole is not grass but a smooth concrete floor or tarmac, the pole will form dihedral reflection on all three bands, showing a strong highlight.

There are two reasons that a van with a roof has different scattering intensity in different wavelength images, as shown in Figure 7.39. On one hand, similar to telegraph pole imaging, due to the high diffuse reflection of the grass background in short wave, the electromagnetic wave cannot generate dihedral reflection on the ground and in the back of the van, while long wave can generate a dihedral angle reflection. On the other hand, the enclosed van roof will lead to electromagnetic wave reflected by the roof, so that unlike the van without a roof, it cannot generate dihedral angle reflection.

These image features can be summarized as the following: under grassland background, L band is more suitable for hard target detection, such as telegraph poles, stone pillars, hidden vehicles, etc., than C and X band.

#### 7.5.3.3 Multiband Fusion Classification Analysis

The pseudocolor synthetic image of multiband SAR can characterize the response of different landforms on different bands, and the feature classification can be performed accordingly. Taking the triband pseudocolor synthetic (red: L band; green: X band; and blue: C band) as an example, red pixels suggest that bottom scattering dominates and the surface in solid structure is hard to penetrate, or there are hard solid targets under vegetation cover. Green pixels indicate that the surface is dominant, and the surface is covered by a shallow layer of vegetation. Blue pixels indicate that the middle layer scattering is dominant, and generally bare land and sand are dominating. When there is a high scattering surface at all three bands, it must be bare solid targets, and the pseudosynthetic color is white.

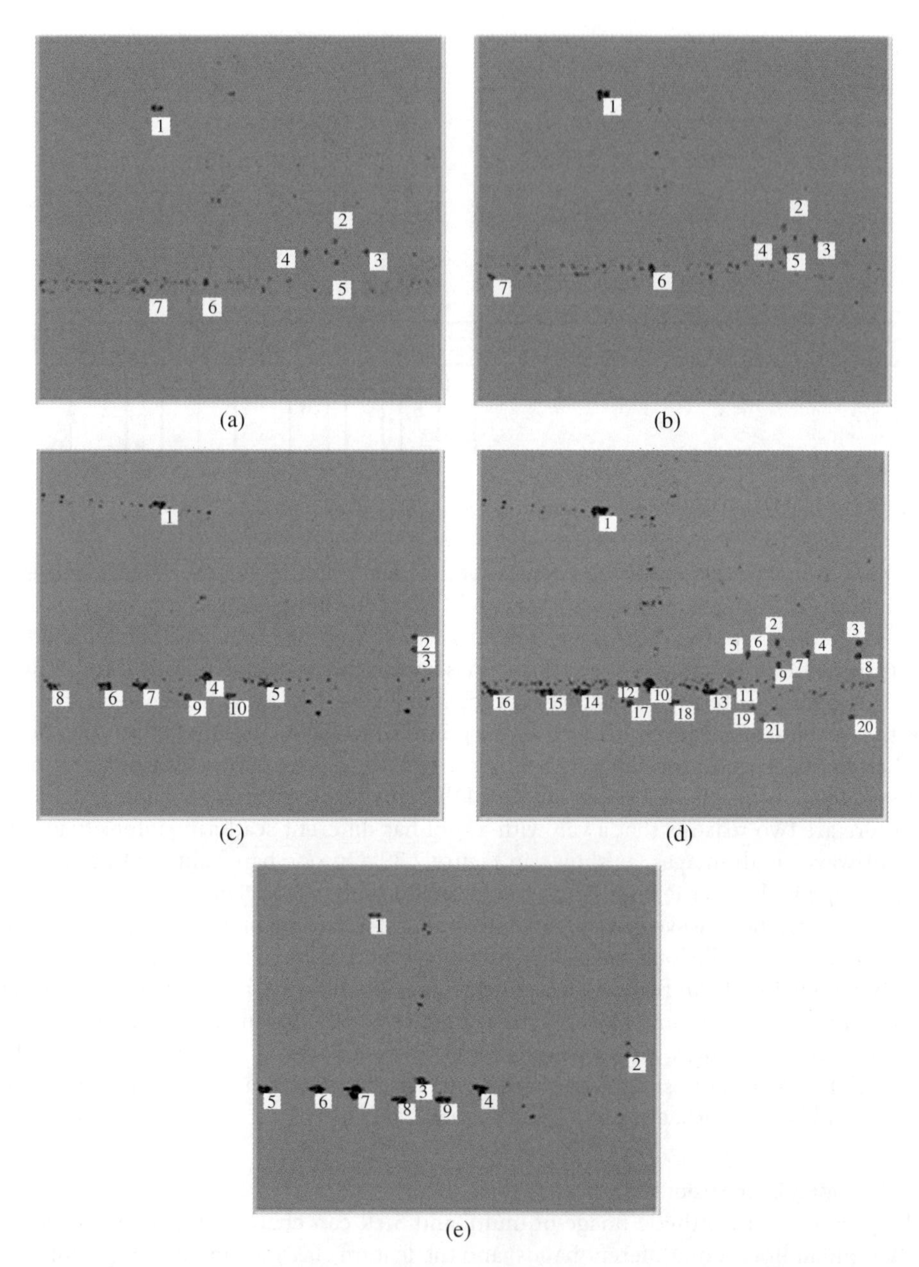


**Figure 7.34** Target detection results. (a) Ku band; (b) X band; (c) L band; (d) Ku, X, and L triband fusion data; (e) P band.

Figure 7.40 is the pseudosynthetic color image of the vegetation area at L, X, and C band. The vegetation in this area is flourishing. It is clear that the X-band echoes in region 1 are weak, C-band echoes are relatively strong, and L-band echoes are strong, which reflects a variety of surface features. X band echoes in region 2 dominate, and foliage reflection of the vegetation/crop has strong echoes in X band. Region 3 is a forest.

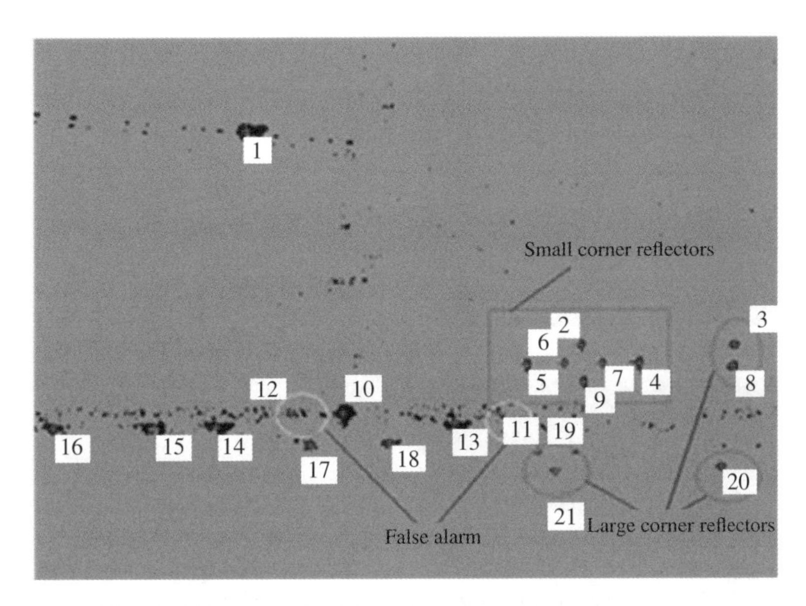


**Figure 7.35** Analysis on multiband synthetic data detection results.

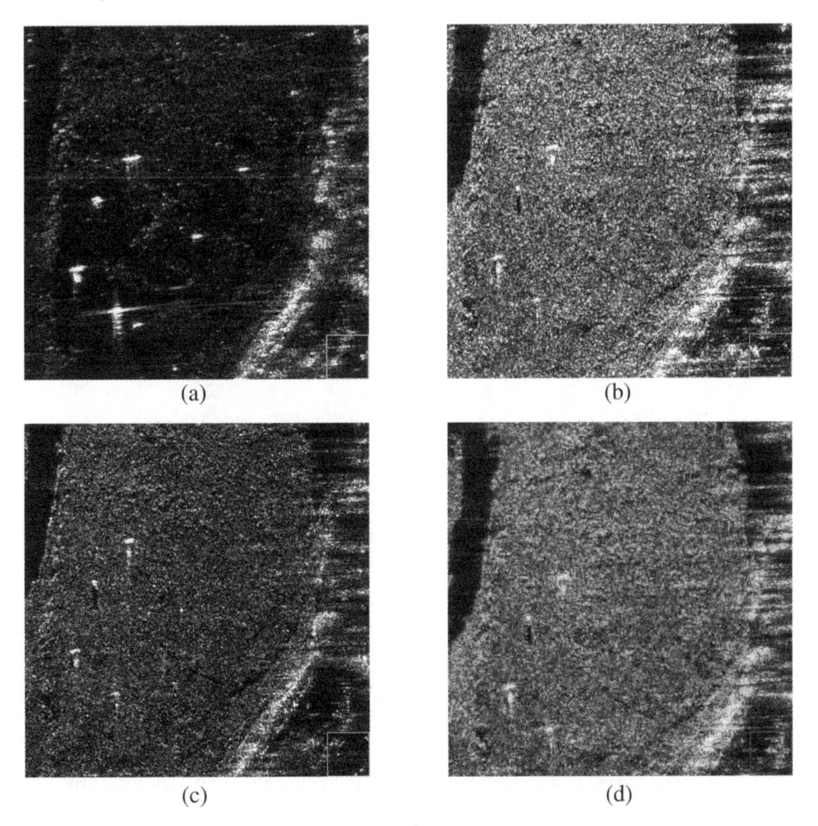


**Figure 7.36** Multiband characteristic analysis and fusion effect. (a) L band; (b) X band; (c) C band; (d) fused image. (See color plate section for the color representation of this figure).

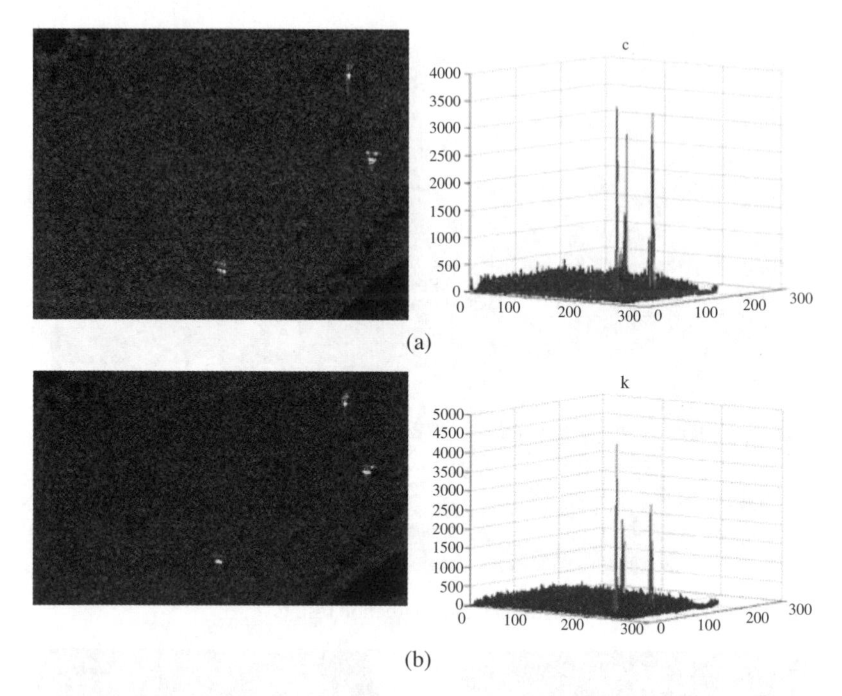


**Figure 7.37** Ground target features comparison of multiband SAR image. (a) C-band target image (left: amplitude; right: 3D view); (b) X-band target image (left: amplitude; right: 3D view).

**Table 7.11** Vehicle SNR.

| No. | L-band SNR/dB | C-band SNR/dB | X-band SNR/dB |
|---|---|---|---|
| 1 | 103.27 | 31.27 | 47.78 |
| 2 | 160.23 | 28.79 | 30.17 |
| 3 | 65.51 | 4.16 | 2.57 |
| 4 | 86.73 | 24.83 | 22.84 |

Remarks: The SNR at each band is the ratio of the peak to average amplitude of the same background region.

At X/C band it shows a round canopy characteristic. Both X and C bands have strong echoes. L band penetrates the canopy. In region 4, hidden objectives under trees are present at L band. Figure 7.41 shows the single-band SAR images at Ku, L, and X band and multiband SAR image after triband fusion. Pseudocolor fusion of multiband data can be used to fuse SAR images, as shown in Figure 7.42. The classification results of single-band SAR images and fused SAR images are shown in Figure 7.43.

The classification analysis of experimental SAR image data shows that the Ku, L, and X single-band are inferior to the multiband SAR classification. The reason is that multiband data can offer more classification information, whereas multiband data ($N$ bands) can offer $N$-dimensional characteristic information, with the corresponding value of

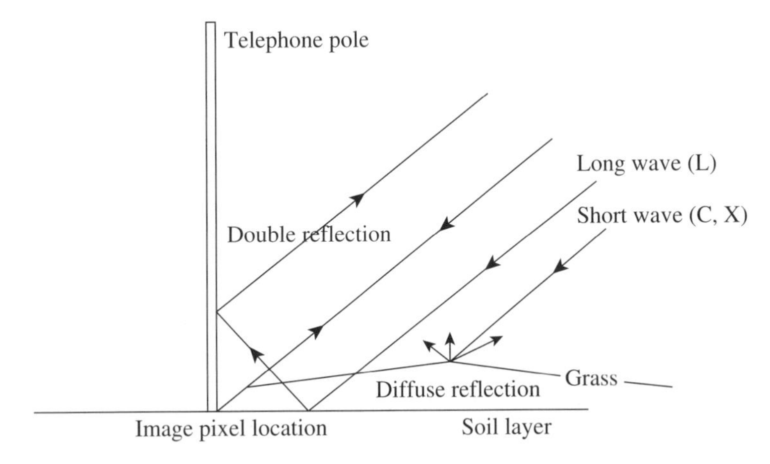


**Figure 7.38** Geometry sketch of a telegraph pole imaging in grassland background at different bands.

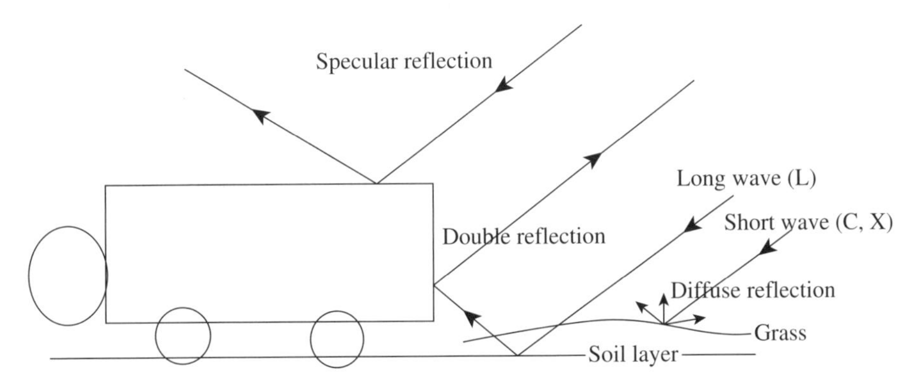


**Figure 7.39** Geometry sketch of the imaging of a van with a roof in grassland background in different bands.

single band of 1. At the same time, multiband data can offer more image texture information.

#### 7.5.3.4 Multiband Marine Target Detection

The multiband SAR data is at L, C, and X band, as shown in Figures 7.44 and 7.45. It can be seen from images of the sea area taken at the same time but at different bands that a small boat shown in the L-band image is not clearly shown in the X-band image. Compared with C and X band, L band can highlight the ship target better and improve the target detection rate.

From parameters of the ship targets, it is known that the size of the same target is different in images acquired at different bands. The size is the largest in the L-band image, the smallest in the X-band image, and the second in the C-band image. Due to images generated by synthetic aperture of different antenna beam width, the focused pulse number of point targets is different. The focusing time is the longest in L band and the shortest in X band. This means that the target size in the X-band image is mostly approaching the actual size of the target.

**Figure 7.40** Multiband pseudocolor synthesis of vegetation area. (See color plate section for the color representation of this figure).

Through analysis of ship target detection in three bands, it is known that the use of both high and low bands not only can improve detection rate but also can more accurately estimate target parameters.

An experiment on ship detection has been performed in five bands over the sea. The multiband and multipolarization data is used for detection. The ship detection results after statistical global CFAR amplitude correction are shown in Table 7.12. The table shows that L band has the highest detection rate and the lowest false alarm rate, while X band has more false alarms. This is mainly due to the noise introduced by the ship highly scattering on side lobe.

#### 7.5.3.5 Multipolarization Building Detection

Figure 7.46 shows airborne full-polarization SAR data at X band with 1 m resolution against a land scene. First, the pixel-level fusion of the multipolarization image is performed; acquiring the full-power diagram, the CFAR detection technology is used to detect the highly scattering targets. From the detection results we can see, there is large number of missing detections in the target area, and only a few parts are detected with single-channel detection. However, the merging of multipolarization can make almost all target areas detected and is beneficial to complete performance of the target shape, greatly improving detection rate and solution of the target attribute parameters.

#### 7.5.3.6 Multipolarization Ship Detection

Ship detection has been performed by using the airborne full-polarization data (2 m resolution). Figure 7.47 shows the HH, HV, VH, and VV multipolarization images of the experiment area. The resolution of high-definition airborne polarization SAR is more sensitive to sea surface waving, which easily increases the false alarm rate especially in the VV image. If the target detection is performed after polarization fusion of full-power images, the ship target can be detected accurately, as shown in Figure 7.48.

(a)

(b)

(c)

**Figure 7.41** Multiband SAR image. (a) SAR image at Ku band; (b) SAR image at L band; (c) SAR image at X band.

## 7.6 Technology Outlook

The SAR image information processing system is discussed in this chapter. The target detection, change detection, target recognition, and multisource image fusion are introduced as well as the corresponding experiment results. Chinese universities and research institutes have also made a lot of efforts in this area. However, the SAR image intelligence is far from practical application, as can be seen in both China and overseas.

**Figure 7.42** Postfusion pseudocolor SAR image. (See color plate section for the color representation of this figure).

Long-term exploration and research are still required. The current application bottlenecks are summarized as follows:

(1) *The technology is still in the research stage and is far beyond engineering application.* Some algorithms have drawbacks in the design. For example, some can only process images with some data types and some resolution, without considering processing of large-size images. Target detection is performed under simple background. Some algorithms are too complicated and innovative, while stability of the algorithm is ignored. Complex algorithms cause more computation or may cause calculation divergence, which cannot acquire the desired results. For practical application, the

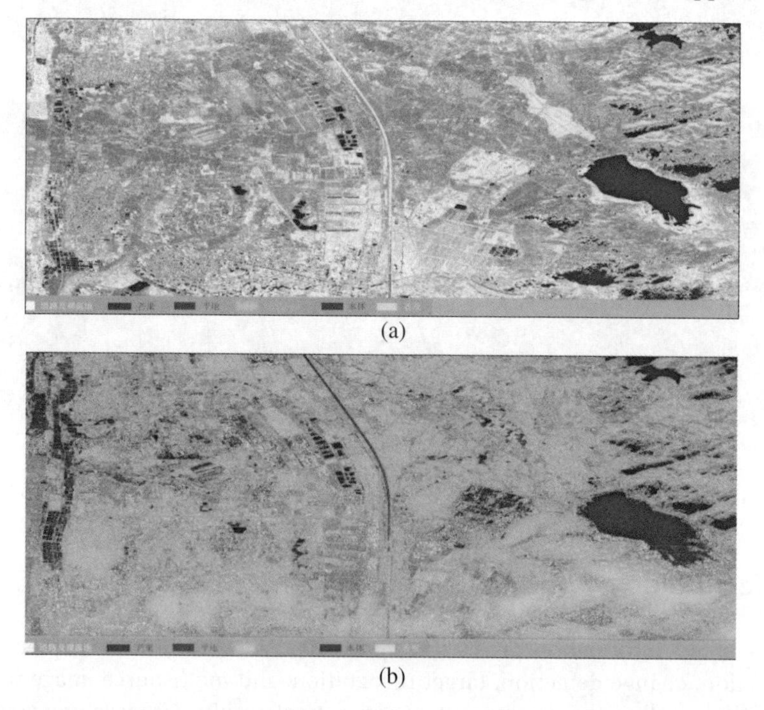

(a)

(b)

**Figure 7.43** Classification results of single-band SAR images and fused SAR image. (a) Classification image at Ku band; (b) classification image at L band; (c) classification image at X band; (d) classification image after multiband fusion. (See color plate section for the color representation of this figure).

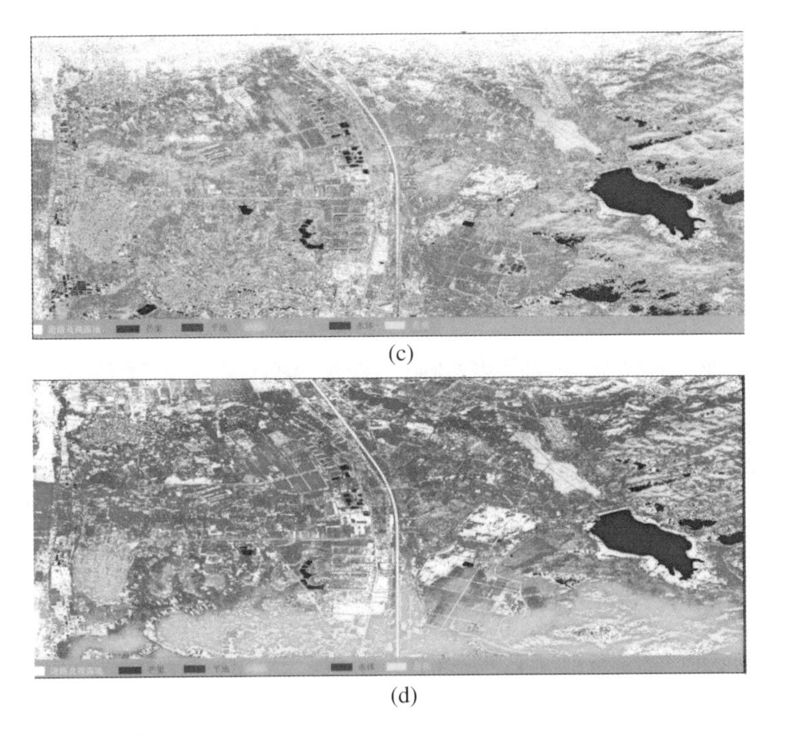

(c)

(d)

**Figure 7.43** *(Continued)*

capability to automatically deal with massive real-time data is very important. Currently, most algorithms only consider single-image processing. Parallel optimization is not considered, which leads to high time complexity. Obviously, the technology algorithm engineering applicable to SAR practice is one of the important developing trends.

(2) *Lack of data source for target recognition.* Target recognition is an important goal of a high-resolution SAR system. Although each recognition technology has a relatively mature theory, the main difficulty of target recognition in the field of SAR is lack of target sample database. An effective target recognition system must be supported by sufficient samples. In practice, it is difficult to acquire the target image, due to high flight cost and low efficiency, and the training requirements of the target recognition system cannot be met. Therefore, how to solve the problem of target data source needs to be one of the prominent studies in the future.

(3) *Insufficient study on information application system.* At present, there is much research on the algorithm. However, it is not enough to perform the research from the application system point of view, such as management of massive data, calculation of big data, display of findings, etc. With the development of the SAR sensor, geographic information system, big data, and computer and artificial intelligence technology, the user needs a comprehensive solution that integrates various SAR image processing technologies from image to information to meet the requirements for different application scenarios.

Based on the above bottlenecks, three important development directions are briefly described: algorithm engineering application, electromagnetic simulation and

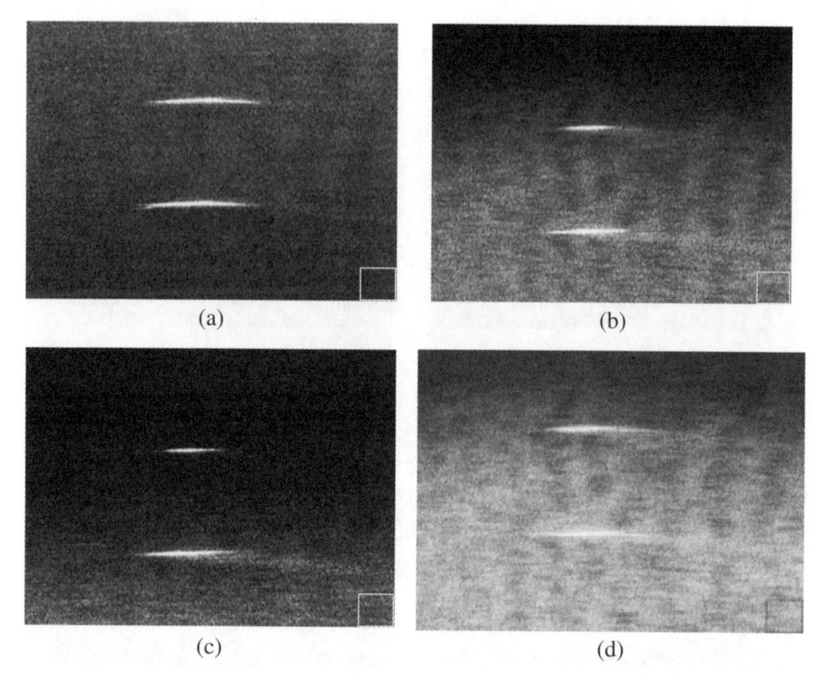


**Figure 7.44** Ship target comparison 1 in multiband SAR image. (a) L band; (b) X band; (c) C band; (d) fusion. (See color plate section for the color representation of this figure).

intelligent target recognition of SAR target image, and the SAR image processing system.

### 7.6.1 Research on Algorithm Engineering Application

The purpose of the research on algorithm engineering application is to improve the practicability of the SAR image processing algorithm, which includes three aspects, namely, adaptability, stability, and real-time improvement. Therefore, it is needed to perform the corresponding improvement design, as shown in Figure 7.49.

Improving adaptability mainly refers to improvement of the processing algorithm to adapt to image data processing with different data types, image sizes, and scenarios to ensure normal calculation. Improving stability mainly refers to improvement of the simplified processing flow and usage of a stable algorithm, so that the target detection algorithm can have stable output for different types and background images. Real-time improvement mainly is to take appropriate means to improve efficiency of the algorithm and to ensure continuous processing of massive data. The major ways to improve real-time performance include parallel algorithm design and computation reduction design.

### 7.6.2 Research on Electromagnetic Simulation and Intelligent Target Recognition of the Target Image

The electromagnetic simulation of the SAR target image is an effective approach to solve the lack of target data in current target recognition. Based on high-precision 3D models

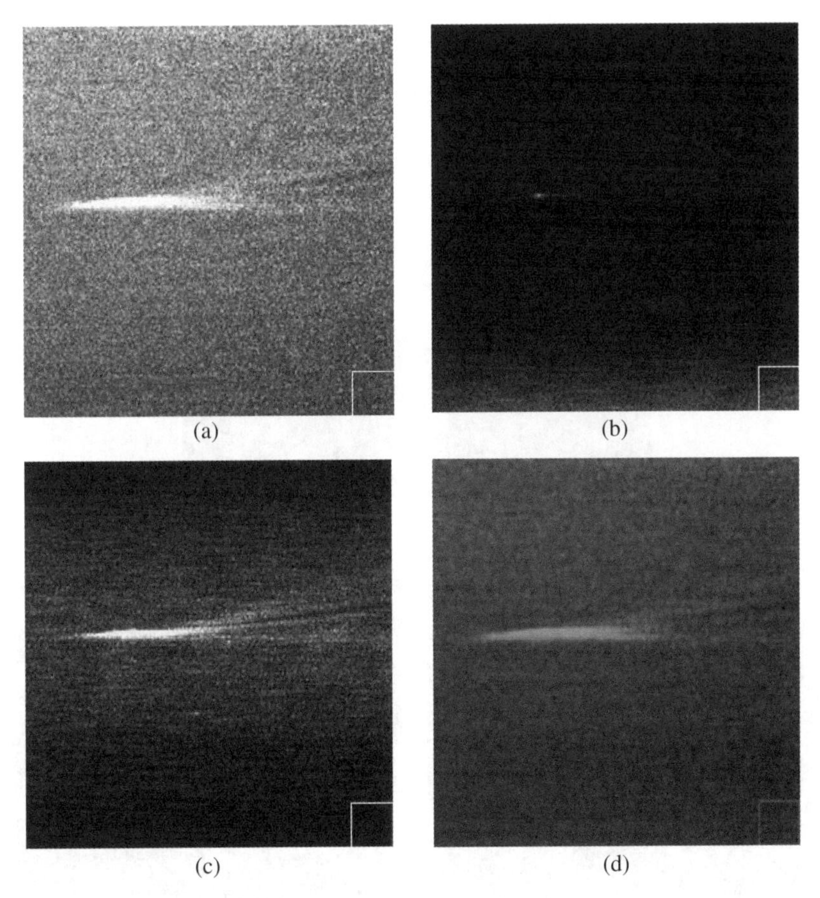


**Figure 7.45** Ship target comparison 2 in multiband SAR images. (a) L band; (b) X band; (c) C band; (d) fusion. (See color plate section for the color representation of this figure).

**Table 7.12** Statistic of ship detection parameters.

| Items | C (HH) | L (HH) | X (VV) |
|---|---|---|---|
| Actual ship targets | 42 | 42 | 42 |
| Detected number | 41 | 42 | 46 |
| Correct target number/detection rate (%) | 40/95.24 | 41/97.62 | 40/95.24 |
| False alarm number/false alarm rate (%) | 1/2.44 | 1/2.38 | 6/13.04 |

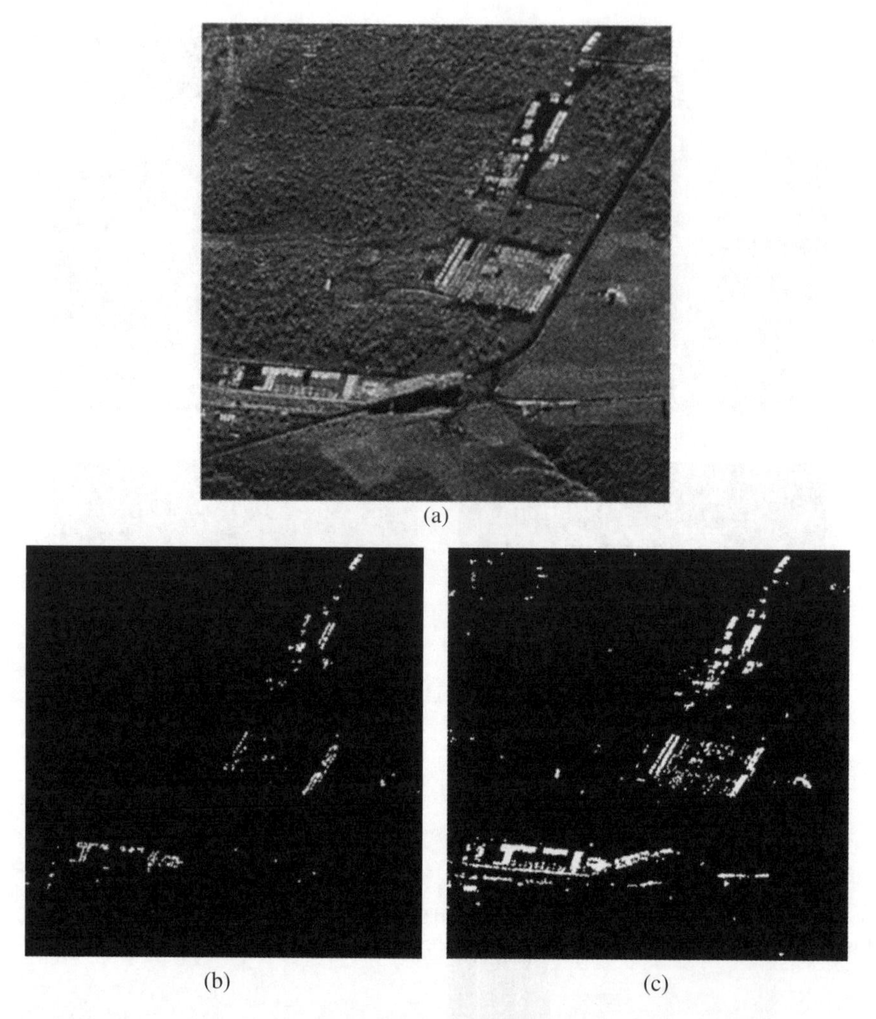

(a)

(b) (c)

**Figure 7.46** Detection results of buildings with multipolarization fusion method. (a) Primitive HH channel image; (b) the detection results of bright area in HH channel; (c) the detection results of bright area in multipolarization fusion.

of the target, high-frequency electromagnetic calculation approaches, such as the ray method and physical optics, are used to simulate the target echo. Then the image is simulated by the appropriate algorithm. Mature SAR target image simulation tools are already available abroad, such as Xpatch, GRECO, OKTAL-SE, and RadBase. In the meantime, relevant universities and research institutes are researching and developing electromagnetic simulation technology, trying to generate an electromagnetic simulation tool with a SAR target image.

The general electromagnetic simulation flow of a SAR target image is shown in Figure 7.50.The main input parameters are target profile and imaging parameters. First, a 3D model of target and scene is established. Then, the ray distribution is estimated under the support of the imaging geometry, the scene is composed by ray tracker, and the ray path is obtained by the acceleration algorithm. According to the scattering and

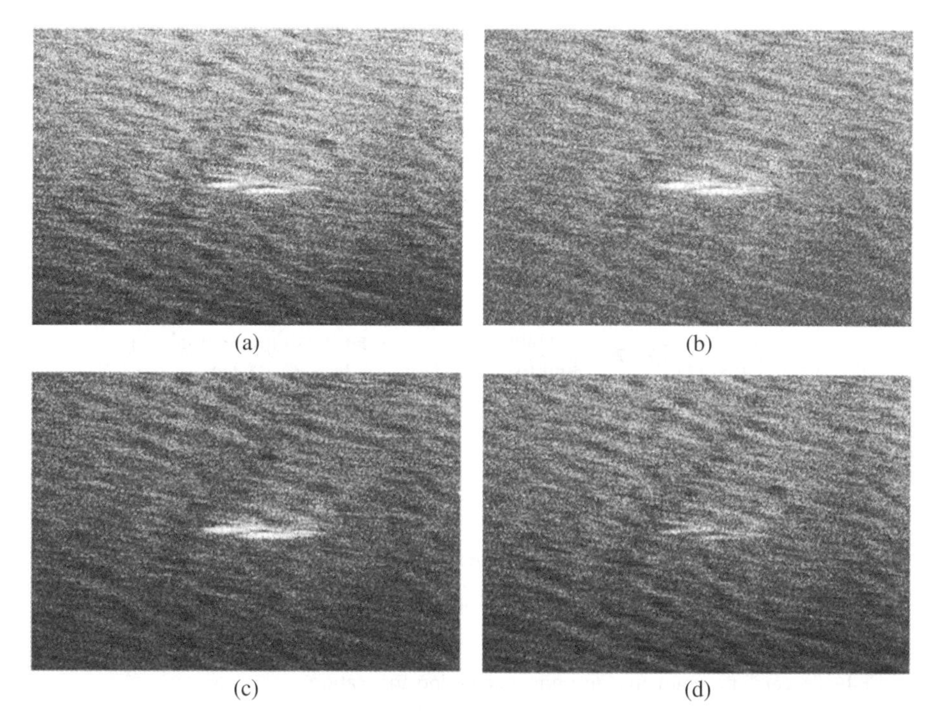


**Figure 7.47** Full-polarization SAR images of ships. (a) HH; (b) HV; (c) VH; (d) VV.

**Figure 7.48** Ship detection results of multiple-polarization fusion.

attenuation characteristics of the target, the scattering calculation module calculates the scattering contribution of the ray path. Finally, the image is calculated by the imaging module.

When the data source is available, various algorithms of target recognition can be fully tested. Among them, the intelligent target recognition technology based on deep learning is one of hot topics in current studies. Deep learning is a branch of machine

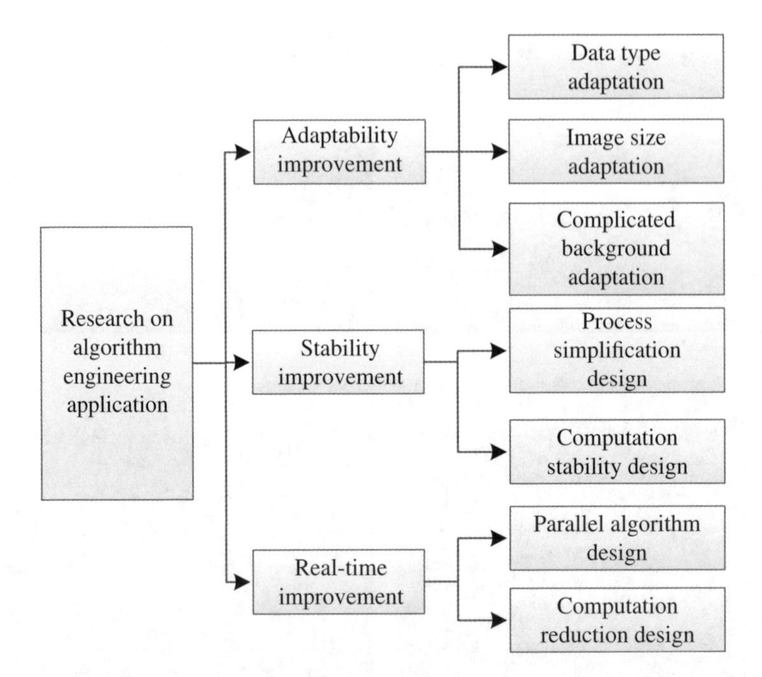


**Figure 7.49** Research program on algorithm engineering application.

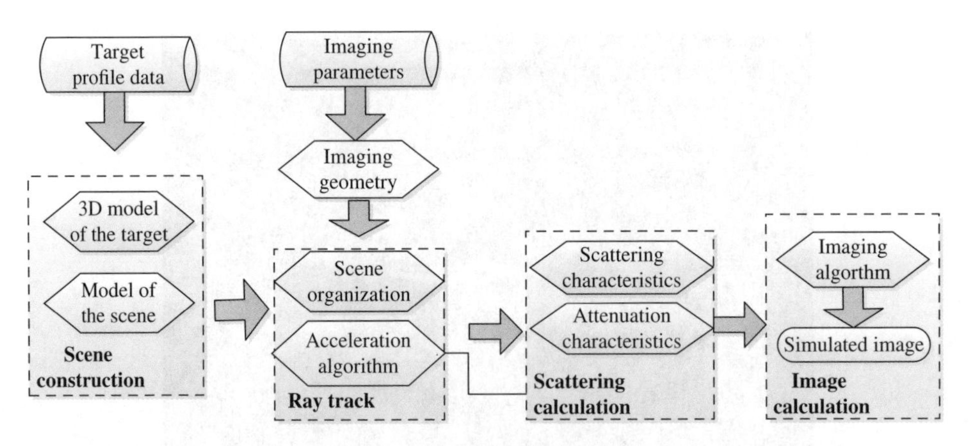


**Figure 7.50** Electromagnetic simulation flow of SAR target image.

learning. By means of a multilayer artificial neural network model, it is possible for computers to learn potential data characteristics from a large number of data samples, which can then be used to identify new samples intelligently. At present, deep learning technology has been successfully applied in voice, text, and image recognition. With the continuous improvement of SAR image resolution, deep learning technology will play an important role in SAR image classification and recognition.

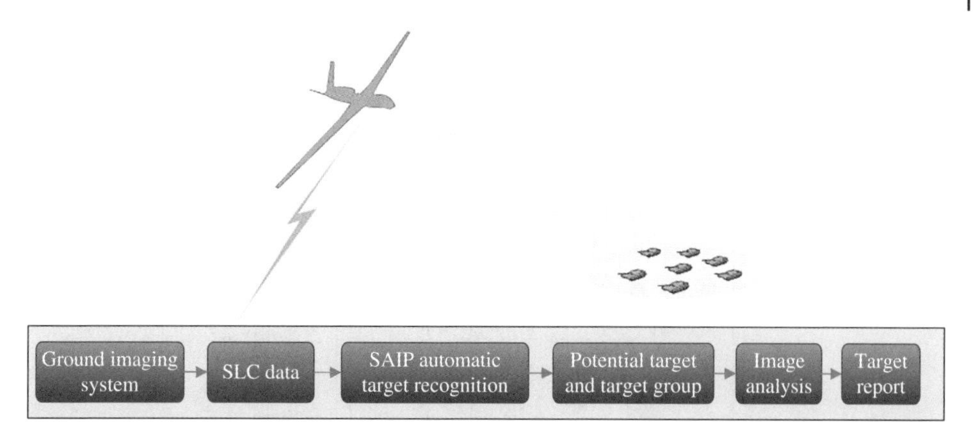


**Figure 7.51** Processing flow of SAIP SAR image information. SLC: single-look complex. (See color plate section for the color representation of this figure).

### 7.6.3 Research on SAR Image Information Processing System

Beginning in the middle of the 1980s, the Lincoln Laboratory of the Massachusetts Institute of Technology performed a study on SAR automatic target recognition. The Semi-Automated Image Intelligence Processing (SAIP) system was published in 1997, as shown in Figure 7.51. This is a joint research project led by Lincoln Laboratory along with a number of other laboratories (ERIM space laboratory, Sandia National Laboratories, and Wright Laboratory). The goal of the project is to provide intelligence within 5 min after two image analysts and one administrator receive a SAR image. At present, many countries have offered mature SAR image information processing systems. When combined with the radar system, an operation mission can be quickly completed, such as the U.S. Joint Surveillance Target Attack Radar System and the British airborne standoff radar. With the development of the SAR system in China, the SAR image information processing system has gradually matured.

The SAR image information application system is normally composed of three layers. The top layer is the user-oriented application layer. Next is the service layer for information processing and management. The bottom layer is the core data layer of the database and the supporting layer of the network platform. The software architecture is shown in Figure 7.52. It is based on service-oriented architecture and standard service interface specifications to realize the efficient operation of data layer, service component layer, and application layer.

The data layer realizes the storage and management of massive amounts of image data and geographic data mainly through a distributed relational database, distributed NoSQL database, and distributed file system.

The service layer includes a variety of data processing services, such as display, image processing, data management, and basic map. The service layer completes the core operations, reduces the pressure of the client, and improves the use efficiency of the server resources through parallel computing. With the development of computer technology, parallel computing has been widely used in the image processing industry. For massive and large-size SAR image data processing, the parallel design is mainly done from three aspects, namely, multicore parallel central processing unit (CPU) and

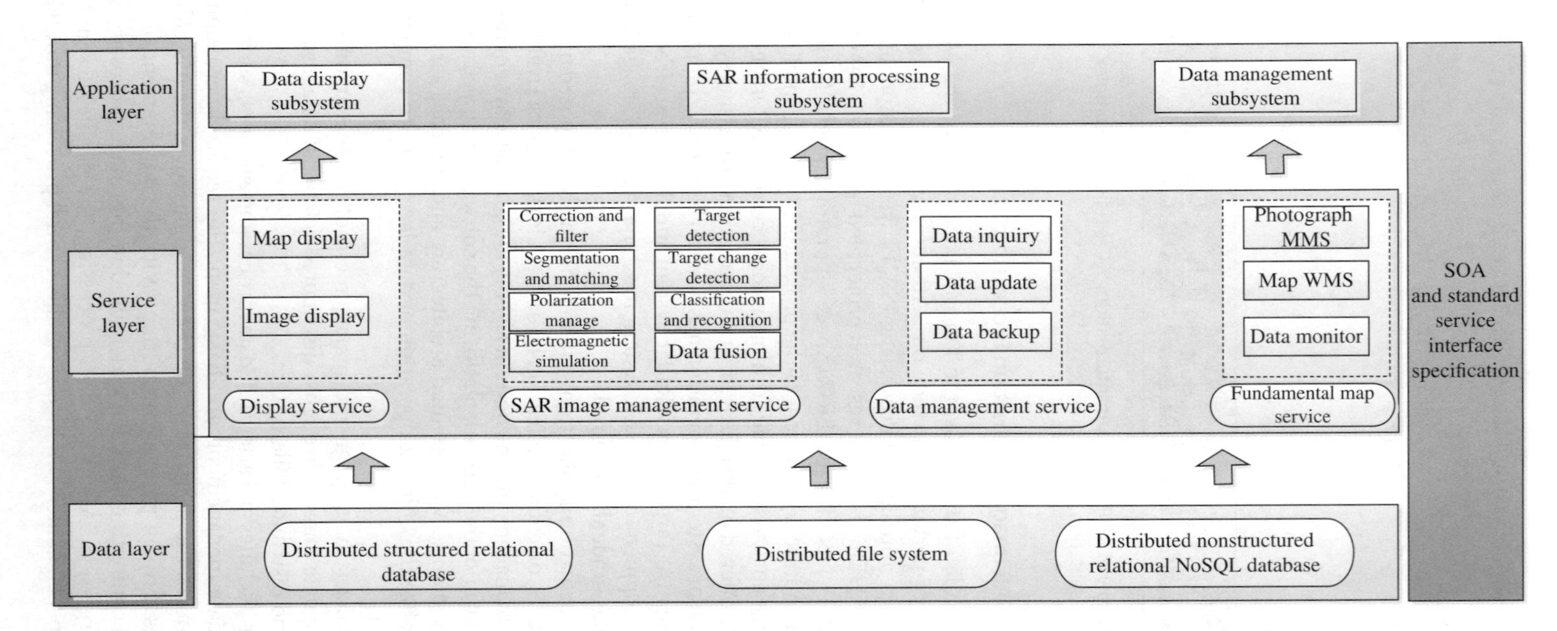


**Figure 7.52** SAR image information application system logic diagram. SOA: service-oriented architecture. (See color plate section for the color representation of this figure).

parallel and distributed parallel graphics processing unit (GPU). Multicore architecture is the mainstream of current CPUs. The computing efficiency can be greatly improved by making full use of multicore architecture for local operations of image data. Parallel GPU makes full use of the GPU computing resources of the graphics card. The image data is transferred into the GPU, where parallel computing is performed by using hundreds or even thousands of computing cores of the GPU. Then results are fed back to the CPU for synthesis. Parallel GPU computing has become a common way to solve big data computing. For distributed stored image data, the distributed parallel computation is used to process the big data, such as MapReduce parallel architecture.

The application layer includes a variety of client-oriented software systems, such as the data display subsystem, SAR information processing subsystem, data management subsystem, etc.

Through research on SAR image information application systems, the actual operation capability of the SAR system can be improved, and the SAR system can be used to meet the requirements of its users.

## References


1 Oliver, C. and Quegan, S. (1998). *Understanding Synthetic Aperture Radar Images.* Norwood, MA: Artech House.
2 Kuang, G., Gao, G., Jiang, Y. et al. (2007). *SAR Target Detection Theory, Algorithm and Application.* National Defense Science and Technology University Press.
3 Wu, T., Chen, X., Niu, L. et al. (2013). Non Supervised SAR Image Change Detection Research Progress. *Remote Sensing Information* 1: 110–118.
4 Wu, T., Xiangwei, R., and Jianbo, T. (2009). Review of support vector machine of SAR image interpretation. *Remote Sensing Information* 5: 90–95.
5 W. Tao, C. Xi, R. Xiangwei, and N. Lei (2009). Study on SAR target recognition based on support vector machine. *Proceedings of the 2009 2nd Asian-Pacific Conference on Synthetic Aperture Radar*, Xian, China (October 26–30, 2009). Piscataway, NJ: IEEE.
6 Novak, L.M., Owirka, G.J., Brower, W.S. et al. (1997). The automatic target recognition system in SAIP. *The Lincoln Labroratory Journal* 10 (2): 187–202.
7 Ross, T., Worrel, W., Velten, V. et al. (1998). Standard SAR ATR evaluation experiments using the MSTAR public release data set. *Proceedings of SPIE* 3370: 566–573.
8 Wu, T. et al. (2011). Segmentation process form SAR imagery based on graph-cuts algorithm. In: *Proc. 2011, IEEE CIE, International Conference on Radar*, 1627–1630. IEEE.
9 Wu, T., Ruan, X., and Chen, X. (2009). A modified method for estimation of SAR target aspect angle based on MRF segmentation. *Proceedings of SPIE* 7495: 74953U.
10 Vapnik, V.N. (2013). *The Nature of Statistical Lerning Theory.* New York: Springer-Berlag.
11 Patnaik, R. and Casasent, D. (2007). SAR classification and confuser and clutter rejection tests on MSTARten-class data using Minace filters. *Proceedings of SPIE* 6574: 657402.

# Index



*Design Technology of Synthetic Aperture Radar,* First Edition. Jiaguo Lu.

## t